Fisheries and Aquaculture

Fisheries and Aquaculture

Edited by **Roger Creed**

New York

Published by Syrawood Publishing House,
750 Third Avenue, 9th Floor,
New York, NY 10017, USA
www.syrawoodpublishinghouse.com

Fisheries and Aquaculture
Edited by Roger Creed

International Standard Book Number: 978-1-68286-031-1 (Hardback)

Printed in the United States of America.

Contents

Preface

Over the recent decade, advancements and applications have progressed exponentially. This has led to the increased interest in this field and projects are being conducted to enhance knowledge. The main objective of this book is to present some of the critical challenges and provide insights into possible solutions. This book will answer the varied questions that arise in the field and also provide an increased scope for furthering studies.

Aquaculture and fisheries have a great significance in global economy. This book aims to understand the different methods to cultivate aquatic organisms for commercial purposes. It details the tools and techniques that are already being practiced worldwide. It also presents some case studies which shed light on the present scenario in fisheries and aquaculture. It also sets base for new research and suggests some innovative methods that can be put to practice. It discusses some crucial aspects of aquatic farming such as feeding, sustainability, quality of seawater, protection from predators, etc. The contributors of this book hail from the top universities of the world. They have contributed their research and their pool of knowledge to help students and professionals in the field of fisheries and aquaculture.

I hope that this book, with its visionary approach, will be a valuable addition and will promote interest among readers. Each of the authors has provided their extraordinary competence in their specific fields by providing different perspectives as they come from diverse nations and regions. I thank them for their contributions.

Editor

Theoretical and experimental study for rectangular spraying

Junping Liu[1]*, Xingye Zhu[1,2] and Shouqi Yuan[1]

[1]Research Center of Fluid Machinery Engineering and Technology,
Jiangsu University, P. R. China, 212013.
[2]Key Laboratory of Modern Agricultural Equipment and Technology, Ministry of Education and Jiangsu Province,
Jiangsu University, P. R. China, 212013.

The use of rectangular spraying can reduce the water and energy requirements in variable pressurized irrigation system. The water application cost can be minimized in rectangular spraying after theoretical analysis. The aim of this study was to achieve rectangular spraying by variable pressures of the pump. Experiments were carried out on a turbo-type whirling sprinkler to determine the spraying effect. Water distribution by the sprinkler was evaluated in an indoor facility. The three-dimensional water distribution figure was drawn using Matrix Laboratory (MATLAB). The result demonstrated that rectangular spraying can be achieved.

Key words: Rectangular spraying, sprinkler, wetted radius.

INTRODUCTION

Sprinkler irrigation is characterized by high potential irrigation efficiency (Clemmens and Dedrick, 1994). In the course of development of sprinkler irrigation technology, a broad range of solutions has been applied to improve irrigation processes from the technical, organizational, and economic point of view. A sprinkler irrigation system represents one of the most common types of irrigation system. In many areas, the water and energy required for irrigation are scarce; hence, sprinkler irrigation systems must apply water with less energy consumption. This generally requires an improvement in the application of water (Martin-Benito et al., 1992).

One of the most important factors in the operation of sprinkler irrigation systems is the spaying effect of the distribution of water over a field. The most common shape is the circular spraying. Non-circular spraying sprinkler was discovered in 1927 by Donald (Donald, 1927). James and La (1952) developed a reaction sprinkler with adjustable rotational speed. Charles (1977)

invented a center pivot irrigation system of rectangle spraying. For an historical note on Edwin's discovery (Edwin, 1982). Since Edwin's discovery, the phenomenon has been studied widely, both experimentally and theoretically. Ohayon (2000) realized the automatic control function of flow-rate and pressure in impact sprinkler. Zhu et al. (2002) studied the variable irrigation flow controls. Han (2003) studied the variable rate watering and contour controlled precision sprinkler. Wallender (2007) suggested water-related research on sustainability.

The research of non-circular spraying began 1920s abroad and 1990s in China. It is a new type sprinkler which was developed for avoiding overlapping or leaking in the irregular parcel. However, when the sprinkler was working, either restricting flow or changing the elevation of pipeline was used to change the irrigated wetted radius. There were a few drops for pressure head, and the energy was wasteful. In order to attach the target of saving water and saving energy together, the frequency control technology was brought into the sprinkler irrigation system. One of the main tasks of irrigation research is to find new ways to minimize energy and water use in irrigated agriculture through experimentation.

*Corresponding author. E-mail: liujunping401@163.com.

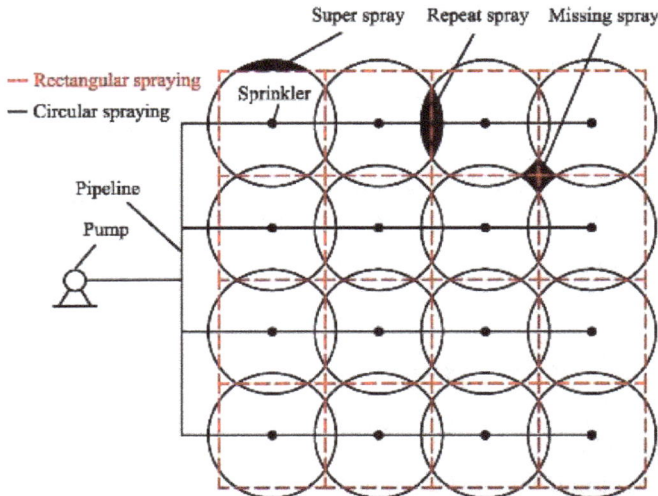

Figure 1. Water applications in circular and rectangular spraying sprinklers.

Figure 2. Frequency inverter.

The objective of this study was to achieve rectangular spraying.

Analysis of water application cost

The precision agriculture has been become the important way of modern agricultural production. According to the characteristics of the agricultural development, the technological system of water-saving precision agriculture should be developed in the near future. Under the conditions of assuring the correct hydraulic performance, the water application cost with sprinkler irrigation includes investment costs (pumping, pipes, and sprinklers), energy, labor, maintenance, and the water cost. Neither labor nor maintenance is taken into account, because they are similar for all the different subunit designs. Some

research (Ortega et al., 2004) shows the great influence of the number of sectors and their size on investments costs. On the other hand, they also indicate the small effect of the application rate or uniformity on these costs.

The water application cost in a sprinkler irrigation subunit, understood as the cost of cubic meter of water added to soil for crop use, have been determined as the sum of the investment costs, energy costs, and the water costs. Water application costs in a sprinkler irrigation subunit are conditioned by factors such as sprinkler layouts, subunit size, and working pressure (Tarjuelo et al., 1999). To analyze the design of subunits to minimize the water application cost, it is necessary to consider the influence of those factors (Lunk et al., 2010).

A subunit formed by a pump at one end, some pipelines, and several laterals (Figure 1) has been designed. Circular and rectangular spraying sprinklers were proposed in this work. For the comparisons, the cost of taking water to the inlet of the subunit has been estimated. Pumping costs from the water source, transport water costs for the network of pipes. For the same subunit, the investment and energy costs are similar both for them.

Figure 1 shows the water applications in circular and rectangular spraying sprinklers. As shown in Figure 1, combination of circular spraying sprinklers will arise the problems of super spray, repeat spray, and missing spray. The whole field can be sprayed well in the combination of rectangular spraying sprinklers, and the water application cost can be minimized.

EXPERIMENTS

The experimental setup was designed and constructed at the Indoor Sprinkling Laboratory, Jiangsu University, China (Fraisse et al., 1995). The laboratory is circular with a diameter of 44 m. A centrifugal pump supplied water to the irrigation system from a reservoir maintained at a constant level. The pump was driven by a variable speed motor with control panel (King and Wall, 2000). Figure 2 shows the photo of the common Mitsubishi frequency inverter, typed FR-F740-0.75K~55K-CHT1. The speed of pump was changed by the instruction of the inverter. To minimize pressure variation by the pump, an air vessel and subsequent settling chamber were put in the line between the delivery side of the pump and the bottom of the riser. The frequency inverter was controlled by the virtual instruments (Figure 3).

Figure 3 shows the photo of the virtual instruments, sending out the orders the frequency inverter needed. For the instruments, combined with the hardware, the software was developed in the program of LabView. It contains the front and back panels of data acquisition module, which is shown in Figures 4 and 5, respectively. The sprinkler being tested was the turbo-type whirling sprinkler. The sprinkler had one nozzle 4 mm. The nozzle size was chosen because it is the standard one according to the manufacturer. The collectors were cylindrical and had a sharp-edged round opening of 200 mm diameter and 600 mm height. The single-leg test was conducted, and the center-to-center spacing of the collectors was 1 m in the direction of the radius of the wetted circle. The sprinkler being tested was placed on a riser with an inner diameter of 20 mm and a 1.4 m height above ground and approximately 800 mm above the collector tops. The test duration was 1 h for every pressure (Sharda et al., 2010; Fulton et al., 2005a). The flow rate through the riser (that is, through the nozzle of

Figure 3. Virtual Instruments.

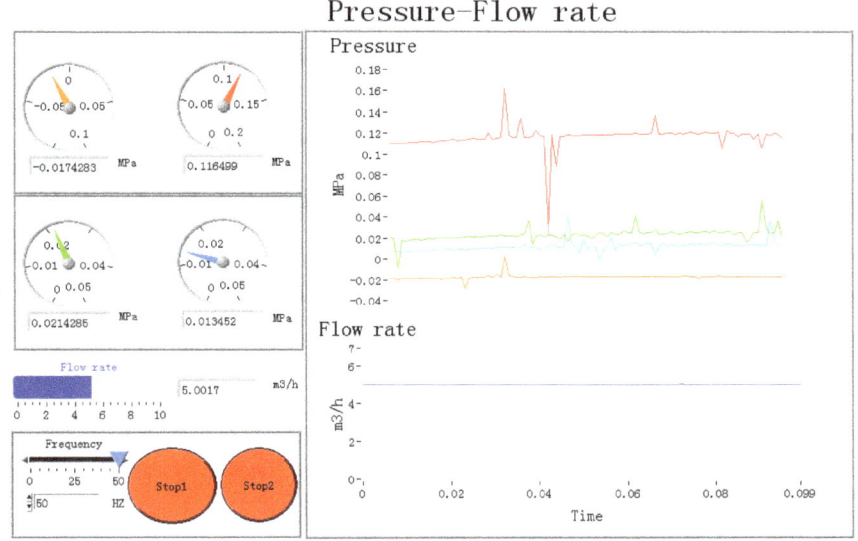

Figure 4. Front panel of data acquisition module.

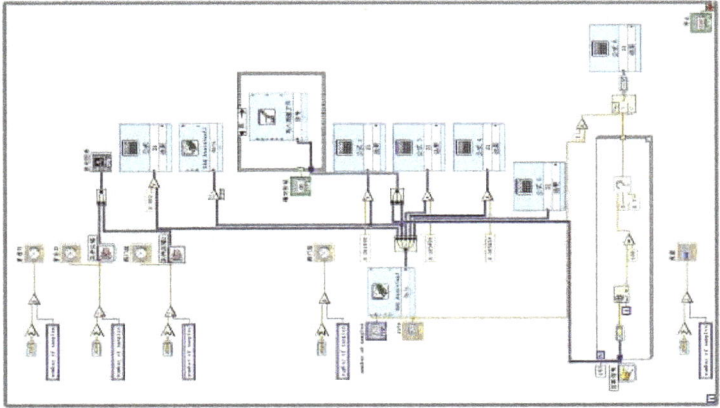

Figure 5. Back panel of data acquisition module

Table 1. Measured data of water distribution.

Degree	Working pressure (kPa)	Flow rate (m³h⁻¹)	Distance from the sprinkler (m)	0	1	2	3	4	5	6	7	8	9	10	11
0°	132	0.61		1.3	1.7	2.6	2.5	2.2	2.1	3.8	0.3	0	0	0	0
25°	146	0.78	Depth of water distribution/mm	1.9	2.3	3	2.5	2.2	1.9	2	3.3	1.4	0	0	0
45°	195	0.87		3.0	3.5	3.8	3	2.7	2.5	2.2	2.3	2.8	2.1	0.3	0

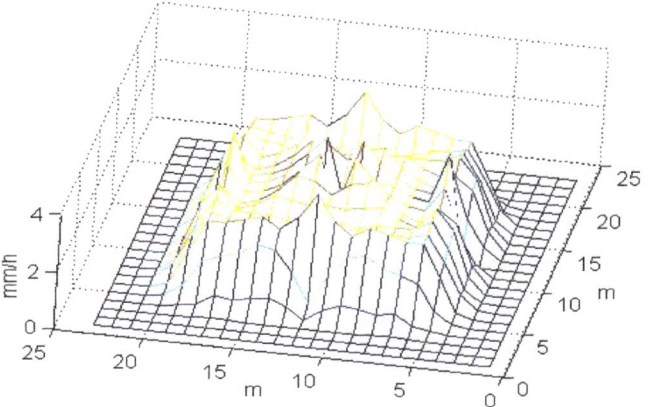

Figure 6. Three-dimensional water distribution of the sprinkler.

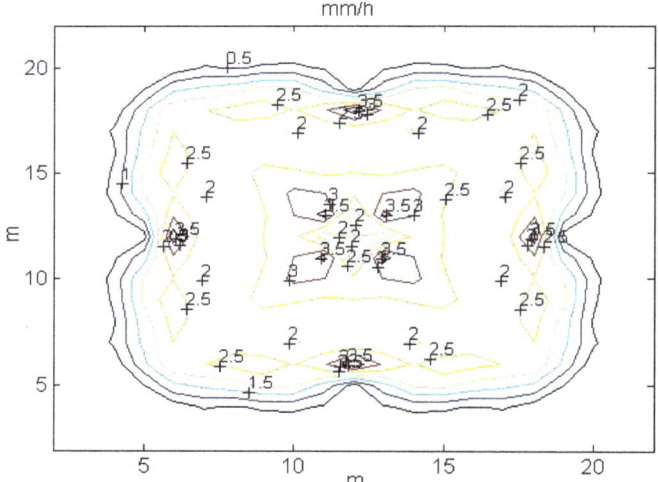

Figure 7. Water distribution contour of the sprinkler.

the sprinkler head) was measured using a calibration tank and stopwatch. The water temperature was measured during the course of the test by a thermometer placed in the main tank. The relative humidity and ambient temperature were also measured during the course of the test. This procedure was chosen to decrease the effect of liquid splashing. To determine the sprinkler water distribution radial profile in no-wind conditions, the following standards were adopted: American Society of Agricultural Engineers (ASAE) S.330.1 (1985); ASAE S.398.1 (1985). A total of 14 useful tests were conducted to study the water distribution of the sprinkler.

RESULTS AND DISCUSSION

The rectangular spraying can be achieved by the experimental setup mentioned above. In the experiment, the maximal wetted diameter for the sprinkler was 10 m, as well as the working pressure for 0, 22.5 and 45° was 132, 146, and 195 kPa, respectively. The catch-can tests were conducted in a radial data for these three working pressure and the tested data was considered to be the radial data of 0, 22.5 and 45°, respectively. Table 1 shows the water depths along these pressures by the test. According to the symmetry, the data for 45 to 90° were calculated out using the data for 0 to 45°. At last the data for 90 to 180° and 180 to 360° can be got using the same method. According to the calculating data, three-dimensional water distribution picture was drawn using Matrix Laboratory (MATLAB). Figure 6 shows three-dimensional water distribution of the sprinkler. Figure 7 shows water distribution contour of the sprinkler.

As can be seen from Figures 6 and 7, it can fulfill rectangular irrigation effectively. Any point around the sprinkler can be got easily from Figure 6. The irrigated intensity is shown in Figure 7. A case study shows that MATLAB is reliable for analyzing water distribution in sprinkler irrigation.

Conclusion

The frequency control technology was brought into the sprinkler irrigation system for the first time. Rectangular irrigated area was chosen to illustrate the quantitative result of this study. After experiment, it can fulfill rectangular spraying effectively. Mathematical program was established using MATLAB. Three-dimensional and contour of water distribution were figured out. It supplied a new way for achieving non-circular spraying of the sprinkler.

ACKNOWLEDGEMENTS

Many thanks to the Jiangsu University for helping in this research. Partial funding in support of this work was provided by National Natural Science Foundation of China (No. 51109098), Program for National Hi-tech Research and Development of China (863 Program, No.2011AA100506), Program for China Postdoctoral

Science Foundation (No. 20110491357) and Open Fund for Key Laboratory of Modern Agricultural Equipment and Technology, Ministry of Education and Jiangsu Province, Jiangsu University (No.NZ201008) and is gratefully appreciated.

REFERENCES

ASAE Standards, 32nd Ed. (1985). S 330.1. Procedure for sprinkler distribution testing for research purposes [S]. St. Joseph, Mich.: ASAE.

ASAE Standards, 32nd Ed. (1985). S 398.1. Procedure for sprinkler testing and performance reporting [S]. St. Joseph, Mich.: ASAE.

Charles HG (1977), Center pivot irrigation system[P]. United Stated Patent Office: US4011990. pp. 03-15

Clemmens AJ, Dedrick AR (1994). Irrigation techniques and evaluations [M]. *Advanced series in agricultural sciences*, K.K. Tanji and B. Yaron, eds., Springer, Berlin. pp. 64-103.

Donald E (1927). Apparatus for Watering Areas of Land [P]. United Stated Patent Office: US1637413, 1927-08-02.

Edwin J (1982). Rotating sprinkler [P]. GB 2094181A. pp. 09-15.

Fraisse CW, Duke HR, Heermann D.F (1995). Laboratory evaluation of variable water application with pulse irrigation[J]. Trans. ASAE. 38 (5):1363-1369

Fulton JP, Shearer SA, Higgins SF (2005a). Distribution pattern variability of granular VRT applicators [J]. Trans. ASABE. 48(6):2053-2064.

Han WT (2003). Variable range watering and contour controlled precision sprinkler and sprinkler irrigation [D]. Yangling: North-West Science-Technology University Of Agriculture And Forestry.

James T, La M (1952). Sprinkler Device[P]. United Stated Patent Office: US2582158. pp. 01-08.

King BA, Wall RW (2000). Distributed instrumentation for optimum control of variable speed electric pumping plants with center pivots[J]. Appl. Eng. Agric. 16(1):45-50.

Lunk JD, Pitla SK, shearer SA (2010). Potential for pesticide and nutrient saving via map-based automatic boom section control of spray nozzles[J]. Comp. Elec. Agric. 70(1):19-26.

Martin-Benito JMT, Gomez MV, Pardo JL.(1992). Working conditions of sprinkler to optimize application of water[J]. J. Irrig. Drain. Eng. 118(6):895-913.

Ohayon S (2000). Automatic Adjustable Sprinkler for Precision Irrigation[P]. United Stated Patent Office: US6079637. pp. 06-27.

Ortega JF, Juan JA. de, Tarjuelo JM (2004). Evaluation of the water cost effect on water resource management: application to typical crops in a semiarid region[J]. Agric. Water Manag. 66:125-144.

Sharda A, Fulton P, Mcdonald TP (2010). Real-time pressure and flow dynamics due to boom section and individual nozzle control on agricultural [J]. Trans. ASABE 53(5):1363-1371.

Tarjuelo JM, Montero J, Honrubia FT (1999). Analysis of uniformity of sprinkle irrigation in a semi-arid area[J]. Agric. Water. Manag. 40:315-331.

Wallender WW (2007). Scales, scaling, and sustainability of irrigated agriculture [J].Trans. ASABE 50(5):1733-1738.

Zhu HP, Sorensen RB, Butts CL (2002). A Pressure Regulating System for Variable Irrigation Flow Controls [J]. Appl. Eng. Agric. 18(5):533-540.

2

Age and growth of *Gerres filamentosus* (Cuvier, 1829) from Kodungallur, Azhikode Estuary, Kerala

Megha Aziz, V. Ambily and S. Bijoy Nandan

Department of Marine Biology, Microbiology and Biochemistry, Cochin University of Science and Technology, Fine Arts Avenue, Cochin 682016, Kerala, India.

The age and growth, length – weight relationship and relative condition factor of *Gerres filamentosus* (Cuvier, 1829) from Kodungallur, Azhikode Estuary were studied by examination of 396 specimens collected between May 2008 to October 2008. Here, length frequency method was used to study age and growth in fishes. $L\infty$, K and t_0 obtained from seasonal and non - seasonal growth curves. *Gerres filamentosus* showed a low mortality rate (Z) $3.702\ y^{-1}$. *G. filamentosus* has moderately low K value and long life span. The relation between the total length and weight of *G. filamentosus* was described as Log W = 1.321+2.5868 log L for males, Log W = 1.467 + 2.7227 log L for females and Log W = 1.481 + 2.7316 log L for sexes combined. The mean relative condition factor (Kn) values ranged from 0.9 to 1.14 for males, 0.89 to 1.11 for females and 0.73 to 1.08 for sexes combined. The length weight relationship and relative condition factor showed that the wellbeing of *G. filamentosus* were good. The morphometric measurements of various body parts were recorded. The morphometric measurements were found to be nonlinear and there is no significant difference observed between the two sexes.

Key words: Age and growth, length weight relationship, condition factor, morphometry, *Gerres filamentosus*.

INTRODUCTION

Studies on age and growth are important in fisheries research. Most of the methods employed for assessing the state of exploited fish stocks rely on the availability of age composition data (Ricker, 1975). Information on growth rate, natural and fishing mortality, age at maturity and spawning, age composition of the exploited populations etc. can be generated from age data of fish populations. Length – weight relationship studies of any fish species is a pre requisite for the study of its population (Le Cren, 1951). The ponderal index or condition factor or the 'fatness' (K) was worked out to assess the well-being of the population with the assumption that the growth of fish in ideal conditions maintain an equilibrium in length and weight (Hile, 1936).

The data on length – weight relationship and the associated condition factor also enables to compare the population of the same species from different environments. The study of morphometric characters in fishes is important because they can be used for the differentiation of taxonomic units. The present study provides comprehensive information on the age and growth, length weight relationship, relative condition factor and morphometry of *Gerres filamentosus* (Cuvier, 1829) from Kodungallur, Azhikode Estuary, Kerala.

G. filamentosus is an economically important food fish species, also known as silver – biddies of the family Gerreidae. The importance of the species of the genus *Gerres* as food and ornamental fish is significant in the context of its sustainable use of the resource and conservation of the endemic fish germplasm. The study on the biology of food and feeding habits of *Gerres oblongus* (Abeyrami and Sivashanthini, 2008), *Gerres macracanthus* (Badrudeen and Mahadevan, 1996) and population dynamics of *Gerres abbreviates* (Kuganathan, 2006) and *Gerres setifer* (Sivashanthini et al., 2004) were the works done on *Gerres*. A survey on the literature of *Gerres* sp.; showed that, there was very little information available on the age and growth of *G. filamentosus* from this sub continent. It was in view of this, that the age and growth of *G. filamentosus* from a wetland system in Kerala is presented in this study.

MATERIALS AND METHODS

Fresh fish samples were collected weekly during May 2008 to October 2008 from the fishermen of the Anappuzha region, Azhikode estuary, Kodungallur (Latitude 10° 11'53" N and Longitude 76°12'13" E). The backwater is known for fishing activities. A total of 396 specimens of *G. filamentosus* ranging in size from 6.8 to 24.2 cm in total length (TL) were used for the age and growth, length- weight analysis and various morphometric measurements. The relationship between various parameters was determined by the method of least square.

Length of fish was measured to the nearest mm and weight up to 0.1 g. The fishes were then sexed by observing the gonads after dissecting the abdomen. The total length was measured from the tip of snout to tip of upper caudal lobe to the nearest mm and weight to 0.5 accuracy. The growth parameters were estimated using Von Bertalanffy's (1938) equation based on the length data on weekly modal progression. The data on males, females and both combined were subjected to analysis using the length frequency data analysis (LFDA) version 5.0 (Kirkwood and Hoggarth, 2006). The total mortality rate (Z) was estimated following Beverton and Holt (1956) method. The Von Bertalanffy growth function of the length frequency data was estimated base on the following:

1. Shepherd's length composition analysis (SLCA, Shepherd, 1987).
2. The projection matrix method (PROJMAT, Rosenberg et al., 1986).
3. ELEFAN method (Pauly, 1987).

The Von Bertalanffy growth function (VBGF) was used to describe the growth the simplest version of VBGF has the form

$$Lt = L\infty (1 - e^{-k(t-t_0)})$$

Where, Lt is the length of fish at age t, $L\infty$ is the average maximum length, K is a measure of the growth rate (the rate at which $L\infty$ is approached) and t_0 is the time (age) at which length is zero.
The length – weight relationship of the form $W = aL^b$ was calculated for male, female and pooled, which was transformed in logarithmic form as Log W = Log a + b Log L. Ponderal index (Kn) was observed separately for males and females of different length groups. It was calculated for each 3 cm length interval. The smoothed mean weights W, for each length group have been computed from this Log formula. LeCren's (1951) modified formula, Kn = W/aL^n was used for calculation of the relative condition factor. Nineteen morphometric characters were studied following the standard procedures described by Appa (1966) and Dwivedi and

Menezes (1974).

RESULTS

Age and growth

In practice, it has been found that the SLCA method does not perform well when estimating seasonal growth parameters, and so LFDA allows only the PROJMAT and ELEFAN methods to be used for fitting seasonal growth curves (Tables 2 and 3). Total mortality was estimated by using the methods of Beverton and Holt (1956). Z was calculated from the mean length derived from the Von Bertalanffy parameters. All three of the mortality estimators available in LFDA are based on non-seasonal von Bertalanffy growth curves. The von – Bertalanffy growth equations for the different sets of parameters can be written as: From nonseasonal growth curves,

$$Lt = 335.00(1 - e^{-1(t + 0.47)})$$ for combined and male

$$Lt = 335.00(1 - e^{-1(t + 0.53)})$$ for female (Table 1)

Here the values obtained by SLCA method. By this method values of K, L∞ and t zero were same (Table 1). From Hoeing seasonal growth curves both pooled and males the same values were obtained from PROJMAT and ELEFAN method. And they can be written as:

$$Lt = 467.79(1 - e^{-0.30(t + 0.16)})$$ pooled (ELEFAN METHOD)

$$Lt = 414.96 (1 - e^{-0.87(t + 0.41)})$$ pooled (PROJMAT METHOD)

$$Lt = 639.9 (1 - e^{-0.47(t + 0.47)})$$ male (ELEFAN METHOD)

$$Lt = 414.9(1 - e^{-0.41(t + 0.86)})$$ male (PROJMAT METHOD)

$$Lt = 412.3(1 - e^{-0.99(t + 0.39)})$$ female (PROJMAT METHOD)

$$Lt = 399.9(1 - e^{-0.60(t + 0.80)})$$ female (ELEFAN METHOD) (Table 2)

From Pauly seasonal growth curves different values were obtained for male, female and pooled.

$$Lt = 374.73(1 - e^{-0.96(t + 0.09)})$$ pooled (PROJMAT)

$$Lt = 299.33(1 - e^{-1(t + 0.58)})$$ pooled (ELEFAN)

$$Lt = 693.66(1 - e^{-0.25(t + 0.78)})$$ male (ELEFAN)

$$Lt = 515(1 - e^{-0.39(t + 0.75)})$$ male (PROJMAT)

$$Lt = 391.35(1 - e^{-0.86(t - 0.54)})$$ female (ELEFAN)

Table 1. The different growth parameters using non seasonal von Bertalanffy growth model in *G. filamentosus.*

Sex	Method	K	L ∞	T0
Combined	SLCA	1	335	-0.47
Male	ELEFAN	1	335	-0.47
Female	SLCA	1	335	-0.53

Table 2. The different growth parameters using Hoenic seasonal von Bertalanffy growth model in *G. filamentosus.*

Sex	Method	K	L ∞	T0	Ts	C
Combined	PROJMAT	0.87	414.96	-0.41	0.07	0.54
Combined	ELEFAN	0.30	467.79	-0.16	-0.4	0.44
Male	PROJMAT	0.86	414.9	-0.41	0.07	0.58
Male	ELEFAN	0.47	639.9	-0.47	-0.4	0.48
Female	PROJMAT	0.99	412.3	-0.39	0.13	0.87
Female	ELEFAN	0.60	399.9	-0.80	0.07	1.30

Table 3. The different growth parameters using Pauly seasonal Von Bertalanffy growth model in *G. filamentosus*

Sex	Method	K	L ∞	T0	Ts	NGT
Combined	PROJMAT	0.96	374.73	-0.09	-0.78	0.86
Combined	ELEFAN	1	299.33	-0.58	-0.48	0.01
Male	PROJMAT	0.39	515	-0.75	-0.26	0.67
Male	ELEFAN	0.25	693.66	-0.78	-0.47	0.68
Female	PROJMAT	0.93	408.48	-0.12	-0.16	0.61
Female	ELEFAN	0.86	391.35	-0.54	-0.38	0.67

$Lt=408.48(1-e^{-0.93(t-0.12)})$ female (PROJMAT) (Table 3 and Figures 1, 2, 3).

The L∞ values obtained were found to range from 374.73 to 693.66 mm from seasonal growth curves (Tables 2 and 3) and 335 mm in non seasonal growth curves (Table 1). The range of estimated K value was 0.25 to 1 from seasonal growth curves and 1 from seasonal growth curves. According to, von Bertalanffy plot was created to estimate the growth coefficient K. The K was determined from non seasonal curves as 1 and t zero as -0.47 years. The total mortality rates (Z) obtained from a length converted catch curve is $3.702y^{-1}$.

Length-weight relationship

Length – weight equations were calculated separately for males, females and sexes combined. The fish samples were divided into 3 cm length groups (Table 4 and Figures 4, 5, 6). When empirical values of lengths were plotted against their respective weight on an arithmetic scale, smooth curves were obtained. The regression coefficients were calculated using the methods of least squares for male and female, *G. filamentosus* in the size range 6.8 to 24.2 cm gave the following equations:

Male $W = 0.0478 L^{2.5868}$
               Log W = 1.321+2.5868 log L

Female $W = 0.0341 L^{3.7227}$
               Log W = 1.467 + 2.7227 log L

Pooled $W = 0.0330 L^{2.7316}$
               Log W = 1.481 + 2.7316 log L

As may be seen from the equations, the exponential values for males, females and sexes combined were practically identical.

Growth Curve = Pauly Seasonal. Linf = 408.48. K = 0.93. Tzero = -0.12. Ts = 0.16. NGT = 0.6

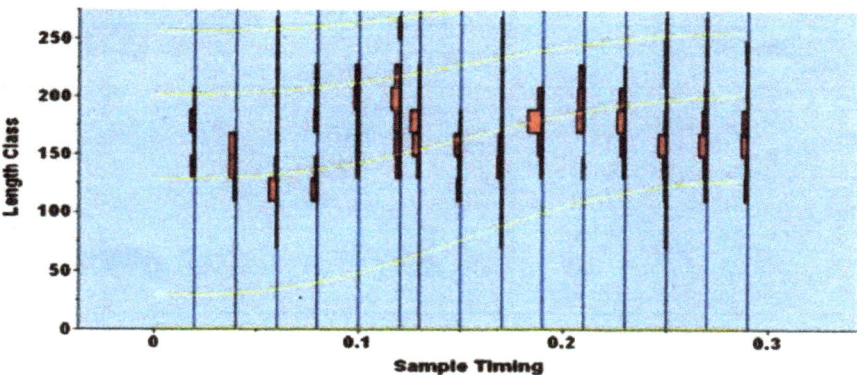

Figure 1. Growth curve (Pauly seasonal), combined sex.

Growth Curve = Pauly Seasonal. Linf = 408.48. K = 0.93. Tzero = -0.12. Ts = 0.16. NGT = 0.6

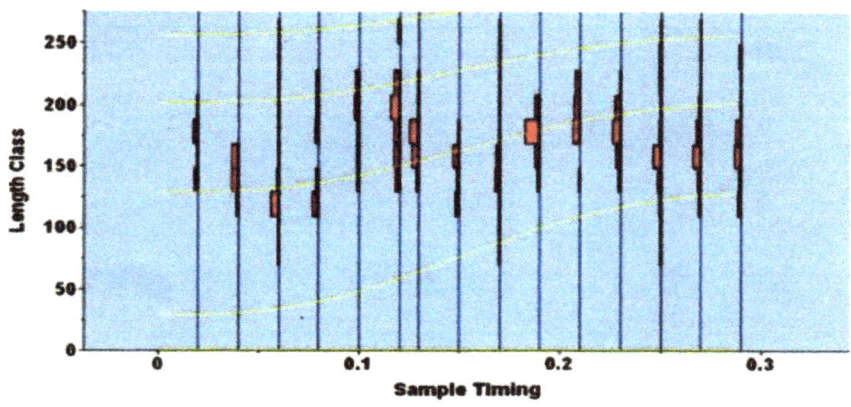

Figure 2. Growth curve (Pauly seasonal), female.

Growth Curve = Pauly Seasonal. Linf = 374.73. K = 0.96. Tzero = -0.09. Ts = -0.78. NGT = 0.86

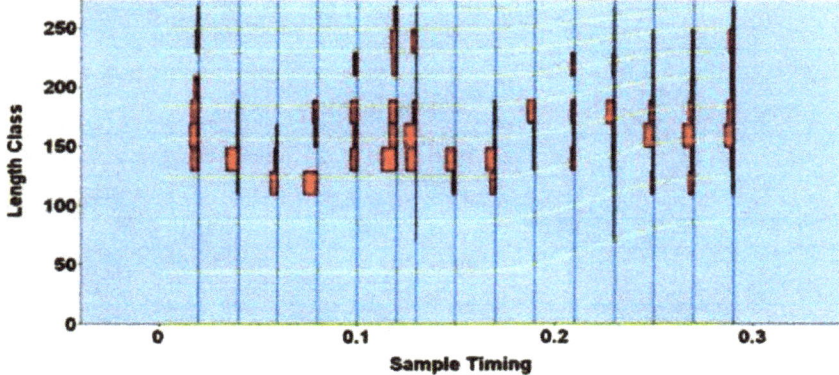

Figure 3. Growth curve (Pauly seasonal), male.

Table 4. Data on length and weight of *G. filamentosus* from Anappuzha.

Length group (cm)	No of fishes combined	Mean length (cm)	Mean weight (g)	No of male	Mean length (cm)	Mean weight (g)	No of female	Mean length (cm)	Mean weight (g)
6-9	37	8.86	9.22	0	0	0	0	0	0
9-12	61	10.21	18.72	28	9.92	21.71	18	10.69	20
12-15	111	13.64	41.11	65	13.22	38.97	46	13.61	42.83
15-18	123	16.53	70.21	46	16.58	69.37	77	16.51	68.87
18-21	48	19.21	11.16	6	19.2	114.66	42	19.67	110.79
21-24	17	21.84	130	10	22	134.5	7	21.6	123.57
24-27	3	24.23	198.33	1	24.1	195	2	24.3	200

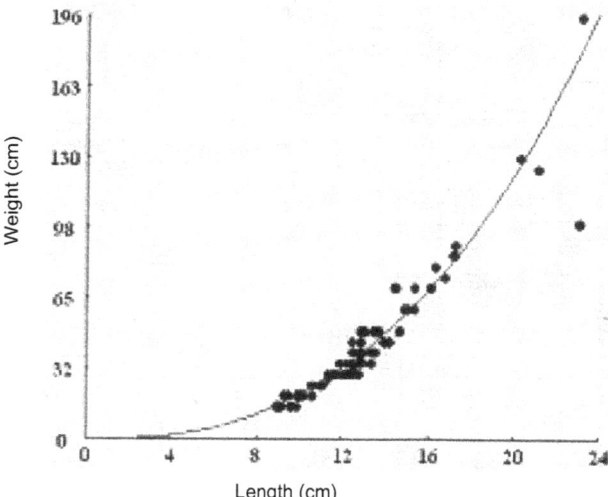

Figure 4. Length weight relationship of *Gerres filamentosus* in Azhikode estuary (male).

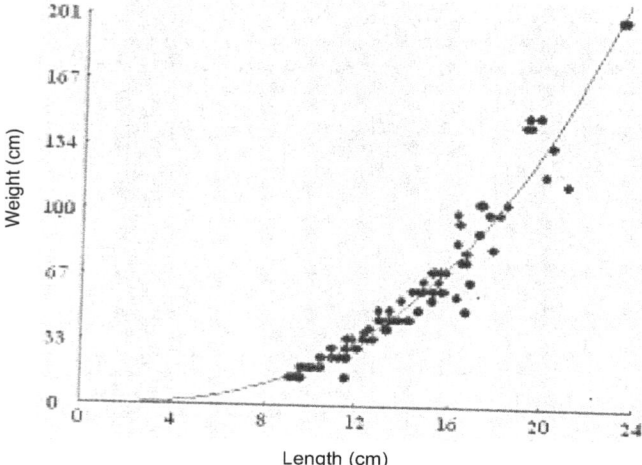

Figure 5. Length weight relationship of *Gerres filamentosus* in Azhikode estuary (Female).

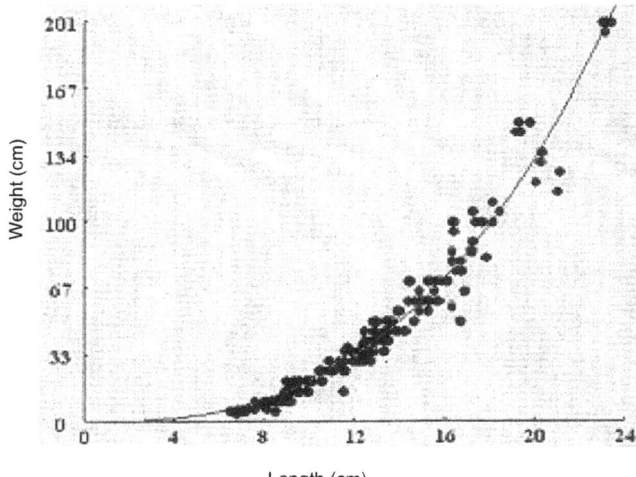

Figure 6. Length weight relationship of *Gerres filamentosus* in Azhikode estuary (combined).

Relative condition factor

The relative condition factor (Kn) can be computed using the formula:

Kn = W / a L^n (Le Cren, 1951)

This can be expressed as Kn = W/W'; Where W is observed weight and W' the calculated weight as determined from the length – weight equations. The relative condition factor (Kn) for all fish samples were determined from the average lengths and weights of 3 cm interval of total length (Table 5). The values of Kn showed fluctuation in all size groups of both males females and sexes combined. The weekly Kn values were calculated for various length groups. Values of Kn for different size groups ranged from 0.9 to 1.14 in males, 0.89 to 1.11 in females and from 0.73 to 1.08 in sexes combined (Tables 5 and 6).

Table 5. Relative condition factor (Kn) values of *G. filamentosus* from Anappuzha.

Length group (cm)	Male			Female			Combined		
	Observed Wt (g)	Calculated Wt (g)	Kn	Observed Wt (g)	Calculated Wt (g)	Kn	Observed Wt (g)	Calculated Wt (g)	Kn
6 - 9	0	0	0	0	0	0	9.22	12.63	0.73
9 - 12	21.71	19.04	1.14	20	19.76	1.01	18.7	18.54	1.01
12 - 15	38.92	39.97	0.973	42.82	38.61	1.11	41.11	40.60	1.01
15 - 18	69.38	71.72	0.97	68.87	65.97	1.04	70.21	68.28	1.03
18 - 21	114.6	104.85	1.09	110.79	107.23	1.03	11.16	102.53	1.08
21 - 24	134.5	148.85	0.90	123.57	139.01	0.89	130	145.08	0.89
24 - 27	195	188.35	1.04	200	192.71	1.04	198.3	192.14	1.03

Table 6. K and Kn values of *G. filamentosus* for different length groups.

Male			Female			Combined		
Length (cm)	K	Kn	Length (cm)	K	Kn	Length (cm)	K	Kn
0	0	0	0	0	0	8.286	1.62	0.73
9.917	2.23	1.14	10.69	1.64	1.01	10.209	1.76	1.01
13.22	1.69	0.97	13.61	1.69	1.11	13.635	1.62	1.01
16.58	1.52	0.97	16.514	1.53	1.04	16.529	1.55	1.03
19.2	1.62	1.09	19.67	1.46	1.03	19.206	1.57	1.08
22	1.26	0.9	21.6	1.23	0.89	21.835	1.25	0.89
24.1	1.39	1.04	24.3	1.39	1.04	24.233	1.39	1.03

Morphometric characters

Morphometric measurements of various parts of the body and their percentage ratio in relation to TL for males and females of 75 fishes (36 males and 39 females) ranging from 7.2 to 22.3 cm, TL are given in Table 7. As may be seen from the tables, fork length, standard length, body depth, pre anal length, pre dorsal, pre pectoral, base of dorsal, base of pelvic, length of pectoral, length of pelvic, length of anal, length of caudal, least width of caudal peduncle, head length, snout length, post orbital length, inter orbital length and gape width are highly correlated with TL and the relationship between body measurements are found to be non linear.

DISCUSSION

Quantification of age and growth is a vital component for understanding the ecology and life history of any fish species (Thomas et al., 2006). Age and growth of fishes can be estimated by indirect methods such as length frequency analysis and rings appearing on the scales. In the present study L∞ was estimated by non seasonal and seasonal growth curves. In non seasonal methods L∞ and K obtained in male, female and sexes combined are equal. In seasonal growth curves L∞ obtained are higher in males than females (Tables 1, 2 and 3).

In the present study the K value range from 0.25 to 1. Pauly (1984) reported that species having shorter life have high K value and reach their L∞ within one or two years. Similar observations were seen in the study on *P. hamrur*, where K value was 0.69 and their L∞ 360 mm (Yassar et al., 2009). On the other hand, those having flat growth rates have lower K values and take many years to reach their L∞. Most of the fish species have a growth rate, K of between 0.1 and 1.0 per year. With a K of 0.1 per year fish grow 9.5% closer to L∞ each year. With a K of 1 per year, they grow 63% closer to L∞ each year. The parameter t zero is the theoretical age (t) at which the fish would have had zero length if growth had followed the VBGF from birth. *G. filamentosus* has moderately low K value and long life span. All three of the mortality estimators available in LFDA are based on non-seasonal von Bertalanffy growth curves, so it cannot be used to estimate mortality for a stock displaying strongly seasonal growth. *G. filamentosus* showed a low mortality rate (Z) $3.702y^{-1}$. Low mortality rates showed by catfish like *Pylodictis divaris* (Das, 1994).

It is universal that growth of fishes or any other animal increases with the increase in body length. Thus, it can be said that length and growth are interrelated. Length weight relationship is expressed by the cube formula $W = aL^3$ by earlier workers (Brody, 1945; Largler, 1952;

Table 7. Regression values for various morphometric characteristics as function of total length.

Parameter	a	b	R	R^2	Y = b X + a
Fork length	5.488	11.880	0.924	0.854	11.880TL + 5.488
Standard length	7.173	15.527	0.870	0.756	15.527TL + 7.173
Body depth	2.556	5.534	0.916	0.839	5.534TL + 2.556
Pre anal length	2.631	5.695	0.710	0.505	5.695TL + 2.631
Pre dorsal length	1.669	3.614	0.904	0.818	3.614TL + 1.669
Pre pectoral length	1.227	2.657	0.941	0.885	2.657TL + 1.227
Pre pelvic length	1.449	3.138	0.925	0.855	3.138TL + 1.449
Base of dorsal	1.948	4.218	0.949	0.901	4.218TL +1.948
Base of pelvic	0.431	0.933	0.695	0.484	0.933TL +0.431
Length of dorsal	2.606	5.642	0.943	0.888	5.642TL + 2.606
Length of pectoral	2.030	4.395	0.871	0.758	4.395TL + 2.030
Length of pelvic	1.533	3.378	0.914	0.836	3.378TL + 1.533
Length of anal	1.732	3.750	0.761	0.580	3.750TL + 1.732
Pre orbital length	0.789	1.707	0.700	0.490	0.789TL + 0.789
Inter orbital width	0.520	1.126	0.710	0.505	1.126TL+ 0.520

Brown, 1957). In the present study the value of 'b' in *G. filamentosus* was found to range between 2.5860 to 2.7316. The highest 'b' value was arrived in females followed by males. The exponential value of 2.7227 implies that the female gain weight at a faster rate in relation to the length than male (2.5868). Le Cren (1951) reported that females are heavier than the males of the same length probably because of the difference in fatness and gonadal development. All the earlier reports (Hile, 1936; Tesch, 1968; Narejo et al., 2000) are in compliance with the present findings on the length-weight relationship in *G. filamentosus* in which the 'b' values were very close to the isometric value of 3. This indicated that *G. filamentosus* in the present study showed an isometric growth.

In the present study, sex - wise analysis of Kn values in males (1.14) was higher than that of females (0.927) (Tables 2 and 3 and Figure 4). In sexes combined, the mean value was 1.03. According to Le Cren (1951), Kn values greater than 1 indicates good general condition of the fish whereas values less than 1 denotes the reverse condition. High Kn values were recorded in *Labeo rohita* (1.0129) and *Catla catla* (0.9967) by Pandey and Sharma (1998) from Uttar pradesh. In the present study also males showed the highest value (1.14) when compared to females. This indicates that males are in better condition when compared to females. The values of K showed significant fluctuation in both males and females which may be due to difference in the weight of food contents in the stomach. This result supports the results of Kader and Rahman (1978); Umesh et al. (1996) and Das et al. (1997). In males, Kn values remained high in size group up to 9 to 12 cm and then gradually declined by 12 – 15 and 15 – 18 values showing major fluctuations

which may be due to several factors like spawning activity, feeding condition, maturation cycle and several other unknown factors of the species. In females Kn value remained almost equal in size group up to 12 to 15 cm followed by 15 to 18 cm indicating attainment of maturity in this size group.

Thus from the study, it is observed that, the body parameters grew symmetrically when observed in different length groups. Similar observations were reported in *Mahseer* spp by Mann (1976), Talwar and Jhingran (1992) and Muhammad et al. (2002). The morphometric measurements were found to be linear and there is no significant difference observed between the two sexes. From the regression results of morphometric characters, the coefficient of determination (r^2) was noted to varying strength relationships between the total length against other measurements. As such the relative growth of the morphometric characters in relation to the total length was noted to be the least in the base of pelvic (b= 0.933) and the highest in the standard length (b= 15.527). Thus from the present investigation on age and growth of *G. filamentosus* indicated that the wellbeing of the species studied were good and showed a long life span.

ACKNOWLEDGMENT

We are thankful to the Head, Department of Marine Biology, Microbiology and Biochemistry, Cochin University of Science and Technology, Fine Arts Avenue, Cochin, Kerala, India for the facilities for undertaking this work.

REFERENCES

Abeyrami B, Sivashanthini K (2008). Some aspects on the feeding of *Gerres oblongus* dwelling from the Jaffna lagoon. Pak. J. Boil. Sci. 11(9):1252-1257.

Appa RT (1966). On some aspects of biology of *Lactarius lactarius* (Schn). Indian J. Fish. 13:334-349.

Badrudeen M, Mahadevan-Pillai PK (1996). Food and feeding habits of the bog – eyed mojara, *Gerres macracanthus* (Bleeker) of the Palk Bay and the Gulf of Mannar. J. Mar. Biol. Ass. India 38(1,2):58-62.

Beverton RJH, Holt SJ (1956). A review of methods for estimating mortality rates in fish populations, with special reference to sources of bias in catch sampling. Rapp. P-v. Reun. CIEM. 17A:1-153.

Brody S (1945). Bioenegetics and growth. Reinhold publishing Corporation, New-York.

Brown ME (1957). The physiology of fishes. 1. Metabolism. Academic press. Inc., New York. P. 371.

Das M (1994). Age Determination and Longetevity in Fishes. Gerontology 40(24):70-96.

Das NG, Majumder AA, Sarwar SMM (1997). Length – weight relationship and condition factor of catfish *Arius tenuispinis*, Day, 1877. Indian J. Fish. 44(1):81-85.

Dwivedi SN, Menezes MR (1974). A note on morphometry and ecology of *Brachiunius orientalis* (Bloch & Schenider) in the estuary of Goa. Geobios. 1:80-83.

Hile R (1936). Age and growth of Cisco *Leucichthys artedi* (Le sucur) in the lakes of the north eastern highlands, Wisconsin. Bull. Bur. Fish. 48:211-317.

Kader MA, Rahman MM (1978). The length – weight relationship and condition factor of tilapia (*Tilapia mossambica*, Peters) J. Asiatic. Soc. Bangladesh Sci. 3(2):1-17.

Kirkwood GP, Hoggarth DP (2006). Stock assessment for fishery management, FAO. P. 127.

Kuganathan S (2006). Population dynamics of *Gerres abbreviatus* from the Parangipettai waters, Southeast coast of India. Srilanka J. Aquat. Sci. 11:1-19.

Largler KF (1952). Freshwater Fishery Biology. Wm.C. Brown co., Dubuque, Lowa. P. 317.

Le cren E D (1951). The length – weight relationship and seasonal cycle in gonad weight and condition in the perch (*Perca fluviatilis*). J. Anim. Ecol. 20:2-19.

Mann RHK (1976). Observation on the age, growth, reproduction and food of the pike (*Esox lucius* L.) in two Rivers in Southern England. J. Fish. Biol. 8:179-197.

Muhammad Z, Abdul N, Nasim A, Mechdi SMH, Naqvi, Zia- Ur Rehman M (2002). Studies on Meristic counts and Morphometric Measurements of Mahseer (*Tor putitora*) from a spawning ground of Himalayan Foot- hill River Korang Islamabad, Pakistan. Pak. J. Biolog. Sci. 5(6):733-735.

Narejo NT, Jafri SIH, Shaikh SA (2000). Studies on the age and growth of palri, *Gudusia chapra* (Clupeidae: Teleostei) from the Keenjhar Lake (District: Thatta), Sindh, Pakistan. Pak. J. Zool. 32(4):307-312.

Pandey AC, Sharma MK (1998). Bionomics of the Indian major carps cultivated on sodic soil pond conditions in U.P., India. Indian J. Fish. 45(2):207-210.

Pauly D (1984). Fish population dynamics in tropical waters: a manual for use with programmable calculators. ICLARM Stud. Rev. 8:325.

Pauly D (1987). A review of the ELEFAN system for analysis of length-frequency data in fish and aquatic vertebrates. Pages 7-34 in Pauly D and Morgan, G. R. (Eds) Length-based methods in fisheries research. ICLARM Conference proceedings 13. ICLARM. Manila, Philippines and Kisr, Safat, Kuwait. P. 468.

Ricker WE (1975). Computation and interpretation of biological statistics of fish populations. Bull. Fish. Res. Board Can. 191:382.

Rosenberg AA, Beddington JR, Basson M (1986). The growth and longevity of krill during the first decade of pelagic whaling. Nature 324:152-154.

Shepherd JG (1987). A weakly parametric method for the analysis of length compositiondata. Pages 113-120 in Pauly, D. and Morgan, G. R. (Eds) Length-based methods in fisheries research. ICLARM, Manila, Philippines and KISR, Safat, Kuwait. P. 468.

Sivashanthini K, Khan K, Ajmal S (2004). Population dynamics of silver biddy *Gerres setifer* (pisces: perciformes) in the Parangipettai waters, Southeast coast of India. Indian J. Mar. Sci. 33(4).

Talwar PK, Jhingran AG (1992). Inland fishes of India. Rec. Ind. J. 3:19-24.

Tesch FW (1968). Age and growth. In: Methods for the Assessment of Fish Production in Freshwater, W. R. Ricker (Ed.), IBP Hand book. 3:98-130.

Thomas JK, William EP, Waters DS (2006). Age, Growth and Mortality of introduced flathead catfish in Atlantic rivers and a review of other populations. North Am. J. Fish. Manage. 26:73-87.

Umesh CG, Nripendra NS, Mahadev C (1996). Studies on the relative condition factor (Kn) in *Clarias batrachus* (Linn), an endemic catfish of Assam from the Brahmaputra River system. Indian J. Fish. 43(4):355-360.

Von Bertalanffy L (1938). A quantitative theory of organic growth. Hum. Biol. 10:181-213.

Yassar S, Chakraborty SK, Jaiswar AK, Panda D (2009). Age, growth, mortality and stock assessment of *Priacanthus hamrurs* from Mumbai waters. J. Mar. Biol. Ass. India 51(2):184-188.

Phytoremediation of polluted water by trees: A review

Muhammad Luqman[1], Tahir Munir Butt[2], Ayub Tanvir[3], Muhammad Atiq[4], Muhammad Zakaria Yousuf Hussan[5] , Muhammad Yaseen[1]

[1]University College of Agriculture, University of Sargodha, Pakistan.
[2]University of Agriculture, Faisalabad, Sub-Campus Toba Tek Singh-Pakistan.
[3]Department of Forestry, University of Agriculture, Faisalabad-Pakistan.
[4]Department of Plant Pathology, University of Agriculture, Faisalabad-Pakistan.
[5]Department of Agriculture, Government of the Punjab, Pakistan.

Presence of heavy metals and other pollutants in the aquatic systems has become a serious problem in many developing countries for environmental scientists and also for agencies engaged in environmental production. In this regard, there has been a great deal of attention given to new technologies for removal of heavy metals from contaminated water because conventional technologies to provide safe and clean water to living beings are not so far implemented. In this manner, the use of plants to remove heavy metals and other pollutants known as "phytoremediation" from the water is relatively cheaper as compared to other expensive engineering operations as plants remove pollutants from water and render them harmless. Five main subgroups of phytoremediation have been identified by the environmental scientists as "Phytoextraction, Phytodegradation, Rhizofiltration, Phytostabilisation and Phytovolatilisation". The identification and selection of plants that are suitable for successful remediation of water pollution is a matter of great concern. It is recommended that plants that have long and extensive root system should be planted at sites which are polluted due to industrial and sewage water.

Key words: Water pollution, heavy metals, phytoremediation.

INTRODUCTION

Water availability and pollution

Water is an essential element for life and is considered as most important and beneficial natural resource. According to an estimate, about 70% of the earth's surface is covered by water, approximately 97.5% of that amount is in the oceans and generally not available for daily use. Major portion of the remaining 2.5% is found in icecaps present in the Polar Regions or mountain peak and is similarly unavailable. Less than 1% of the earth's water is fresh water on the land surface, as groundwater, in the atmosphere and of this amount, only eight 10,000 of 1% is both readily available and renewable in lakes and streams for use by the earth's population (ODI, 2002). Some research studies reported that only a small percentage (0.01%) of the fresh water is only available for human use (Hinrichsen and Tacio, 2002). While this water volume remains generally constant, the population using this water continues to rise, stressing this supply more critically each year (USDA, 2000). The above mentioned water crises and availability of safe and fresh water becomes a greatest challenge for development agencies in the global world because all the ground water gets polluted due to rapid urbanization and industrialization revolution in the developed and developing world (World Bank, 1998). It has been reported by the press release of UNO Secretary General on world water day 2002 that about 1.1 billion people lack access to safe drinking water, 2.5 billion people have no access to proper sanitation, and more than 5 million people die each year from water-related diseases.

IMPACTS OF WATER POLLUTION

The extent of anthropogenic environmental pollution in the developing world is well documented (Mattina et al., 2003). Among overall environmental pollution, water pollution is one of the major threat to public health especially in developing and under developed countries as drinking water quality in these countries is poorly managed and monitored (Mwegoha, 2008; Azizullah et al., 2011). Both surface and ground drinking water get contaminated with coli forms, toxic metals and pesticides. About 2.3 billion peoples are suffering from water related diseases worldwide (UNESCO, 2003). The presence of heavy metals (elements with an atomic density greater than 6 g/cm^3) is one of the most persistent pollutants present in water. Unlike other pollutants, they are difficult to degrade, but can accumulate throughout the food chain, producing potential human health risks and ecological disturbances (Akpor and Muchie, 2010). In developing countries, more than 2.2 million people die every year due to drinking of contaminated water and inadequate sanitation (WHO and UNISEF, 2000). In general, water pollution has served impacts on the quality of fresh water and aquatic system. Water pollution also has negative impacts on food production, heath and social development and economic activities. Poor quality of surface and groundwater has become a threat to supplies of drinking water throughout the world (World Bank, 1998). In general, the decreasing availability of safe and healthy drinking water due to pollution, in terms of quality and quantity has been a major health concern in South Asia.

FACTORS RESPONSIBLE FOR WATER POLLUTION

There are so many factors which are responsible for water pollution, but it is most often due to human activities. Increasing population, geological factors, rapid urbanization, agricultural developments, global markets, industrial development, industrialization and poor wastewater regulation have affected the quantity and the quality of water (Saleem, 2001; Farooq et al., 2006). Besides the indiscriminate disposal of industrial, municipal and domestic wastes in water channels, rivers, streams and lakes etc. are regarded as the documented source of water pollution (Kahlown and Majeed, 2003). Kampa et al. (2001) reported that untreated domestic

waste, discharges from industries, rapid deforestation and poor agricultural practices result in the soil erosion and leaching down of nutrients, pesticides and insecticides. An estimated 2 million tons of sewage and other effluents are discharged into the world's waters every day. In developing countries, the situation is worse where over 90% of raw sewage and 70% of untreated industrial wastes are dumped into surface water sources (Anonymous, 2010). Rapid industrialization in urban and Peri-urban areas and high living standards are mainly responsible for discharge of wastewater in the rivers and streams (Minareci et al., 2009). Other sources of water pollution are sewage and waste water, marine dumping, industrial waste, radioactive waste, oil pollution, underground storage leakages, atmospheric deposition, global warming and eutrophication. The Global Environmental Monitoring System (GEMS) of the United Nations Environmental Program (UNEP) have reported heavy pollution in several rivers around the World (Bichi and Anyata, 1999).

PHYTOREMEDIATION OF WATER POLLUTION

It is universally accepted that trees as a suitable vegetation cover improve the quality of life as they absorb dangerous pollutants from the environment (Aronsson and Perttu, 1994; Glimmerveen, 1996; Beckett et al., 1998; EPA, 2000). Literature shows that a healthy, well managed forest can provide many ecological benefits (Yang et al., 2005). If water flows quickly over the surface of land, many of the pollutants present on the surface, the run-off carries will reach the main body of water. If the water flows more slowly due to the presence of vegetation on land more of the pollutants will be filtered out, either by adhering to plants and soil, or by being absorbed through the root systems of plants. Trees act as water filters and improve water quality. They utilize waste water and uptake heavy metals due to their extensive root system (Bose et al., 2008).

Trees have been suggested as a low cost, sustainable and ecological sound solution to the remediation of heavy metals contaminated water as trees uptake of these metals and dangerous pollutants from soil and water. The main characteristics of trees are to make them suitable for phytoremediation by their large biomass both below and above ground (Ghosh and Singh, 2005; Coder, 1996). Salt et al. (1998) described this process to remove pollutants from environment including natural aquatic system as phytoremediation. Five main subgroups of phytoremediation have been identified:

(1) Phytoextraction: Plants remove heavy metals and other pollutants from the soil as well as groundwater and concentrate them into their harvestable parts (Kumar et al., 1995).
(2) Phytodegradation: Plants and associated microbes

degrade organic pollutants (Burken and Schnoor, 1997).
(3) Rhizofiltration: Plant roots absorb metals from waste streams (Dushenkov et al., 1995).
(4) Phytostabilisation: Plants reduce the mobility and bioavailability of pollutants in the environment either by immobilization or by prevention of migration (Vangronsveld et al., 1995).
(5) Phytovolatilisation: Volatilization of pollutants into the atmosphere via plants (Burken and Schnoor, 1999; Banuelos et al., 1997).

Plantation and vegetation can filter and immobilize sediment and other water contaminants, such as fertilizer and pesticide run-off, reducing water pollution (Schnoor, 2002). It has long been recognized that natural lands such as forests, parks and wetlands can help to slow and filter the water before it gets to rivers, reservoirs or aquifers, keeping those drinking water sources cleaner and making treatment cheaper (Crompton, 2008). Some woody species have the capacity to accumulate heavy metals as pollutants present in the ground water (Unterbrunner et al., 2007). A study of 27 water suppliers found that water treatment costs for utilities using primarily surface water supplies varied depending on the amount of forest cover in the watershed. For every 10% increase in forest cover in the source area (up to about 60% forest cover), treatment and chemical costs decreased by approximately 20%. Approximately 50 to 55% of the variation in operating treatment costs could be explained by the percent of forest cover in the source area (Ernst et al., 2007).

Plants, especially woody plants, are very good at removing nutrients (nitrates and phosphates) and contaminates (such as metals, pesticides, solvents, oils and hydrocarbons) from soil and water. These pollutants are either used for growth (nutrients) or are stored in wood. In one study, a single sugar maple growing roadside removed a considerable quantity of Cadmium, Chromium, Nickel, and Lead in a single growing season. Studies in Maryland showed reductions of up to 88% of Nitrate and 76% of Phosphorus after agricultural run-off passed through a forest buffer (Cotron n.d.). Natural forests and planted trees play an important role in protecting water quality as pointed out by many engineers, planners and community leaders as forests are the most beneficial land use for protecting water quality, due to their ability to capture, filter, and retain water (Singh et al., 2010).

Forests are also essential to the provision of clean drinking water to over 10 million residents of the watershed and provide valuable ecological services and economic benefits including carbon sequestration, flood control, wildlife habitat, and forest products. Another research study shows that trees play a crucial role in protecting water quality. Leaves and needles break the force of rain, slowing the movement of water and reducing water pollution, run-off and flooding (Kuchelmeister, 2000). Keeping in view the importance of natural and planted vegetation to remediate and restoration of hazardous polluted water due to extensive anthropogenic activities also known as phytoremediation, has gained increasing attention to environmental scientists as it is cost effective and non-intrusive means of remediation from contaminated ground water (Ouyang, 2002). It is an emerging natural and environmental friendly technology that can be considered for remediation of contaminated groundwater because of its aesthetic advantages, and long-term applicability (Chaney et al., 2005; Huang et al., 2004; Susarla et al., 2002; Pivetz, 2001). There are several advantages of phytoremediation, some of them are reported by Morikawa and Erkin (2003) as (1) it is an aesthetically gratifying, solar-energy motivated cleanup technology; (2) there is minimal environmental distraction and in situ treatment conserve earth; (3) it is most useful at sites with low levels of contamination; (4) it is useful for treating a broad range of environmental contaminants; and (5) it is inexpensive (60 to 80% or even less costly) than conventional physicochemical and other conventional methods (Schnoor, 1997). The use of natural and artificial planted vegetation as phytoremediation of polluted has its limitations. It is a time consuming process, and it may take at least several growing seasons to cleanup a site. Plants that absorb toxic heavy metals or persistent chemicals may pose a risk to wildlife and contaminate the food chain (Mwegoha, 2008).

In this way, the potential use and selection of suitable plant species for phytoremediation research and implementation is one of the challenges that need to be met and a pre-requisite for successful phytoremediation research. Phytoremediation of different types of contaminants requires different general plant characteristics for optimum effectiveness. Aquatic plants for example, duckweed and pennywort, also Brassica and sunflower remove contaminants like metals, radionuclide's, hydrophobic organics from groundwater. The cultivation of Dalbergia sissoo as woody species may be extended to industrial and urban areas where industrial and municipal wastewater is the only source of irrigation (Farooq et al., 2006). On the other hand, Popular and Willow trees remove inorganic, nutrients, and other chlorinated solvents present in the groundwater (Schooner, 2002). A special characteristic of Willow, which makes it a very suitable tree for use in phytoremediation, is that it can be frequently harvested by coppicing, yielding as much as 10 to 15 dry t ha^{-1} year^{-1} (Riddell-Black, 1993; Punshon et al., 1995; Pulford and Watson, 2003). The concentration of heavy metal pollutants in the bark and wood of 20 different Willow varieties were determined by Pulford et al. (2002). Wetland plants generally are not "hyperaccumulator", they store metals in the below ground organ than above ground organ (Weis and Weis, 2004).

CONCLUSION AND RECOMMENDATIONS

There is a growing demand of groundwater for drinking and an irrigation purpose since it is the most readily available low cost source of water supplies to low income countries. The problem of water pollution as a result of contamination of groundwater is constantly increasing especially in developing and low income countries due to the fact that there are limited financial and technological resources to remediate polluted water sources. In this situation, the use of trees to remediate polluted water is considered as the new emerging technology which is relatively cheaper than the conventional technologies. The technology of Phytoremediation offers viable solution to water pollution. It offers restoration of sites, limited decontamination, preservation of the biological activity and physical structure of soils, and is potentially cheap, visually inconspicuous. It is also reported that trees can withstand good in heavy metal contamination than agricultural crops. The critical point in this technology is the selection of appropriate plant species that is suitable in the prevailing environmental conditions. The emphasis is given on the plantation of terrestrial plants than aquatic plants due to their larger root system. It is recommended that there must be multi disciplinary calls for collaboration between universities, research institutes and other environmental protection agencies to create voluntary teams to address questions like agronomic practices needed for successful establishment of flora; identification of locally available plant species for specific remediation requirements and expansion of these plant species at local and national level.

REFERENCES

Akpor OB, Muchie M (2010). Remediation of heavy metals in drinking water and wastewater treatment systems: Processes and applications. Int. J. Phys. Sci. 5(12):1807-1817.

Anonymous (2010). World Water Day 22.03.2010. United Nations.

Aronsson P, Perttu K (1994). Willow vegetation filters for municipal wastewaters and sludges: A biological purification system. Uppsala: Swedish University of Agricultural Science, P. 230.

Azizullah A, Khattak MN, Richter P, Hader DP (2011). Water pollution in Pakistan and its impact on public health- A review. Environ. Int. 37(02):479-97.

Banuelos GS, Ajwa HA, Mackey LL, Wu C, Cook S, Akohoue S (1997). Evaluation of different plant species used for phytoremediation of high soil selenium. J. Environ. Qual. 26:639-646.

Beckett KP, Freer-Smith PH, Taylor G (1998). Urban Woodlands: Their role in reducing the effects of particulate pollution. Environ. Pollut. 99:347-360.

Bichi MH, Anyata BU (1999). Industrial waste pollution in the Kano River Basin. Environ. Manag. Health 10:112-116.

Bose S, Vedamati J, Rai V, Ramanathan AL (2008). Metal uptake and transport by Tyaha angustata L. grown on metal contaminated waste amended soil: An implication of phytoremediation. Geoderma 145:136-142.

Burken JG, Schnoor JL (1997). Uptake and metabolism of atrazine by poplar trees. Environ. Sci. Technol. 31:1399-406.

Burken JG, Schnoor JL (1999). Distribution and volatilisation of organic compounds following uptake by hybrid poplar trees. Int. J. Phytoremediat. 1:139-51.

Chaney RL, Angle JS, McIntosh MS, Reeves RD, Li YM, Brewer EP, Chen KY, Roseberg RJ, Perner H, Synkowski EC, Broadhurst CL, Wang S, Baker AJ (2005). Using hyperaccumulator plants to phytoextract soil Ni and Cd. Z. Naturforsch [C] 60:190-198.

Coder RD (1996). Identified Benefits of Community Trees and Forests. University of Georgia.

Crompton JL (2008). Empirical Evidence of the Contributions of Park and Conservation Lands to Environmental Sustainability: The Key to Repositioning the Parks Field. World Leisure No. 3.

Dushenkov V, Kumar PBAN, Motto H, Raskin I (1995). Rhizofiltration: The use of plants to remove heavy metals from aqueous streams. Environ. Sci. Technol. 29:1239-45.

EPA (2000). Introduction to phytoremediation. Washington: U.S. Environmental Protection Agency. EPA/600/R-99/107.

Ernst C, Gullick R, Nixon K (2007). Protecting the source: Conserving forests to protect water. In C.T.F. de Brun (editor). The economic benefits of land conservation. San Francisco: Trust Public Land, pp. 24-27.

Farooq H, Siddiqui MT, Farooq M, Qadir E, Hussain Z (2006). Growth, Nutrient Homeostatis and Heavy Metal Accumulation in Azadirachta indica and Dalbrgia sissoo Seedlings Raised from Waste Water. Int. J. Agric. Biol. 8(4):504-507.

Ghosh M, Singh SP (2005). A review on phytoremediation of heavy metals and utilization of its byproducts. Appl. Ecol. Environ. Res. 3(1):1-18.

Glimmerveen I (1996). Heavy metals and trees. Edinburgh: Institute of Chartered Foresters, P. 206.

Hinrichsen D, Tacio H (2002). The coming freshwater crisis is already here: The linkages between population and water. Washington, DC: Woodrow Wilson International for Scholars.

Huang XD, El-Alawi Y, Penrose DM, Glick BR, Greenberg BM. (2004). A multi-process phytoremediation system for removal of polycyclic aromatic hydrocarbons from contaminated soils. Environ. Pollut. 130:465-476.

Kahlown MA, Majeed A (2003). Water-resources situation in Pakistan: challenges and future strategies. Water resources in the south: Present scenario and future prospects in Islamabad, Pakistan. Comm. Sci. Technol. Sustain. Dev. South (COMSATS) pp. 21–39.

Kampa E, Choudhury K, Kraemer RA (2001). Protecting water resources: Pollution Prevention, Thematic background paper. International conference on Fresh Water, December, 2001, Bonn.

Kuchelmeister G (2000). Trees for the urban millennium: Urban forestry update. Unasylva, 200(51). Germany.

Kumar PBAN, Dushenkov V, Motto H, Rasakin I (1995). Phytoextraction: The use of plants to remove heavy metals from soils. Environ. Sci. Technol. 29:1232-1238.

Mattina MJI, Berger WL, Musante C, White C (2003). Concurrent plant uptake of heavy metals and persistent organic pollutants from soil. Environ. Pollut. 124:375-378.

Minareci O, Ozturk M, Egemen O, Minareci E (2009). Detergent and phosphate pollution in Gediz River, Turkey. A. J. Biotechnol. 8(15):3568-3575.

Morikawa, H, Erkin OC (2003). Basic processes in phytoremediation and some application to air pollution control. Chemosphere 52:1553-1558.

Mwegoha WJS (2008). The use of phytoremediation technology for abatement soil and groundwater pollution in Tanzania: opportunities and challenges. J. Sustain. Dev. Afr. 10(01):140-156.

ODI (2002). The water crises: Faultiness in Global debates.ODI Briefing paper.

Ouyang Y (2002). Phytoremediation: Modeling plant uptake and contaminant transport in the soil-plant-atmosphere continuum. J. Hydro. 266:66-82.

Pivetz RE (2001). Phytoremediation of Contaminated Soil and Ground Water at Hazardous Waste Sites. EP A/540/S-0 1/500.

Pulford ID, Watson C (2003). Phytoremediation of heavy metal-contaminated land by trees - A review. Environ. Int. 29:529-540.

Pulford ID, Riddell-Black D, Stewart C (2002). Heavy metal uptake by willow clones from sewage sludge-treated soil: The potential for phytoremediation. Int. J. Phytoremediat. 4:59-72.

Punshon T, Lepp NW, Dickinson NM (1995). Resistance to copper toxicity in some British Willows. J. Geochem. Explor. 52:259-66.

Riddell-Black D (1993). A review of the potential for the use of trees in the rehabilitation of contaminated land. WRC Report CO 3467. Water Research Centre, Medmenham.

Saleem MA (2001). Industrialization and Water Pollution. Daily Dawn, Lahore. 15th April, 2001. Page Agriculture and Technology.

Salt DE, Smith RD, Raskin I (1998). Phytoremediation. Annu. Rev. Plant Physiol. 49:643-68.

Schnoor JL (1997). Phytoremediation: Technology Evaluation Report, GWRTAC Series TE-98-0 1.

Schnoor JL (2002). Technology Evaluation Report: Phytoremediation of Soil and Groundwater. GWRTAC Series TE-02-01.

Singh G, Bhati, M, Rathod T (2010). Use of tree seedlings for the phytoremediation of a municipal effluent used in dry areas of north-western India: Plant growth and nutrient uptake. Ecol. Eng. 36:1299-1306.

Susarla S, Medina VF, McCutcheon SC (2002). Phytoremediation: An ecological solution to organic chemical contamination. Ecol. Eng. 18:647-658.

UNESCO (2003). Water for people water for life. The United Nations World Water Development Report. United Nations Educational, Scientific and Cultural Organization (UNESCO) and Berghahn Books.

Unterbrunner R, Puschenreiter M, Sommer P, Wieshammer G, Tlustos P, Zupan M, Wenzel WW (2007). Heavy metal accumulation in trees growing on contaminated sites in Central Europe. Environ Poll. 148:107-114.

USDA Forest Service (2000). Water and the Forest Service. FS-660. Washington DC, P. 26.

Vangronsveld J, van AF, Clijsters H (1995). Reclamation of a bare industrial area contaminated by non-ferrous metals: *In situ* metal immobilization and revegetation. Environ. Pollut. 87:51-9.

Weis SJ, Weis P (2004). Metal uptake, transport and release by wetland plants: Implications for phytoremediation and restoration. Environ. Int. 30:685-700.

WHO, UNICEF (2000). Global water supply and sanitation assessment 2000 Report. USA: World Health Organization and United Nations Children's Fund.

World Bank (1998). World Resources 1998-99, New York, Oxford, Oxford University Press.

Yang J, McBride J, Zhou J, Sun Z (2005). The urban forest in Beijing and its role in air pollution reduction. Urban For. Urban Greening 3:65-78.

4

Distribution and size structure of comb jellyfish, *Mnemiopsis leidyi* (Ctenophora) in the southwestern Caspian Sea

Siamak Bagheri[1,2]*, Mashhor Mansor[2] , Azemat Ghandi[1] and Esmaeil Yosefzad[1]

[1]Inland Waters Aquaculture Institute, Iranian Fisheries Research Organization, 66 Anzali, Iran.
[2]School of Biological Sciences, Universiti Sains Malaysia, 11800 USM, Penang, Malaysia.

Temporal and vertical distributions of *Mnemiopsis leidyi* were studied along three transects namely Lisar, Anzali and Sephidrood in the southwestern Caspian Sea during 2001 to 2010. The maximum lengths of the ctenophore were 60 to 70 mm, and bulk of individuals (90.6%) were <5 mm in length. The means of abundance and biomass during the whole period were 3032 ind/m^2 and 293.54 g/m^2, respectively. The highest average abundance value (1017 ind/m^2) was measured in summer 2006 and the lowest abundance value (54 ind/m^2) was in spring 2010. The average biomass of *M. leidyi* ranged between 1175.40 and 0.85 g/m^2 in summer 2005 and winter 2008, respectively. The highest annual of abundance and biomass were obtained in 2005 and 2006. The dominant of *M. leidyi* bulk occurred above 20 m depth, whilst the ctenophore population sharply decreased below 20 m. The seasonal pattern of *M. leidyi* is related to water temperature, as evidenced from the positive correlation between the water temperature and *M. leidyi* population.

Key words: *Mnemiopsis leidyi*, abundance, biomass, size, Caspian Sea.

INTRODUCTION

The Caspian Sea is the largest inland water body on earth; it is located at the far end of southeastern Europe, bordering Asia (Kosarev and Yablonskaya, 1994). Approximately, 130 rivers with various sizes drain into the Caspian Sea. With an average annual input of about 300 km^3, the river Volga contributes up to 82% of the inflow (Dumont, 1998). The south of Caspian Sea receives more than 100 rivers, the Sephidrood is the largest river with 67,000 km^2 catchment area and discharge of 4,037 million m^3 (Lahijani et al., 2008; Bagheri et al., 2012a). The Caspian Sea is known for its traditional sturgeon fishery and in particular the caviar industries, it also supports a large scale pelagic fishery that is mainly made up of three small pelagic fish species of the kilka (*Clupeonella cultriventris*, *C. engrauliformis* and *C. grimmi*); the biological diversity of the Caspian Sea and its coastal zone makes the region one of the most valuable cosystems in the world (Shiganova, 2002). The native habitats of the ctenophore, *M. leidyi*, are temperate to subtropical estuaries along the Atlantic coast of North and South America. In the early 1980s, it was accidentally introduced in the Black Sea, where it flourished and expanded into the Azov, Marmara,Mediterranean seas (Purcell et al., 2001). The possibility of *M. leidyi* introduction into other sensitive, neighboring ecosystems, notably the Caspian Sea, had been mentioned during the GESAMP meeting in 1994. As expected, this ctenophore was reported to be present in the Caspian Sea by November 1999 (Esmaeili et al., 1999; Ivanov et al., 2000). Ivanov et al. (2000) suggested that this ctenophore was transported via ballast water

*Corresponding author. E-mail: siamakbp@gmail.com

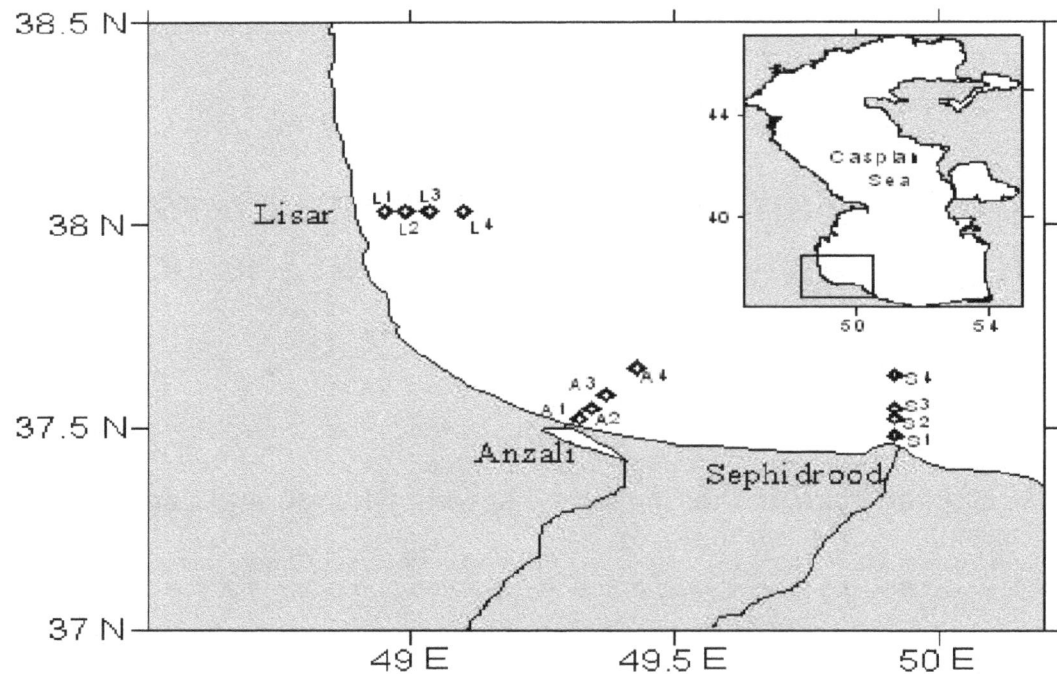

Figure 1. Sampling transects in the southwestern Caspian Sea during 2001 to 2010.

taken aboard either in the Black Sea or the Sea of Azov (where *M. leidyi* occurs in the warm months) and released after ballast-loaded ships passed through the Volga Don Canal and the shallow freshwater areas of the northern Caspian Sea into the saltier central or southern Caspian waters. This species invasion was the start of one of the most important anthropogenic problems the Caspian Sea ecosystem has ever experienced (Bagheri et al., 2012a).

Following the introduction, a few basin-wide surveys were undertaken to understand the distribution of *M. leidyi* in the Caspian Sea (Shiganova et al., 2001a; Kideys, 2002). Investigations of the new invader to the Caspian Sea in 2000 to 2001 showed that it was found almost everywhere, including the northwestern Caspian, where salinity exceeded 4 ppt (Shiganova et al., 2003).

There was an increasing trend in the abundance of *M. leidyi* in 2001 compared to 2000. The average and maximum biomass of *M. leidyi* over the entire middle and southern Caspian Sea were as high as 120 and 351 g wet weight (WW) m^{-2}, respectively, compared to a mean value of 60 g/m^2 in the summer of 2000 (Shiganova et al., 2001a; Kideys and Moghim, 2003). Furthermore, Roohi et al. (2008, 2010) documented the seasonal fluctuation of *M. leidyi* in the southern Caspian Sea from 2001 to 2006. The authors concluded that the total number and biomass of *M. leidyi* decreased to a certain extent in the years after 2003. Recently, Kideys et al. (2008) and Roohi et al. (2010) reported, the introduction *M. leidyi* have played important role in the variation of phytoplankton composition and decrease of zooplankton population after 2000.

Although a few *M. leidyi* studies have been conducted on the Caspian Sea in recent years (Ivanov et al., 2000; Shiganova et al., 2001a, b; Kideys and Moghim, 2003; Roohi et al., 2008, 2010), there is no adequate survey on the *M. leidyi* community in the southwestern Caspian Sea and at present, only a few publications on the *M. leidyi* are found (Bagheri and Kideys, 2003; Bagheri and Sabkara, 2003; Bagheri, 2006; Bagheri et al., 2010,2012a) for this region. This study intended to evaluate the distributions of *M. leidyi* abundance and biomass, and size structure in the southwestern Caspian Sea during July 2001 to October 2010, furthermore this study also intended to evaluate interactions between *M. leidyi* density, size structure and water parameters in the southwestern Caspian Sea.

MATERIALS AND METHODS

In this study, spatial and temporal distributions of *M. leidyi* were studied along three transects namely Lisar, Anzali and Sephidrood in the southwestern of Caspian Sea during 2001 to 2010; each transect had four stations located at depth 5 m (L1, A1 and S1), 10 m (L2, A2 and S2), 20 m (L3, A3 and S3) and 50 m (L4, A4 and S4) contours (Figure 1). The distance between stations differed among transects (Table 1).

Sample collections at all transects were accomplished on the same day using a speed boat. Sampling was planned at monthly intervals, however, due to certain metrological problems, no sampling could be done for some months during the period of 2001 to 2010, as shown in Table 2.

Table 1. Sampling region and stations in the southwestern Caspian Sea during 2001 to 2010.

Region	Station	Depth (m)	Latitude	Longitude	Distance from shore (km)
Lisar	L1	5	48° 51' 42"	38° 02' 21"	2
	L2	10	48° 58' 30"	38° 04' 51"	9
	L3	20	49° 04' 21"	38° 03' 40"	16
	L4	50	49° 11' 30"	37° 59' 34"	26
Anzali	A1	5	49° 29' 31"	37° 29' 00"	1
	A2	10	49° 28' 59"	37° 29' 20"	3
	A3	20	49° 29' 43"	37° 30' 30"	6
	A4	50	49° 28' 37"	37° 35' 07"	15
Sephidrood	S1	5	49° 56' 00"	37° 28' 08"	2
	S2	10	49° 55' 20"	37° 29' 42"	4
	S3	20	49° 54' 59"	37° 30' 31"	6
	S4	50	49° 55' 16"	37° 31' 29"	10

Table 2. Sampling frequency in the southwestern Caspian Sea during 2001-2010.

Year	Jan	Feb	Mar	Apr	May	Jun	Jul	Aug	Sep	Oct	Nov	Dec
2001							*	*	*	*	*	*
2002	*			*	*	*	*	*	*			
2003	*			*		*	*	*		*	*	*
2004		*		*	*		*		*	*	*	
2005		*		*			*	*	*		*	*
2006		*		*			*	*	*	*		*
2008		*		*			*				*	
2009							*				*	
2010	*		*							*		

Water samples were collected using a 1.71 L Nansen water sampler (Hydro–Bios, Germany; TPN; Transparent Plastic Nansen water sampler, No: 436201), and water temperature levels of the seawater at 5, 10, 20 and 50 m depth were measured in situ by using a reverse thermometer (Hydro–Bios, TPN). Salinity was estimated using a salinometer (Beckman; RS-7B, U.S. Patent, No: 2542057). At each station, water transparency was measured with a Secchi disk at each depth. M. leidyi populations were sampled using a 500 µm mesh sized closing plankton net having a diameter of 50 cm with a large cod-end (volume 1000 ml) suitable for the collection of ctenophores (Kideys et al., 2001). Samples were obtained via vertical towing from the bottom to the surface for all stations except the deepest stations.

At the deepest stations, due to the existence of thermocline, two vertical wings were carried out: from 50 m to 20 m and from 20 m to the surface. At the end of each tow, the net was washed from the exterior, and the cod end was passed into a container immediately to enumerate ctenophores by naked eye. The abundance of M. leidyi per unit area (that is, m²) was calculated from the diameter of the net and the tow depth. The ctenophores were sorted based on their length groups of 0 to 5, 6 to 10, 11 to 15 mm and so on for determining the abundance of different size groups. A total of 179,000 individuals were measured and grouped in this way. Individual weighting of these animals was not practical at sea.

Therefore, weights of these animals were calculated from an equation which was obtained from individual lengths (with lobe using a ruler) and weight measurements (using a digital balance with a sensitivity of 0.001 g) of 269 individuals obtained from the different depths in July 2001: $W = 0.0013 \times L^{2.33}$ where W is wet weight of M. leidyi in g and L is the length in mm (Bagheri and Kideys, 2003).

Statistical comparisons between months were made by using statistical (Statsoft) software SPSS version 15 for Windows. Analysis of variance comparisons (One-way ANOVA) for water parameters and nonparametric test (Kruskal–Wallis) for M. leidyi abundance and biomass were used to identify the importance of variables between different seasons and years. Spearman rank correlation coefficients (r) were calculated to evaluate the relationships between M. leidyi number biomass and water parameters.

RESULTS

No differences were noted in the spatial distribution of M. leidyi and water parameters among the three transects of Lisar, Anzali and Sephidrood during 2001 to 2010

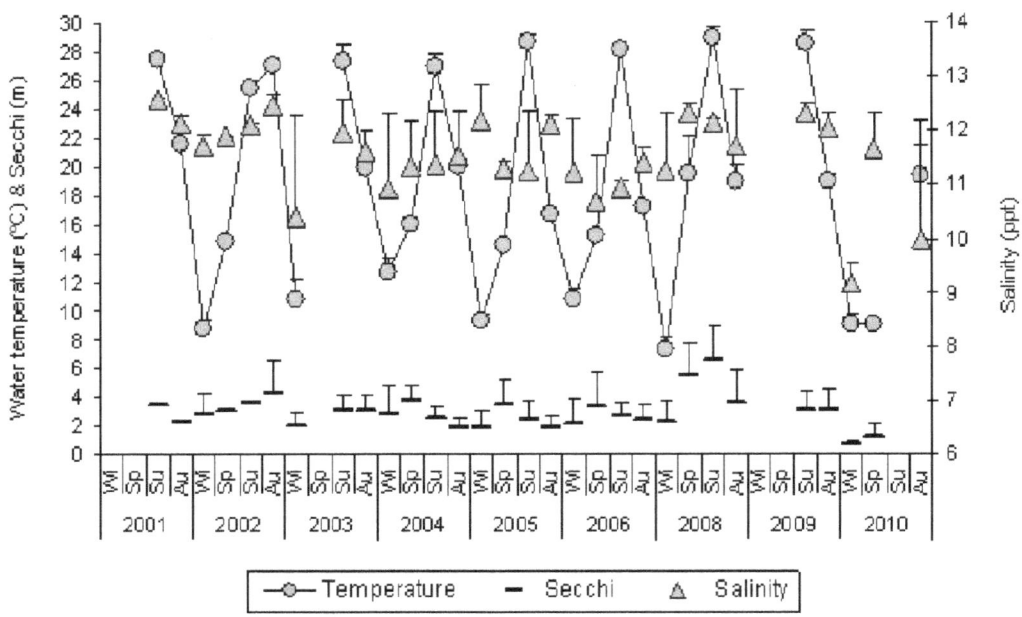

Figure 2. The average (±SD) surface water temperature, salinity and secchi depth at three transects and all stations in the southwestern Caspian Sea during 2001-2010. Wi, Winter; Sp, spring; Su, summer; Au, autumn.

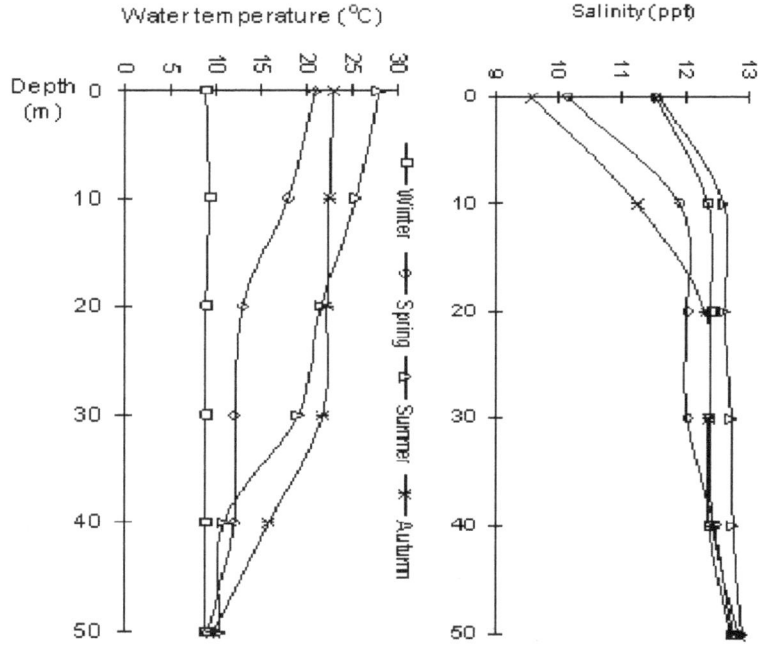

Figure 3. Vertical profiles of water temperature and salinity at Anzali transect (50 m depth) in the southwestern Caspian Sea in 2004.

(abundance: Df=2, p = 0. 285; biomass: Df= 2, p = 0.324. Therefore, the data of the three transects were combined per year and season. The sea-surface temperature distribution during the study period is showin Figure 2. The sea-surface temperature ranged between 7.4 and 29.0°C during 2001 to 2010. Differences in temperature

variations among the years were not significant (p > 0.05) while there was a significant difference among seasons (p < 0.05). Seasonal stratification patterns were similar throughout the study period as shown in Figure 3. Here, different seasons in 2004 were chosen as representative (Figure 3).The thermocline started to form in summer and

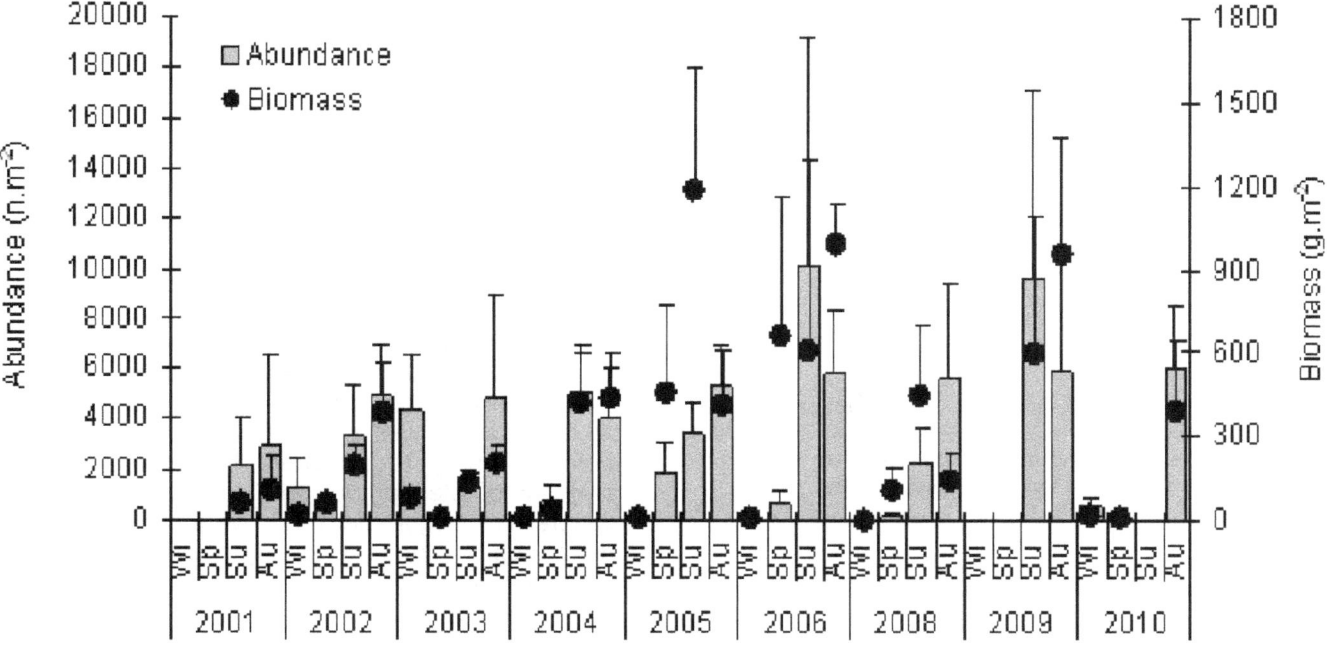

Figure 4. The average (±SD) of *M. leidyi* abundance and biomass at three transects and all stations in the southwestern Caspian Sea during 2001-2010. Wi, winter; Sp, spring; Su, summer; Au, autumn).

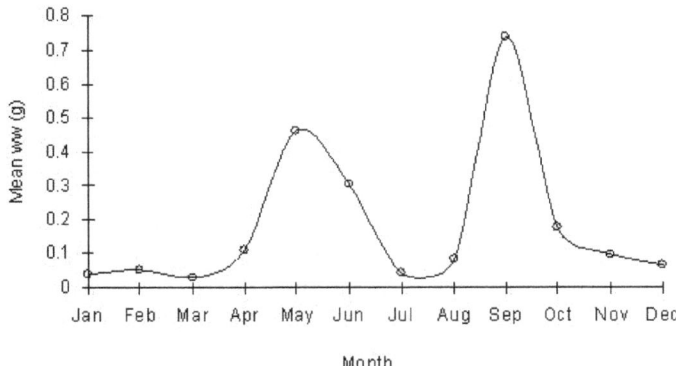

Figure 5. *M. leidyi* mean wet weight in the southwestern Caspian Sea during 2001-2010.

autumn at around 30 m depth. The sharpest thermocline was observed in summer. The halocline present all year round is sharpest in winter. The average salinity of all surface stations (10.68 ± 1.01 ppt) was lower than at 50 m depth (12.79 ± 0.07 ppt; Figure 3). There were increasing and decreasing trend in the average surface salinities from winter to spring (9.1 to 10.7 ppt) and summer to autumn (10.9 to 9.9 ppt), respectively (Figure 2). These variations could be related to fresh water input from rivers during different seasons. However, average salinities were not significantly different between seasons ($p > 0.05$), but there were meaningful difference between years ($p < 0.05$). The Secchi disk depth, being an indicator of water turbidity, changed from 0.7 to 6.6 m

during the study period with an overall average of 2.9 ± 1.7 m. The Secchi disk depth has fluctuations during 2001 and 2010 (Figure 2), and was significantly different only between years ($p < 0.05$). *M. leidyi* was present at all regions, depths and seasons studied. There was a seasonal succession of ctenophore densities every year, with the maximum being observed in summer and the minimum density in winter. A positive correlation was found between water temperature and abundance (r = 0.86) and biomass (r = 0.80) of *M. leidyi* ($p < 0.01$). There was no significant correlation between the biomass of *M. leidyi* and salinity (r = -0.14, $p > 0.05$). Figure 4 shows changes in the abundance and biomass of *M. leidyi*. The highest values of abundance and biomass were obtained in 2006 and 2005, respectively.Statistical nonparametric test (Kruskal–Wallis) showed that *M. leidyi* abundance and biomass were not significantly different between the years ($p > 0.05$), however, the difference between abundance and biomass *M. leidyi* were significant among the sampling seasons ($p < 0.05$). The highest abundance value (10117 ind/m^2) was measured in summer 2006 and the lowest abundance value (54 ind/m^2) was in spring 2010 (Figure 4). The biomass of *M. leidyi* ranged between 0.85 and 1175.4 g/m^2 during the study period in the southwestern Caspian Sea (Figure 4). The abundance and biomass values of ctenophore were low during winter and early spring, and gradually increased during summer and autumn (Figure 4). The means of abundance and biomass during the whole study period were 3032 ind/m^2 and 293.54 g/m^2, respectively. Minimum mean weight of specimens in the population

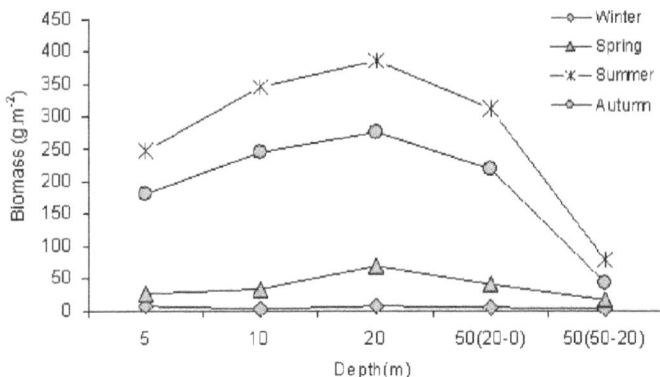

Figure 6. Temporal distribution of *M. leidyi* biomass in different seasons and depths in the southwestern Caspian Sea in 2004. Winter, spring, summer and autumn seasons were represented by February, May, July and November, respectively.

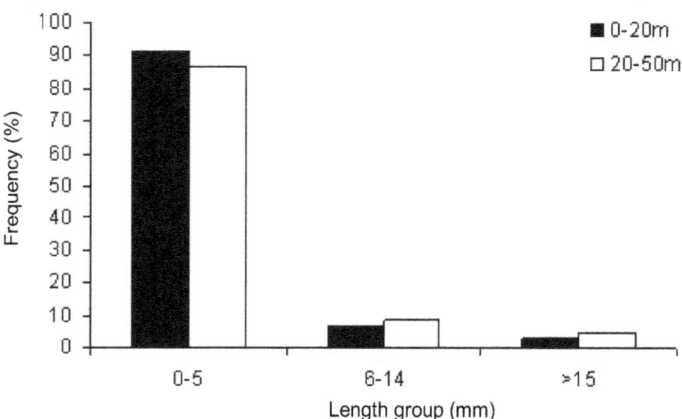

Figure 7. Comparison of size frequencies of the *M. leidyi* population from two layers in the southwestern Caspian Sea during 2001-2010.

was observed in March with a value of 0.03 g. There were two biomass increases after this period. Somatic growth appeared to take place from April to June and from August to September all years (Figure 5), when maximum increase in the weights of specimens were recorded (almost 15-fold). Seasonal stratification patterns were similar in all years and at all stations. Thus, year 2004 was chosen as representative to observe the vertical and seasonal distribution of *M. leidyi*. The highest *M. leidyi* biomass occurred at 20 m depth. The ctenophore population sharply decreased below 20 m depth. The lowest biomass values of *M. leidyi* were observed at the deeper layer (50 to 20 m depth) (Figure 6). *M. leidyi* biomass significantly varied vertically ($p < 0.05$).

During the entire study period, a total of 179,000 specimens were sampled and individually measured for their body length. The length-frequency distribution displayed that whilst 90.6% of the population belonged to the 0 to 5 mm group, only 6.4% were from 6 to 14 mm

length group (Figure 7). Thus, these larvae and juveniles made up 97% of the total population. The largest size that the ctenophore could attain in southwestern Caspian Sea was 60 to 70 mm, which was measured in September 2006. The length-frequency distributions of *M. leidyi* from 0 to 20 and 20 to 50 m depth are presented to understand the vertical distribution of ctenophores with respect to their size. In both layers, small ctenophores (< 5 mm) dominated the *M. leidyi* population (Figure 7), they comprised 91.16% (at 0 to 20 m depth) and 86.62% (at 20 to 50 m depth) of total abundance. However, ctenophores from the deeper layers had larger size compared to those from shallower depths. This indicates that mainly larger individuals penetrate through the thermocline to dwell in deeper waters. Despite the larger animals' ability to penetrate into deeper waters, the majority of ctenophores still remained at the surface waters. The larger specimens were observed more often during April to June (mid spring-early summer) compared to other months (Figure 8).

DISCUSSION

The variations in surface temperatures were between 7.4 and 29.0°C during 2001 to 2010 in the southwestern Caspian Sea (Figure 2). Dumont (1998), Kideys and Moghim (2003), Roohi et al. (2008), and Bagheri et al. (2010,2011,) reported that in the southern Caspian Sea, the maximum and minimum temperatures are 28 to 29 and 7 to 8°C in summer and winter, respectively. Kideys and Moghim (2003), Bagheri et al. (2012 a,b), and Zaker et al. (2007) noted that in the summer and autumn, thermocline was located between 20 and 40 m depths in the southern Caspian Sea; also they reported the depth of the mixed layer was not the same as in the Caspian. In this study, the thermocline started to form in summer and autumn at 30 m depth; the sharpest thermocline was observed in summer (Figure 3).

In additions, the variation of thermocline thickness could be related to meteorological monthly fluctuations during 2001 to 2010 in the southwestern Caspian Sea. Our findings are similar to Shiganova et al. (1998),Zaker et al. (2007) and Bagheri et al. (2012c) who reported stratification is dependent on weather conditions in the Black Sea and Caspian Sea.

There were increasing (winter-spring) and decreasing (summer-autumn) trends in the average surface salinities(Figure 2). This trend is related to fresh water inputs from rivers during seasons. Bagheri et al., (2010, 2011, 2012c) noted there was a strongly negative correlated between salinity and freshwater discharge via rivers in southwestern Caspian Sea. In additions, our findings are similar to the previous findings reported by Kosarev and Yablonskaya (1994), Dumont (1998), Purcell et al. (2001), Kideys and Moghim (2003), Zaker et al. (2007), and Roohi et al. (2008). The variations of fresh water input from Lisar and Sephidrood rivers as well as

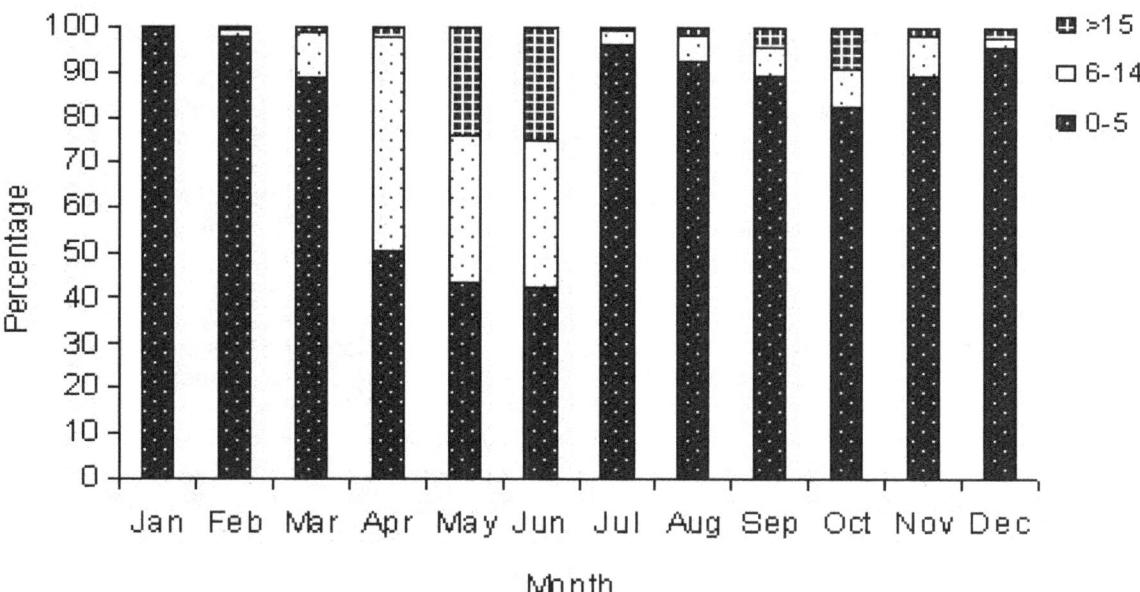

Figure 8. Percentage of *M. leidyi* size (mm) frequency in different months in the southwestern Caspian Sea during 2001-2010.

Anzali wetlands varied; the primary productivity in the southwestern Caspian, subsequently affected the Secchi depth readings in our study (Figure 2). Khodaparast (2006) and Bagheri et al. (2012b) reported that near the southwestern Caspian coasts, chlorophyll a levels increased from 0.56 to 1.34 µg/L in 1994 and 2.71 to 35.25 µg/L in 2006. Recently, Kideys et al. (2008) and Bagheri et al. (2012c) also noted that after 1999, satellite derived chlorophyll a levels gradually increased and reached extremely high levels of 9.26 µg/L in 2008.In the Black Sea, *M. leidyi* was usually found in the upper mixed layer, or in and above the seasonal thermocline, with only a few individuals found in deeper layers with low oxygen concentrations (Vinogradov et al., 1989; Bogdanova and Konsoulov, 1993; Mutlu, 1999; Kideys and Romanova, 2001). The reasons for the scarcity of ctenophores below the thermocline (of which the lower boundary is 25 to 50 m) in the Black Sea have been suggested to be low concentrations of food and temperature (Kideys and Romanova, 2001). Despite the fact that *M. leidyi* is known to display wide salinity and temperature tolerance (Kideys and Niermann, 1994). Bagheri and Kideys (2003), Kideys and Moghim (2003), Roohi et al. (2008) and Bagheri et al. (2012a) observed that *M. leidyi* is distributed generally above the thermocline in the Caspian Sea. *M. leidyi* inhabits mainly the surface layer from 0 to 15 to 25 m, and above seasonal thermocline (Purcell et al., 2001). The present study confirms that *M. leidyi* prefers surface waters or above the thermocline in the Caspian Sea (Figures 3 and 6).

In our study, the abundance and biomass values of *M. leidyi* were low during winter and spring, but gradually increased during summer and autumn. The maximum abundances and biomasses were observed in summer (Figure 4). There was a strong seasonality in the abundance of *M. leidyi* throughout the year; this is similar to the findings of Kideys and Moghim (2003), whilst up to 1200 specimens m^{-2} were recorded during the summer months; the population fell to very low levels of about 50 specimens m^{-2} during the colder period. This observation could be due to mass mortality during the colder months and renewal of the population in the following summer. So, the ctenophore population shrinks and expands again every year from small number of specimens. The seasonal pattern of *M. leidyi* is also related to water temperature, as evidenced from the positive correlation between the water temperature and abundance of *M. leidyi* (Figures 2 and 4), the ctenophore reached its highest abundances in summer when the water temperature was high, similar to the pattern observed by Shiganova (1998) and Finenko et al. (2006) in the Black Sea, Costello et al. (2006) in the Narrgansett Bay, Purcell (2005) in the Chesapeake Bay, and Javidpour et al. (2009) in the Baltic Sea.

The highest abundance and biomass were obtained in 2005 and 2006, respectively, showing the seasonal peaks of abundance and biomass of *M. leidyi* are very variable in the southwestern Caspian Sea from year to year (Figure 4). According to Roohi et al. (2008) during 2001 to 2006, the highest abundance and biomass of *M. leidyi* were observed during summer-autumn months coincident with warm temperatures. Besides, Bagheri and Kideys (2003) reported the biomass mean values of 164±79.8 g/m^2 in June, and 221±91 g/m^2 in October

2001, while Shiganova et al. (2001b) documented mean values of 372 g/m^2 in July and 556 g/m^2 in October 2001 from the basin wide surveys. Kideys and Moghim (2003) reported that the biomass of *M. leidyi* ranged between 3.5 and 351 g/m^2, with an average value of 120 g/m^2 from a basin-wide survey in August 2001. In additions, the highest *M. leidyi* population is showed during warm seasons in the Black Sea, Caspian Sea, Baltic Sea, Narragansett Bay and Chesapeake Bay (Purcell et al., 2001; Shiganova 2002; Kideys and Romanova, 2001; Bagheri and Kideys, 2003; Kideys et al., 2005; Purcell, 2005; Javidpour et al., 2009 ; Bagheri et al., 2012a).

Although there appears to be an increasing trend in the biomass of *M. leidyi* from 2001 to 2010 (Figure 4), the values are lower than the respective values for the Black Sea in the early 1990s. Despite the fact that the highest average biomass values for the Caspian Sea are, as yet, lower than those obtained in the Black Sea, the abundance of *M. leidyi* is much higher in the Caspian Sea due to the dominance of small sized (0 to 5 mm) animals in the population (Figure 8). Whilst maximum abundance of *M. leidyi* were 5976 ind/m^2 in October 2010, 9686 ind/m^2 in July 2009 and 10116 ind/m^2 in August 2006 in our study (Figure 4), respective values were 47 ind/m^2 in June 1991 and 408 ind/m^2 in August 1993 in the Black Sea (Mutlu, 1996). In all years, the mean of somatic growth appeared to take place from April to June and from August to September (Figure 5), when maximum weights of specimens were recorded. The mean weight of the ctenophore was 0.18 g in the southwestern Caspian Sea. Kideys and Moghim (2003) reported the average weight of *M. leidyi* as 0.24 g in the Caspian Sea for summer 2001, whereas for the Black Sea were 5.3 g in August 1993 and 4.2 g in July 1992. Based on the same sampling period, this denotes a 26-fold difference between the mean weight of *M. leidyi* from the two seas.

The size of surviving ctenophores and resulting population size increase with raise the water temperatures. From March to June, the biomass increased due to the intensive somatic growth of the animals that are wintering over; with the mean body weight of the ctenophores in the population been increased to 3.3 to 3.7 g, but the abundance remained low. In July, the abundance as well as the biomass increased due to the start of intensive reproduction; whilst the mean weight of animals in the population decreased by 0.26 g.

During that period, small ctenophores (< 5 mm) contributed 50 to 87% to the total population abundance. During subsequent months, the reproduction continued and the biomass decreased due to the elimination of the large animals that wintered over. The percentage of young animals remained was very high contributing 60 to 80% of the total abundance.

The mean wet weight of the animals in the population achieved its minimum in the first half of August as 0.05 g, when young animals <5 mm in size contributed 87% to the total abundance. *M. leidyi* from the Caspian Sea is smaller than that from the Black Sea. The maximal size of *M. leidyi* in the Caspian Sea (51 mm in Bagheri and Kideys (2003), 45 mm in Kideys and Moghim (2003), 65 mm in Finenko et al. (2006), and 70 mm in the present study) is also smaller than that recorded in the Black Sea, where this ctenophore could attain a length of 180 mm (Shiganova, 1997). The length-frequency distribution based on measurements of a total of 179,000 specimens, displayed that 90.6% of the population belonged to the 0 to 5 mm group and 6.4% were from 6 to 14 mm length group (Figure 7). A characteristic feature of the size composition of the *M. leidyi* population in shallow waters of the Caspian Sea (especially in the south) is the predominance of small ctenophores of <10 mm, similar to the results of Kideys and Moghim (2003), and Roohi et al. (2008), these small ctenophores made up 90% of the total abundance in summer. The low salinity in the Caspian Sea has been suggested to be the reason for the smaller size of ctenophores (Finenko et al., 2006) although Purcell et al. (2001) reported there was no significant relationship between salinity and ctenophore body size in the western Atlantic. In our study as well there was no significant correlation between population of *M. leidyi* and salinity (r = - 0.14). The largest length group dominated during April to June (in spring) and the smallest length group dominated other months (Figure 8). *M. leidyi* population sizes in temperate locations are small during cold winter temperatures, and increase with reproduction in the spring (Kremer, 1994). The greatest numbers of ctenophores in Chesapeake Bay occur in the spring (Purcell et al., 2001). In the southern Caspian (that is, Iranian waters), contribution of smaller specimens to the population was highest in summer, due to high water temperature (Figures 2 and 8).

Conclusions

This study documented the seasonal distribution and size structure of *M. leidyi* in the southwestern Caspian Sea during 2001 to 2010 and attempted to estimate fluctuations in abundance and biomass of *M. leidyi* in comparison with the previous literatures. This survey clearly showed that small individual up to 5 mm dominanted the *M. leidyi* population over the years and the abundance and biomass of M. leidyi has not changed much. We suggested, in future studies, the effects of climate change, shifte in meteorological regime and environmental degradation on plankton community in southwestern Caspia Sea should be undertaken.

ACKNOWLEDGEMENTS

The authors are grateful to John Leslie for improving the English of the draft manuscript. They would like to thank the Inland Waters Aquaculture Institute and Iranian

Fisheries Research Organization for financially supporting this project. The Universiti Sains Malaysia is also gratefully acknowledged. The assistance received from M. Sayad Rahim, Y. Zahmatkesh, A. Abedini, H. Mohsenpour, M. Malekshoumali, J. Khoushhal, and M. Iranpour in this study is greatly appreciated.

REFERENCES

Bagheri S, Kideys AE (2003). Distribution and abundance of Mnemiopsis leidyi in the western Iranian coasts of the Caspian Sea. In: Yilmaz A. (ed.). Proceedings of Second International Conference on Oceanography of the eastern Mediterranean and Black Sea:Similarities and differences of two interconnected basins. Ankara, Turkey. TUBITAK. pp. 851-856.

Bagheri S, Sabkara J (2003). Stomach contents of Mnemiopsis leidyi in the Iranian coastal waters of the Caspian Sea. Iran. J. Fish. Sci. 11:1-12.

Bagheri S (2006). An investigation on abundance and distribution of Mnemiopsis leidyi in Guilan waters, Southwestern Caspian Sea. Iran. J. Fish. Sci. 14:1-16.

Bagheri S, Mashhor M, Makaremi M, Mirzajani AR, Babaei H, Negarestan H, Wan Maznah WO (2010). Distribution and Composition of Phytoplankton in the Southwestern Caspian Sea during 2001-2002, a Comparison with Previous Surveys. World J. Fish. Mar. Sci. 2:416-426.

Bagheri S, Mansor M, Makaremi M, Sabkara J, Wan Maznah WO, Mirzajani A, Khodaparast SH, Negarestan H, Ghandi A, Khalilpour A (2011). Fluctuations of phytoplankton community in the coastal waters of Caspian Sea in 2006. Am. J. Applied Sci. 8: 1328-1336.

Bagheri S, Niermann U, Sabkara J, Mirzajani A, Babaei H (2012a). State of Mnemiopsis leidyi (Ctenophora: Lobata) and mesozooplankton in Iranian waters of the Caspian Sea during 2008 in comparison with previous surveys. Iran. J. Fish. Sci. 11: 732-754.

Bagheri S, Mansor M, Turkoglu M, Makaremi M, Babaei H (2012b). Temporal distribution of phytoplankton in the southwestern Caspian Sea during 2009–2010: A comparison with previous surveys. J. Mar. Biol. Assoc. UK. 92: 1243-1255.

Bagheri S, Mansor M, Turkoglu M, Marzieh M, Wan Maznah WO, Negaresatan H (2012c). Phytoplankton composition and abundance in the Southwestern Caspian Sea. Ekoloji. 21:32–43.

Bogdanova DP, Konsoulov AS (1993). On the distribution of the new ctenophore species Mnemia mccradyi in the Black Sea along the Bulgarian voastline in the summer of 1990. Doki Blug Akad Nauk. 46:71-74.

Costello JH, Sullivan BK, Gifford DJ, Van Keuren D, Sullivan LJ (2006). Seasonal refugia, shoreward thermal amplification, and metapopulation dynamics of the ctenophore Mnemiopsis leidyi in Narragansett Bay, Rhode Island. Limnol. Oceanogr. 51:1819-1831.

Dumont HJ (1998). The Caspian Lake: History, biota, structure and function. Limnol Oceanogr. 43:44-52.

Esmaeili A, Abtahi B, Khoda-Bandeh S, Talaeizadeh R, Darvishi Fd, Terershad H (1999). First report on occurrence of a combjelly in the Caspian Sea. Islamic Azad University. J. Env. Sci. Tech. 3:63-69.

Finenko GA, Kideys AE, Anninsky BE, Shiganova TA, Roohi A, Tabari MR, Rostami H, Bagheri S (2006). Invasive ctenophore Mnemiopsis leidyi in the Caspian Sea: feeding, respiration, reproduction and predatory impact on the zooplankton community. Mar. Ecol. Prog. Ser. 314:71-185.

Ivanov PI, Kamakima AM, Ushivtzev VB, Shiganova TA Zhukova O, Aladin N, Wilson SI, Harbison GR, Dumont HJ (2000). Invasion of Caspian Sea by the comb jellyfish, Mnemiopsis leidyi (Ctenophora), Biol. Invasions 2:255-258.

Javidpour J, Molinero JC, Peschutter J, Sommer U (2009). Seasonal changes and population dynamicsnof the ctenophore Mnemiopsis leidyi after its first year of invasion in the Kiel Fjord, Western Baltic Sea. Biol. Invasions 11: 873-882.

Khodaparast H (2006). Harmful Algal Bloom in the southwestern Basin of the Caspian Sea. Tehran, Iran: IFRO publisher. pp. 10-6.

Kideys AE, Niermann U (1994). Occurrence of Mnemiopsis leidyi along the Turkish coast. ICES J. Mar. Sci. 51:423-427.

Kideys AE, Romanova Z (2001). Distribution of gelatinous macrozooplankton in the southern Black Sea during 1996-1999. Mar Biol. 139: 535-547

Kideys AE, Roohi A, Bagheri S (2001). Monitoring Mnemiopsis leidyi in the Caspian Sea waters of Iran. Baku, Azerbaijan: CEP publisher. pp.1-14.

Kideys AE (2002). Fall and rise of the Black Sea ecosystem. Science 297:1482-1484.

Kideys AE, Moghim M (2003). Distribution of the alien ctenophore Mnemiopsis leidyi in the Caspian Sea in August 2001. Mar Biol. 142:163-171.

Kideys AE, Roohi A, Bagheri S, Finenko G, Kamburska L (2005). Impacts of invasive ctenophores on the fisheries of the Black Sea and Caspian Sea. Oceanography 18:76-85.

Kideys AE, Roohi A, Eker E, Melin F, Beare D (2008). Increased chlorophyll a levels in the southern Caspian Sea, following an invasion of jellyfish. Research Lett. Ecol. pp. 1-4

Kosarev AN, Yablonskaya EA (1994). The Caspian Sea. Moscow,Russia: SPB Academic Publication, pp. 224-249.

Kremer P (1994) Patterns of abundance of Mnemiopsis leidyi in U.S. coastal waters: a comparative overview. ICES J. Mar. Sci. 51:347-354.

Lahijani HA, Tavakoli V, Amini AH (2008). South Caspian River Mouth configureuration under Human Impact and Sea level Fluctuations. Environ. Sci. 5:65-86.

Mutlu E (1996) Distribution of Mnemiopsis leidyi, Pleurobrachia pileus (Ctenophora) and Aurelia aurita (Scyphomedusae) in the western and southern Black Sea during 1991-1995 period; net sampling and acoustical application. Ph. D dissertation, Middle East Technical University, Erdemli, Turkey.

Mutlu E (1999). Distribution and abundance of Ctenophora (Mnemiopsis leidyi) and their zooplankton food in the Black Sea. Mar Biol. 135:603-613.

Purcell JE, Shiganova AT, Decker MB, Houde ED (2001). The Ctenophora Mnemiopsis leidyi in native and exotic habitats: U.S. estuaries versus the Black Sea basin. Hydrobiologia 451:145-147.

Purcell JE (2005). Climate effects on formation of jellyfish and ctenophore blooms: a review. J. Mar. Biol. Assoc. UK 85:461-476.

Roohi A, Yasin Z, Kideys AE, Hwai AT, Khanari AG, Eker-Develi E (2008). Impact of a new invasive ctenophore (Mnemiopsis leidyi) on the zooplankton community of the Southern Caspian Sea. Mar. Ecol. 29:421-434.

Roohi A, Kideys AE, Sajjadi A, Hashemian A, Pourgholam R, Fazli H, Khanari A, Eker E (2010). Changes in biodiversity of phytoplankton, zooplankton, fishes and macrobenthos in the Southern Caspian Sea after the invasion of the ctenophore Mnemiopsis leidyi. Biol Invasions 12:2343-2361.

Shiganova TA (1997). Mnemiopsis leidyi abundance in the Black Sea and its impact on the pelagic community. In: Ozsoy E, Mikaelyan A (eds) Sensitivity of North Sea, Baltic Sea and Black Sea to anthropogenic and climatic changes. Dordrecht: Kluwer Academic. pp. 117-130.

Shiganova TA (1998). Invasion of the Black Sea by the ctenophore Mnemiopsis leidyi and recent changes in pelagic community structure. Fish Oceanogr. 7:305-310.

Shiganova TA, Kamakin AM, Zhukova OP, Ushivtsev VB, Dulimov AB, Museava EI (2001a). The Invader into the Caspian Sea Ctenophora Mnemiopsis leidyi and its initial effect on the pelagic ecosystem. Oceanology 41:542-549.

Shiganova TA, Sokolsky AF, Karpyuk MI, Kamakin AM, Tinenkova D, Kuraseva EK (2001b). Investigation of invader ctenophore Mnemiopsis leidyi and its effect on Caspian ecosystem in Russia in 2001. Baku, Azerbaijan: CEP publisher. pp. 36-58.

Shiganova TA (2002) Environmental impact assessment including risk assessment regarding a proposed introduction of Beroe ovata to the Caspian Sea. Moscowa, Russia : Institute of Oceanl. RAS. pp. 25-40

Shiganova TA, Sapognikov VV, Musaeva EI, Domanov MM, Bulgakova YV (2003). Factors that determine pattern of distribution and abundance Mnemiopsis leidyi in the northern Caspian. Oceanology 43:716-733.

Vinogradov ME, Shushkina EA, Musaeva EI, Sorokin PY (1989). A new acclimated species in the Black Sea: The Ctenophore Mnemiopsis leidyi (Ctenophora: Lobata). Oceanology 29:220-224.

Zaker NH, Ghaffari P, Jamshidi S (2007). Physical Study of the Southern Coastal Waters of the Caspian Sea, off Babolsar ,Mazandaran in Iran. J. Coast Res. 50:564-569.

5

Social values of biodiversity conservation for Mediterranean monk seal (*Monachus monachus*)

Matsiori S.[1], Stamkopoulos Z.[1], Aggelopoulos S.[2], Soutsas K.[3], Neofitou Ch.[1] and Vafidis D.[1]

[1]Department of Ichthyology and Aquatic Environment, School of Agricultural Sciences, University of Thessaly, Nea Ionia, Magnesia, Greece.
[2]Department of Agricultural Development and Agribusiness Management, Alexander Technological Educational Institute of Thessaloniki, 57400, Sindos Thessaloniki, Greece.
[3]Department of Forestry and Management of the Environment and Natural Resources, Democritus University of Thrace, N. Orestiada PC 68200, Greece.

This paper presents goods and services approach to determine the economic value of *Monachus monachus* with the aim of clarifying the role of valuation in the management and conservation of marine biodiversity. More specifically, it uses Contingent valuation method to estimate the existence value of the Mediterranean seal population that is predicted to be lost in the future if we not take any measure today to protect them. Our empirical findings suggest that the major variables affecting respondents' willingness to pay were related to respondents' attitude against reasons (values) of monk seals conservations with non use value to play the most important role. Results show that monk seal is highly valued and people are willing to pay for the most endangered seal in Europe. The results would help the choice of management strategies; the economic valuation of monk seal is a key part of any successful management plane informs conservation biologists and policy makers about opportunity costs of protection activities.

Key words: Contingent valuation, economic value, endangered species, Mediterranean monk seal.

INTRODUCTION

The Mediterranean monk seal (*Monachus-monachus*) is the most endangered seal. Greece is an extremely important country for the monk seal as it holds the largest population in the Mediterranean (Johnson, et al., 2006), with best known populations in the National Marine Park of Alonnissos - Northern Sporades (HSSPMS, 1995) and the Ionian islands (Panou et al., 1993).

Understanding the economic value of monk seal and wildlife in general remains a key part of any wildlife management strategy. Environmental economics can inform conservation biologists and policy makers about why species are endangered, the opportunity costs of protection activities, and the economic incentives for

conservation (Shogren et al., 1999) and a fundamental step in conservation (Pearce and Moran, 1994). The economic value of biodiversity can arise at of benefits that are derived from it: both tangible and intangible. The concept that will be used in these discussions is a measure known as the Total Economic Value (TEV) of a biological asset. It is defined as the sum of all service (use and non use) flows that the asset generates (Freeman, 1993) both now and in the future – appropriately discounted.

A number of valuation techniques were used to estimate the TEV of goods and services provided by environmental. One of the most widely used methods for biodiversity valuation is the contingent valuation method (CVM) (Pearce and Moran, 1994) due in part to recent advances in the theory and especially the testing methodology (Hanneman and Kanninen, 1996) and its cost advantage over other methods. According to Loomis and White (1996) "the contingent valuation method can provide meaningful estimates of the anthropocentric benefits of preserving rare and endangered species". A number of valuation studies have attempted to value marine biodiversity (Van Kooten, 1993; Loomis and Larson, 1994; Stevens et al., 1997).

In Greece Langford et al. (1998) used CVM for estimating willingness to pay for protecting the Mediterranean monk seal (*M. monachus*) in the Aegean. Authors estimate a median willingness to pay (WTP) of 11.7€ and income, sex, age and education were found significant explanatory variables of the function. Langford et al. (2001) also used CV method to estimate the WTP of respondents to financially support a public fund for the protection of the Mediterranean monk seal. Kaval et al. (2009) in a CVM research at Zakynthos Island carried out a CV reaches that discovered residents were willing to pay approximately 30 € more than tourists for the turtle, and a bit more for the seal. Others CVM studies for endangered species in Greece were carried out from Kontogianni et al. (2003).

The focus of this paper is the estimation of Mediterranean monk seal conservation value based on public preferences. More over this paper tries to determine the factors influence willingness to pay for Mediterranean monk seal conservation. The specific objectives are as follows:

(1) To investigate the public's awareness, attitudes and behaviors regarding Mediterranean monk seal conservation;
(2) To estimate the public's willingness to pay (WTP) for the conservation of Mediterranean monk seal;
(3) To identify the factors those affect the WTP.

MATERIALS AND METHODS

In present research to investigate the values people have for monk seals in Volos, a survey instrument was developed and tested. Primary data on the WTP were randomly selected from residents of Volos city through personal interviews[1]. Volos' port is the third of

---

[1]The sample size was determined using the Cluster Sampling formula. All others sampling methods require sampling frames which demand a list of the enumeration units (Tryfos, 1996). This is not always feasible, possible or even available. As consequence entire population is divided into groups, or clusters and a random sample of these clusters is selected (Aaker et al., 2009; Shiver and Borders, 1996). The target population of present survey was recreational visitors of coastal zone and it was not feasible or even possible to have a frame list and cluster sampling was the only technique that we can use. So as clusters unit were assumed the coastal beaches and as elementary unit were the days during a summer holiday season. All visitors who were visiting the beaches in random selected unit – days ware included in the sample. The size of the sample is considered sufficient for the performed statistical analysis.

Greece's major commercial ports, the capital of its prefecture, Magnesia and near National Marine Park of Alonnisos Northern Sporades.

The research's questionnaire comprised 32 items and the CV section was constructed according to guidelines established by the NOAA panel (Arrow et al., 1993). Background information was provided and also information on a hypothetical conservation plan for Mediterranean monk seal tried to elicit values through WTP questions. The question format was a voter referendum to approve this effort. Respondents were asked, prior to the WTP question, whether they would be in favor of supporting such a program through a voter referendum question format[2], and the implementation of these program would cost them a specified amount of money (in €) in a one-time payment. In the second phase, the WTP was elicited only from people who had answered positively to the first question, this time by asking if they are willing to pay a specific amount of money to confirm their participation. Specified amounts were randomly assigned to respondents. Bit step amounts were used based on the results obtained in the pre-test and in the pilot study where an open-ended question ranged from 1 € to 65 € (bit step 2 €). Given this information, respondents were asked whether they would vote 'yes' or 'no' to approve these effort. In questionnaire of pilot study an open-ended question format was included with the aim to specify the bit step amounts of final questionnaire due to lack of previous valuation studies for the study area. The results of the pilot study shows that the WTP amounts were fluctuated between 1 and 65 € (Table 1).

Follow-up questions were asked to determine reasons for respondents' answers. As protest responses were considered those reject some feature of the hypothetical CV scenario rather than from an absence of value.

For this CVM study, the dichotomous choice method, which seeks simple 'yes' or 'no' answers to an offered bid, is used. In order to be able to calculate the correct willingness to pay (WTP), a function was formulated which described the relationship between a person's WTP (dependent variable) and a number of socio-economic characteristics (independent variables) that influence this choice (Giraud et al., 2002) and variables are associated with respondents' pro-environmental attitudes and knowledge about monk seal (Kotchen and Reiling, 2000). In cases that our dependent variable (WTP) had a dichotomous format (yes/no), a binary logistic regression model should be used (Hosmer and Lemeshow, 1989).

Yes: No f (BID, INCOME, SEX, AGE, VALUE, ECOLOGIST)

Where Yes:No was the dichotomous-choice response to the WTP question, BID was the specified amount (€) respondents are asked to pay, and INCOME was respondent income, VALUE was the result from the value question and ECOLOGIST was respondents' pro-environmental attitudes.

A wide variety of use and non-use values can be derived from an endangered species conservation program. For Loomis and White (1996) TEV of the conservation is a sum use values, option value, existence value and bequest value. TEV of biodiversity derived from future genetic information for potential uses (medicinal and genetic) of endangered species (Loomis, 1996) satisfaction of existence of a particular species (Freeman, 2003). According the above conservation economic values a set of 5 Lickert scale (value scale) items was used to measure respondents attitude against

---

[2]According NOAA Panel suggestion CVM surveys should use a referendum approach. Employing this question format respondent it faced with a particular program and the possibility to pay for the implementation of them through some means, such as higher taxes (Carson et al., 1998). Referendum format resembled the way people actually make choices regarding public programs (Portney, 1994).

Table 1. Respondents opinions for monk seal protection.

Respondents opinions	Strongly disagree	Somewhat disagree	Unsure	Somewhat agree	Strongly agree	r_{i-t}
Want monk seal protection for visiting its habitants	7.798	19.266	22.936	35,321	14.679	0.383
Want monk seal protection for next generations	1.376	5.963	20.642	37,156	34.862	0.686
Want monk seal protection because its ecological importance	1.835	5.046	19,725	33.486	39.908	0.568
Want monk seal protection because not human beings have rights to exist	1.835	2.752	16.972	24.771	53.670	0.496
Want monk seal protection because of its possible use in the future	6.422	13.303	37.156	22.935	20.183	0.408

total economic value of Mediterranean monk seal. These data for the Mediterranean monk seal enable analysis of the way people realize the benefits from monk seal conservation influence CV responses, elicited values, and non-use motivations. Thus, even numbered items were coded as: "strongly agree" = 5; "somewhat agree" = 4; "unsure"= 3; "somewhat disagree" = 2; and "strongly disagree" = 1. The order is reversed for odd-numbered items, with a possible minimum score of 5 and maximum of 25. All respondents were categorized as having weaker, moderate, or stronger "value attitudes" according to value scale results (Kotchen and Reiling, 2000). Weaker attitudes were those with value scores less than 15, moderate were those greater than 15 and less than 19, and stronger were those 20 or greater. The results from the value scale demonstrate a range of respondents' attitudes against monk seal total economic value.

Mean WTP was calculated by assuming no negative values for species protection and using the formula suggested by Hanemann (1989):

$$E(WTP) = \left(\frac{1}{\beta_1}\right) * \ln\left(1 + \exp^{\beta_o}\right)$$

RESULTS AND DISCUSSION

Knowledge of and attitudes toward Mediterranean monk seal conservation

All respondents knew the Mediterranean monk seal and they did not need any further information about them. Then respondents were asked on their stand on different statements concerning monk seal conservation. Most of them (83.9%) known that Mediterranean monk seal is protected and one of the most endangered seal. They also knew (55.5%) the monk seal conservation status and the existence of National Marine Park of Alonissos Northern Sporades. Finally, the majority of respondents rate are as important as the protection of monk seal.

According to the results from value scale significant majorities express the view that non humans being are having existence rights so Mediterranean monk seal must be protected. Strong "value" attitudes are revealed with non use values of monk seal. On the contrary respondents did not want monk seal conservation expecting direct use benefits (as recreation benefits). The

item–total correlations for each item and all correlations are reasonably strong, ranging from a high of 0.68 to a low of 0.38. Cronbach's coefficient a, which is the mean of all split-half correlations 0.736.

Willingness to pay for monk seal conservation

For present study protest answers ware those that reject some feature of the hypothetical CV scenario and were not included in analysis (Mitchell and Carson, 1989). Responses considered as protests are: "natural protection is a responsibility of the government", "natural environment protection is already funded by national and regional governments", "natural protection are public good and do not pay for them" "I don't think the protection program would work" and "I am opposed to any new taxes".

According to the results of logit model, all coefficients have signs in the expected direction. Respondents were sensitive to the price they were asked to pay. Bid amount (BID) was negative and significant. On the contrary higher annual income encourages support of the CV scenario, as INCOME was positive and significant as in previous CVM studies for monk seal which is consistent with the economic theory.

Sex was found to significantly determine WTP, in that, females had higher probability of responding "yes" to the WTP question. Age had a negative and highly significant effect on the probability of respondents answering "yes" or 'no' to the valuation question. This indicates that as the age increases the tendencies to pay monk seal conservation will decrease. Younger people are more possible to participate and in CVM scenario.

On the other hand respondents with stronger pro-environmental attitudes had higher probabilities of responding "yes" to CVM scenario. Finally, the value scale was positive and significant, indicating that higher "value" scores result in higher probabilities of answering "yes".

The percentage of right predictions for the monk seal was 85.3 and R^2 value was 0.57. Mean WTP was approximately 15.67 €.

Environmental value attitude

Only 0.5% of the sample was NGO members. A significant percentage of respondents (10.5%) had actually paid for other species protection in the past. Comparisons among groups who had actually paid for wild life protection and pro-environmental it showed differences between groups are significant (x^2 = 6.728 and p < 0.05), indicating that environmental attitudes were related to the way respondents actually participate in action for wildlife protection.

According the considerable majority of the sample the Mediterranean monk seal conservation it was very important (40%) and we must intensify any effort for this direction. Environmental attitude was found to be significantly related to the way respondents rate the reasons of conservation (scale value). Significance tests for each conservation reason resulted in x^2 equal to 22.39 which is the 95% critical value for 4 degrees of freedom. The most noteworthy differences were in the way attitudes influence ratings of "very important". Stronger pro-environmental attitudes result in higher percentage ratings of "very important" for every conservation reason. The results showed that different reasons for monk seal conservation ware differentiated according to strength of pro-environmental attitudes, except the reason which was related with quasi option value. Respondents with strong pro-environmental attitude were more sensitive with reasons that were related with non use values. Respondents' attitude against the value scale was significant related to age (x^2 = 17.229 and p < 0.05), education level (x^2 = 27.19 and p < 0.05), prior knowledge of monk seal endangered status (x^2 = 24.824 and p = 0.0) and respondent's rate for importance of monk seal conversation (x^2 = 80.063 and p=0.0).

DISCUSSION

This study investigates relationships among environmental attitudes and socioeconomic characteristics of respondents to economic value of Mediterranean monk seal. Dichotomous-choice models of CV responses show how environmental attitude is one of the most significant determinants of yes/no responses. The most significant determinant is respondents' attitude against reasons (values) of monk seals conservation with non use value to play the most important role. The coefficients of the variables included in the model are all have the expected sign. First of all responses to willingness to pay questions are associated to some degree with socio-demographic factors as income, sex and respondents age. Jacobsen and Hanley (2008) investigate the effect of income in 46 CVM surveys for biodiversity conservation and only in 39% income effect on WTP for biodiversity conservation and such correlation were positive and significant. According Lopez et al. (2007) age influence peoples'

decision to support biodiversity conservation and often has been found to have a negative effect on WTP (Carson et al., 1998). For Spash et al. (2009) social psychology respondent's ethical beliefs help to understand WTP results and more over human attitudes toward these species under valuation influences respondents' willingness (Lopez et al., 2007; Serpell, 2004). Langford et al. (1998) valuing Mediterranean monk seal uses as explanatory variables bit amount, income (natural logarithm of annual income), age and education level. In this study education level had no significant effect to respondents' possibility to answer yes to WTP question.

Respondents rate of monk seal difference types of economic value determine their answer to CVM scenario. More over non use values of monk seal had positive influence to decision for monk seal conservation. These results ensure the opinion that the willingness to pay for species conservation is mainly based to non uses values (Jacobsen and Hanley 2008). Accordingly to previous studies pro-environmental attitudes result in significantly higher probabilities of responding "yes" (Kotchen and Reiling 2000). The results show that two main factors determine peoples' willingness to invest in the conservation of certain species: 1) respondent's degree of familiarity with the specific species and 2) the individual's understanding of the role that the specific species plays in the ecosystem. Moreover, the results shown that peoples' decision to support biodiversity conservation was influenced by the knowledge of its non-tourist value (that is, the ecological value), the origin of the respondents and their age.

The values obtained in this study are quantified indications of the value placed by the people on Mediterranean monk seal, one of the most endangered mammals on earth. As such, they are useful for cost benefit analysis and for debate and decision-making on conservation strategies. The study may contribute to drawing the attention of the policy makers in formulation of appropriate policy mechanisms for monk seal conservation polices. The results also point to the need for a better information programmed about the value an ecological importance of monk seal if we want people are more interesting for their conservation.

REFERENCES

Aaker D, Kumer V, Day G (1995). Marketing Research. Fifth Edition. New York: John Wiley & Sons.

Arrow K, Solow R, Portney PR, Leamer EE, Schuman RR (1993). Report of the NOAA Panel on Contingent Valuation. Report to the General Counsel of the US National Oceanic and Atmospheric Administration. Resources for the Future, Washington, D.C.

Carson RT, Hanemann WM, Kopp RJ, Krosnick JA, Mitchell RC, Presser S, Ruud PA, Smith VK, Conaway M, Martin K (1998). Referendum Design and Contingent Valuation: The NOAA Panel's No-Vote Recommendation. Rev. Econ. Stat. 80(2):335-338.

Freeman AM (1993). The Measurement of Environmental and Resource Values. Resources for the Future, Washington DC.

Freeman AM (2003). The Measurement of Environmental and Resource Values: Theory and Methods. Second Edition. Resources for the Future, Washington. DC.

Giraud K, Turcin B, Loomis J, Cooper J (2002). Economic benefit of the protection program for the Steller sea lion. Mar. Pol. 26:451-458.

Hanemann M (1989). Welfare evaluations in contingent valuation experiments with discrete response data: reply. Am. J. Agric. Econ. 71:1057-1061.

Hanneman M, Kanninen B (1996). The statistical analysis of discrete-response CV data. California Agricultural Experiment Station. Working Paper No. 798.

Hellenic Society for the Study and protection of the Monk Seal. (1995). Continuation of the Monitoring of the Monk Seals in the National Marine Park of Northern Sporades. Final Report for the European Commission Project. 4-3010(92):7829.

Hosmer DW, Lemeshow S (1989). Applied Logistic Regression. John Wiley & Sons. P. 307.

Jacobsen JB, Hanley N (2008). Are there income effects on global willingness to pay for biodiversity conservation? Environ. Resour. Econ. 43(2):137-160.

Johnson WM, Karamanlidis AA, Dendrinos P, De Larrinoa PF, Gazo M, Gonzalez I M, Guecluesoy H, Pires R, Schnellmann M (2006). Monk seal fact files: Biology, behaviour, status and conservation of the Mediterranean monk seal, Monachus monachus. The Monachus Guardian. Retrieved 1 December 2011 from www.monachus-guardian.org/factfiles/medit01.htm.

Kaval P, Stithou M, Scarpa R (2009). Social Values of Biodiversity Conservation for the Endangered Loggerhead Turtle and Monk Seal. Int. J. Ecol. Econ. Stat. 14:67-76.

Kontogianni A, Langford IH, Papandreou A, Skourtos MS (2003). Social preferences for improving water quality: an economic analysis of benefits from wastewater treatment. Water Resour. Manag. 17:317-336.

Kotchen JM, Reiling DS (2000). Environmental attitudes, motivations, and contingent valuation of nonuse values: A case study involving endangered species. Ecol. Econ. 32:93-107.

Langford IH, Skourtos MS, Kontogianni A, Day RJ, Georgiou S, Bateman IJ (2001). Use and nonuse values for conserving endangered species: the case of the Mediterranean Monk Seal. In: Turner RK, Bateman IJ, Adger N, editors. Economics of coastal and water resources: Valuing environmental functions. Dordrecht: Kluwer Academic Publishers.

Langford, IH, Kontogianni A, Skourtos MS, Georgiou S, Bateman, IJ (1998). Multivariate mixed models for open-ended contingent valuation data: willingness to pay for conservation of monk seals. Environ. Resour. Econ. 12(4):443-456.

Loomis J (1996). Measuring the Economic Benefits of Removing Dams and Restoring the Elwha River: Results of a Contingent Valuation Survey. Water Resour. Res. 32(2):441-447.

Loomis JB, Larson DM (1994). Total economic value of increasing Gray Whale populations: Results from a contingent valuation survey of visitors and households, Mar Resour. Econ. 9:275-286.

Loomis JB, White DS (1996). Economic benefits of rare and endangered species: Summary and meta-analysis. Ecol. Econ. (18):197-206.

Lopez BM, Montes C, Benayas J (2007). The non-economic motives behind the willingness to pay for biodiversity conservation. Biol. Conserv. 139:67-82.

Panou A, Jacobs J, Panos D (1993). The Endangered Mediterranean Monk Seal Monachus monachus in the Ionian Sea, Greece. Biol. Conserv. 64:129-140.

Pearce D, Moran D (1994). The Economic Value of Biodiversity. Earthscan Publications Limited, London.

Portney PR (1994). The Contingent Valuation Debate: Why Economists Should Care, J. Econ. Persp. 8:3-17.

Serpell JA (2004). Factors influencing human attitudes to animals and their welfare. Anim. Welfare 13:145-151.

Shiver DB, Borders BE (1996). Sampling Tecniques for Forest Resource Inventory. John Wiley & Sons. Inc. USA.

Shogren JF, Tschirhart J, Anderson T, Ando AW, Beissinger SR, Brookshire D, Brown GMJr, Coursey D, Innes R, Meyer SM (1999). Why economics matters for endangered species protection. Conserv. Biol. 13:1257-1261.

Spash CL, Urama K, Burton R, Kenyon W, Shannon P, Hill G (2009). Motives behind willingness to pay for improving biodiversity in a water ecosystem: Economics, ethics and social psychology. Ecol. Econ. 68(4):955-964.

Stevens TH, DeCoteau NE, Willis CE (1997). Sensitivity of contingent valuation to alternative payment schedules. Land Econ. 73:140-148.

Tryfos P (1996). Sampling Methods for Applied Research text an cases. John Wiley & Sons, Inc. Canada.

Van Kooten GC (1993). Bioeconomic Evaluation of Government Agricultural Programs on Wetland Land Econ. 69(1):27-38.

6

Preservation practices and quality perception of shrimps along the local merchandising chain in Benin

E. Y. Kpoclou[1], V. B. Anihouvi[1], M. L. Scippo [2] and J. D. Hounhouigan [1]

[1]Department of Nutrition and Food Science, Laboratory of Food Microbiology and Biotechnology, Faculty of Agronomic Sciences, University of Abomey-Calavi, 01 BP 526, Cotonou, Benin.
[2]Departement of Food Science, Laboratory of Food Analysis, Faculty of Veterinary Medicine, Centre of Analytical Research and Technology (CART), University of Liège, B43b, Boulevard de colonster 20, Star-tilman, B-4000 Liège, Belgium.

This study, performed through a survey, sought to investigate the quality attributes and the artisanal preservation and processing methods of fresh shrimps along the local merchandising chain in the Ouémé-Plateau and Mono Districts in the South of Benin. The survey was in the form of interviews administrated through a questionnaire and observations of shrimp's processors at work. The information gathered focused on the socio-cultural profile of actors, the types of raw materials used, the processing techniques, the quality attributes and the storage of fresh and processed shrimps as well. The data collected were analysed by means of descriptive statistics and correspondence analysis. Through the results it appeared that shrimps processing and commercialization are female activities of various socio-cultural groups including: Pédah (35%), Sahouè (9%), Xwla (20%), Aïzo (20%), Goun (16%) and Kotafon (1%). On the local markets, shrimps offer is constituted of smoked shrimps powder (40%), entire smoked shrimps (32.6%), fresh shrimps (26.6 %) and fried shrimps (0.7%). Fresh shrimp's quality is assessed through the colour, the odour and the size while the smoked ones are appreciated by the colour, the integrity, the firmness, the size and the taste.

Key words: Sea food, processing, smoking, frying, quality attributes, local market.

INTRODUCTION

Fishery products are significant foodstuffs for millions people in Africa; it constitutes a major source of animal proteins, accessible to the households with low incomes, especially in the regions where the price of meat is out of the reach of an average consumer (FAO, 2004; Lem, 2005; Anonymous, 2005a). In Benin, shrimp fishing plays a significant socio-economical role, since fresh shrimps are one of the most important export products of the country (Horemans, 1998; Anonymous, 2005b). This activity occupies about 21 000 fishermen living in more than 200 villages in the south of the country, who catch wild shrimps in the brackish waters of Ahémé and

Nokoué lakes, and Lagoon of Porto-Novo (Gnimadi et al., 2006). The annual production of fresh shrimps in Benin, estimated to approximately 3000 tons, concerns mainly the Penaeidae of which *Penaeus notialis*, *Penaeus monodon*, *Penaeus duorarum* and *Penaeus kerathurus* constituted the major species (Raux, 2009). In Benin, the fishing of shrimps is a seasonal activity (from February to August). In addition, the shrimps are highly perishable food products, which can be degraded within 6 h at ambient temperature (Laghmari and El Marrakchi, 2005). Previous studies showed a fast increase in histamine content of shellfish from 36.6 to 2123.9 mg/kg after 24 h of storage at 25°C (Kim et al., 2009). Recent work carried out by Anacleto et al. (2011) showed that the shelf-life of halieutic processed products like crabs depends on the raw material quality. In this respect, shrimps preservation is undoubtedly a major problem with the need to constitute reserves for off seasons of fishing and to absorb fresh shrimp unsold in plenty period.

In general, the artisanal processing conditions and distribution of halieutic products in Western Africa countries are proved unsatisfactory (Essuman, 1992). It is well known that chemical compounds generated by wood combustion during smoking are transferred to the smoked products (Duedahl-Olesen et al., 2010). Moreover pathogenic microorganisms and mycotoxins which development is associated with preservation conditions were isolated from halieutic products processed in artisanal conditions and collected from markets in Nigeria (Essien et al., 2005). With the challenge of development, the developing countries must guarantee food security, by which any inhabitant can have a sufficient, healthy, balanced and acceptable food. In this respect, there is a need in Benin to improve the country practices of processing and preservation of foodstuffs in general and the products of animal origin such as shrimps in particular. The present field investigation focused on the characterization of the technical system of shrimps processing and preservation methods used along the local chain of commercialization, with the objective to gather information which allow to improve those practices in order to improve the safety of fresh and processed shrimps commercialized along the local merchandising chain, in consideration of the way these products are currently handled or treated.

Indeed, the environment in which the fresh shrimps are processed is generally unhygienic, paving the way for possible microbial contamination and production of food toxicants such as histamine. The consequence of this is quality defects, with occasional public health hazards. So, despite a lack of information on food poisoning caused by shrimps in Benin, there is a potential for sporadic amine poisoning. Thus, this study aimed to give a better understanding on the preservation and processing methods of fresh shrimps, the types of raw materials and other ingredients associated with shrimp treatment and the definition of the most important quality attributes of fresh and processed shrimps according to the actors. The current survey also investigated the specific problems related to the processing, the preservation and the storage of this product, for a further upgrading of the technologies.

MATERIALS AND METHODS

Choice of survey zones

A survey was carried out in the Atlantic, Mono and Ouémé Districts in the south of Benin (Figure 1). The choice of these areas is justified by the fact that shrimps fishing and processing are specific activities of populations living along the lakes and lagoon (Nokoué and Ahémé Lakes, and Lagoon of Porto-novo) located in that places where fresh shrimps are caught. In the selected areas the field investigations were conducted at market level and in the restaurants and hotels (Table 1).

Sampling of actors

The number of actors surveyed per locality and categories of actors in the survey zones is summarized in Table 1. Shrimps processors were randomly selected according to Dagnelie (1998) on the basis of data obtained from a preliminary census of actors of shrimps industry in Benin (Gnimadi et al., 2006). Shrimps sellers were investigated in the markets. In Cotonou municipality (Atlantic District), the markets were chosen at random whereas in Comè (Mono District) and Porto-Novo municipality (Ouémé District), all the markets were taken into account because of their limited number. In these markets, all the shrimps and by-products sellers listed were surveyed by spot. Restaurants and hotels were selected on the basis of the national directory on hotels and restaurants according to the criterion of 50 places at least. Eligible actors were selected by raking of shrimp processors, sellers and users. A total of 167 processors, 139 traders and 17 restaurants and hotels were interviewed (Table 1).

Collection of data

A validated questionnaire was designed to collect data on shrimps processing and the preservation techniques. The survey was in the form of individual interviews or focus group discussions, and observations of processors at work. Demographic data related to gender, age, religion, marital status and academic qualification were collected. Other information collected included the quality attributes of raw and processed shrimp, the processing and preservation methods, the storage duration and the specific problems related to the methods used, the fresh shrimps and by-products.

Data analysis

Descriptive statistics were calculated from the collected data using SAS software (version 9. 2 2003, Cary, NC: SAS Institute Inc.). Multivariate analyses were also performed on the types of shrimp and quality attributes as perceived by actors, using Statistica 7 (StatSoft, Tulsa, USA).

RESULTS AND DISCUSSION

Socio-cultural status of actors of the local merchandising chain of shrimp

The survey showed that the post-capture circuit of shrimps before reaching the local consumer is constituted of fresh shrimp's sellers, shrimps processors/sellers, retailers, and hotels and restaurants (Tables 2 and 3). The survey also revealed that women are the main actors in the artisanal processing and distribution of shrimps, and these activities constituted for most of them the main source of income. The majority (95%) of these women are married and 85.3% of them are married to fishermen or had a family relationship with them. The processors

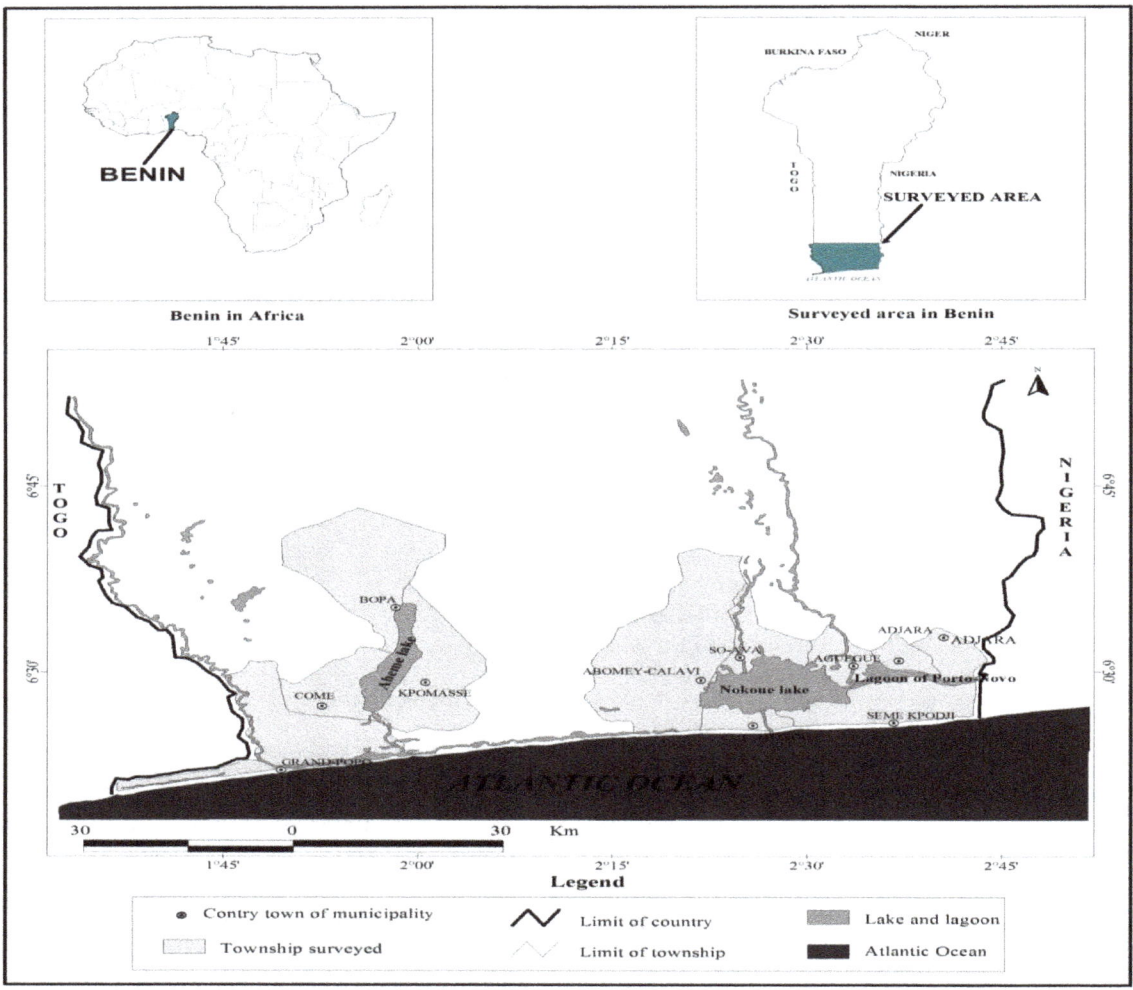

Figure 1. Surveyed zones including fishing and selling areas of shrimps.

Table 1. Number and categories of shrimps sector actors surveyed per locality in the survey zones.

Area surveyed		Number of actors surveyed			Total number of actors surveyed
Department	Township	Number of processors	Number of traders	Number of hotels and restaurants	
Atlantic-District	Abomey-Calavi	5	-	-	5
	Sô-Ava	17	-	-	17
	Cotonou	32	62	6	100
	Sèmè-kpodji	12	-	1	13
	Kpomassè	20	-	-	20
Mono-Couffo District	Bopa	14	-	2	16
	Comè	32	20	-	52
	Grand-Popo	19	-	5	24
Ouémé District	Adjarra	2	-	-	2
	Aguégué	12	-	-	12
	Porto-Novo	2	57	3	62
Total		167	139	17	323

Table 2. Socio-cultural characteristics of shrimps processors and sellers.

Characteristics	Processors (%)	Sellers (%)
Age (years)		
10-20	0	5
21-30	8	24
31-40	25	30
41-50	22	24
>50	45	17
Gender		
Female	100	100
Ethnic groups		
Aïzo	20	5
Goun	16	41
Xwla	20	20
Pedah	35	11
Sahouè	8	0
Kotafon	1	0
Fon	0	18
Nago	0	2
Yorouba	0	3
Academic qualification		
Illiterate (no school)	92	56
Primary school	7	28
Secondary school	1	16
University	0	0
Marital status		
Unmarried	0	5
Married	95	95
Divorcee/widow	5	0
Religion		
Animism	49	5
Christianism	44	87
Islam	3	4
Atheism	4	3

surveyed belonged to the socio-cultural groups of Pédah (35% of them), Sahouè (8%) and Kotafon (1%) living around the Ahémé lake, and Xwla (20%), Aïzo (20%) and Goun (16%) installed along the Nokoué lake and the lagoon of Porto-Novo (Table 2). The majority (92%) of the interviewed processors are aged from 31 to more than 50 years; among them, 44 and 49% are Christians and animists respectively. The majority of processors (92%) had no formal education; a very small number (7%) had secondary education and the remaining (1%) had only primary education (Table 2). All of them have acquired the knowledge in shrimp processing while helping a parent or a member of their husband family. Regarding

the sellers surveyed, all of them are female aged between 21 and 50 years (78%). Most of them (56 %) had no formal education, while 28 and 16% had primary education and secondary education respectively. They belonged to various socio- cultural groups including: Goun (41%), Xwla (20%), Fon (18%), Pédah (11%), Aïzo (5%), Yorouba (3%), and Nagot (2%). In contrary to processors, the majority of them (87%) are Christians. The commercial establishments users of shrimps surveyed are hotels (94%) and restaurants (6%). These establishments are mainly located in the towns: Cotonou (41%), Grand-Popo (29%), Porto-Novo (18%) and Bopa (12%) (Table 3). The employees surveyed in these

Table 3. Characteristics of hotels and restaurants surveyed in shrimps sector.

Characteristics	Percentage
Nature	
Hotels	94
Restaurant	6
Localization (town)	
Cotonou	41
Bopa	12
Grand-Popo	29
Porto-Novo	18
Responsibility surveyed	
Cook	88
Manager	12
Age (years)	
21-30	18
31-40	35
41-50	35
>50	12
Gender	
Male	82
Female	18
Academic qualification	
Primary school	6
Secondary school	88
University	6

establishments are cooks (88%) and managers (12%). These workers are men (82%) and women (18%) having primary education (6%), secondary education (88%) and university education (6%).

Description of the technical system of fresh shrimp preservation

Shrimps' offer in local markets is constituted of smoked shrimps powder (40%), entire smoked shrimps (32.6%), fresh shrimps (26.6%) and fried shrimps (0.7%). Two families of shrimps are commercialized in Bénin: the *Penaeidae* fished in Nokoué and Ahémé lakes, and the lagoon of Porto-Novo, and the *Mysidaceae* caught in the sea; however, the main species of shrimps used in Benin belonged to *Penaeidae*. The reason of such situation is that the fishing of *Mysidaceae* is forbidden by the Benin ministerial order n° 518/MAEP/D-CAB/SGM/DRH/DP/SA, December 2008, related to shrimps fishing. The hotels, restaurants and households bought fresh shrimps from

sellers in the market whereas the processors and sellers purchased or received fresh shrimps from the fishermen and the wholesalers of fresh shrimp. For fresh shrimps preservation, all the restaurants and hotels surveyed usually kept fresh shrimps in a freezer while at the market level, some of the vendors (10% of vendors interviewed) preserve it in a chest filled with ice and the other (7% of respondents) maintain the fresh shrimp at ambient temperature during selling. In addition, the smoking (claimed by 72% of surveyed actors) and the frying (claimed by 4% of surveyed actors) have been identified as shrimp's preservation methods (Figure 2). Fresh shrimps kept at ambient temperature around 30°C during selling may be unsafe for consumers since fresh shrimps are very perishable raw materials.

According to Adams et al. (1987), aquatic food products kept in such conditions may spoil within 12 h. The survey revealed that fresh shrimps are preserved under ice for sale during 3 to 14 days. Works by Laghmari and El Marrakchis (2005) showed that the fresh shrimps exceed quality standard beyond 72 h of preservation under ice. In addition, Anacleto et al. (2011) showed that the merchant quality and the shelf life of the halieutic processed products like crabs depend on the raw materials quality. In this respect, the sanitary quality of fresh shrimps and its by products in the local merchandising chain remains uncertain and consequently needs to be investigated. The survey also showed that the processing of shrimps in the local merchandising chain is essentially an artisanal activity. The processing sites are huts mostly located beside processor residence along the lakes and lagoon, near the fishermen landing areas. One processor transforms a rough average of 26 kg of shrimps per day during the plenty period (from March to August). The processing techniques are simple and rudimentary, equipments such as basins, baskets, metallic grid and artisanal cooker are used. The flow diagram of processing techniques used by the processors is represented in Figure 3.

Smoking method of fresh shrimps

The processing of fresh shrimps into smoked shrimps is the result of a number of steps of unit operations after the purchase of fresh shrimps. The main steps are the sorting, the washing and the draining, the smoking, the cooling and the grinding in the case of smoked shrimps powder.

Fresh shrimps receipt

The fresh shrimps used for processing are essentially (claimed by 76% of processors) received from fishermen. The processors do not impose any criteria during the receiving of shrimps, because of the keen competition with the wholesale fish merchants and the export

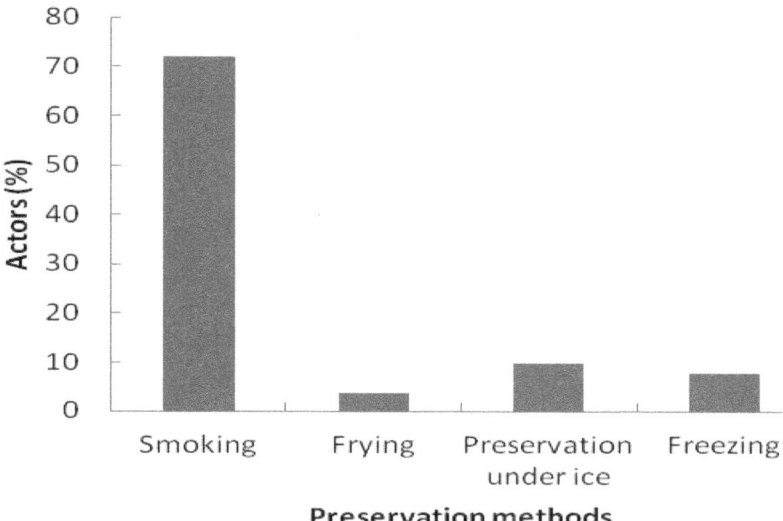

Figure 2. Fresh shrimps preservation methods in the local merchandise chain and categories of actors involved.

factories. For fear of being deprived of next captures, the processors are constrained to receive the entire offer from her fishing partner, whatever the shrimps are in spoilage state.

Sorting, washing and draining of fresh shrimps

The fresh shrimps are then rid of undesirable objects and washed three times with well water or tap water. In order to maximize their profit, the shrimps are sorted, and the big and very fresh ones are resold to the collectors of export factories. The remaining is put in a basket for draining during 10 to 15 min.

Smoking

The smoking is carried out in two successive phases: The cooking phase with intense fire which confers firmness to shrimps and the drying phase with mild fire which assures dehydration of the product. During the two phases, the processors pay attention to the intensity of fire in order to avoid the burnt aspect and get the redcolor of smoked shrimps that some of them (4.40% of processors) seek to improve by using annatto tree (*Bixa orellana*) seeds powder which confers to shrimps more pronounced red color. This powder is only used in the case of smoking. The powder is generally put in water and the shrimps are immerged for about 5 min before to be smoked. The merchant quality of smoked shrimps depends on the conduct of the smoking process. In this respect, a particular attention is granted to the disposal of fresh shrimps on the smoking grid, the wood nature and

the adjustment of smoking fire level. At the beginning of smoking, fresh shrimps are thoroughly spread out over the grid of smoking. Hard woods less smoking and whose embers hold for long time are mostly used. The woods species mostly used include: Pencil tree (*Acacia auriculiformis*, used by 45% of processors surveyed), teak tree (*Tectona grown*, 11%), velvet tamarind tree (*Dialium guineense*, 10%), Australian Pine tree (*Casuarina equisetifolia*, 7%), neem tree (*Azadirachta indica*, 5%), mango tree (*Mangnifera indica*, 5%), red mangrove tree (*Rhizophora racemosa*, 5%), river redgum tree (*Eucalyptus camadulensis*, 2%), ironwood (*Prosopis africana*, 2%) and cashew tree (*Anarcadium occidentale*, 1%). In addition to these fuels, peels of cassava (*Manihot esculenta*, observed with 7% of processors surveyed) and of the coconut raid (*Coconuts nucifera*, observed with 33% of processors) are used to control the fire during the smoking. For the smoking of *mysidaceae* in particular, all the processors interviewed use wood coal to avoid the burn of shrimps because of their small size. The smoking duration is generally ranged between 3 and 4 h, after which the smoked shrimps are cooled at ambient temperature.

Cooling

For cooling, the smoked shrimps are kept on the smoking grid to ambient temperature until total cooling. According to processors, the cooling stage prevents the condensation of water steam during packaging and the loss of the integrity of wetted shrimps during post smoking handling. However, during the cooling period the smoked shrimps are exposed to flies and other types of

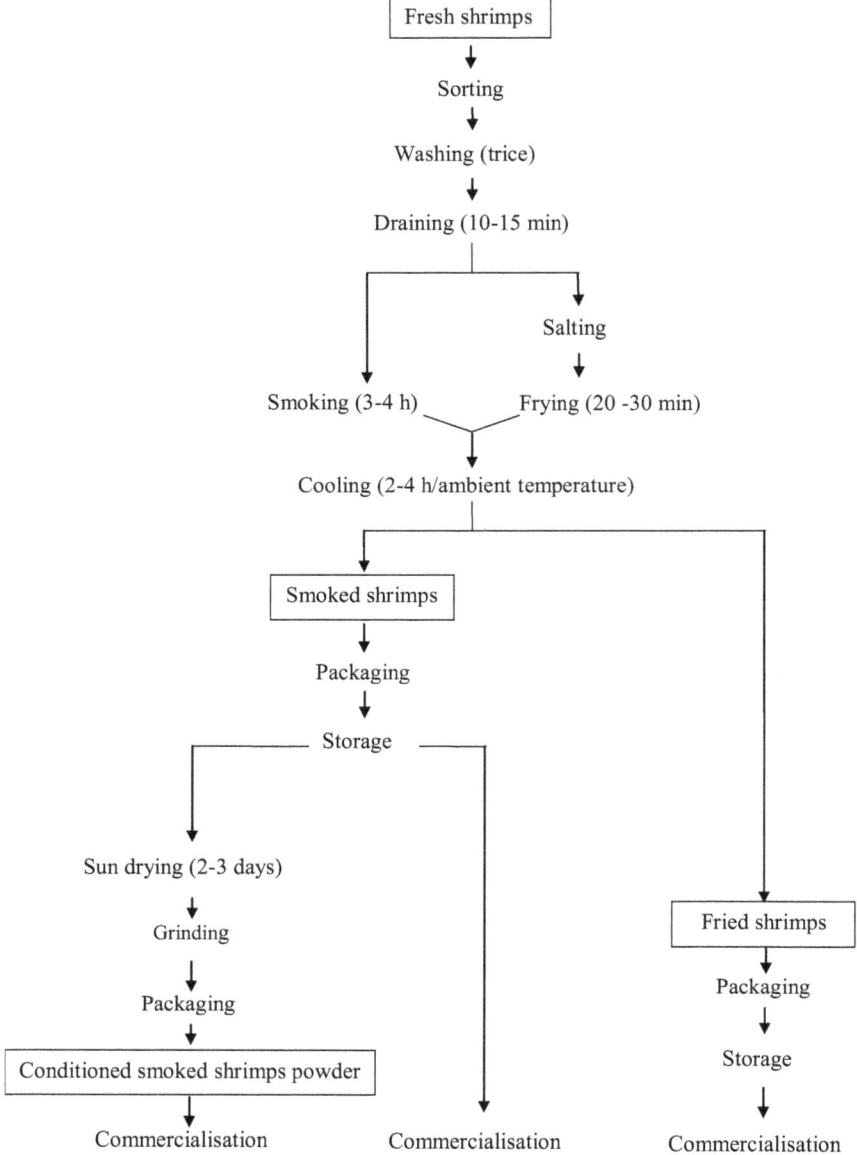

Figure 3. Flow diagram of fresh shrimps processing.

insects. Such practices may lead to post processing contaminations of the end-product. After the cooling stage, the end-product may be sold as entire smoked shrimps or as powder of smoked shrimps. When the end-product is sold as whole smoked shrimps, after cooling, the shrimps are packaged in baskets and covered with cloths for storage. In the case of smoked shrimp powder, the smoked shrimps are ground after the cooling.

Grinding of smoked shrimps

For grinding, the eyes of smoked shrimps are removed and the grinding is done using a mill generally available

in the markets. According to the processors, the grinding contributes to dissimulate the quality defects (colour, broken shrimps) of smoked shrimps (claimed by 8.82% of processors); thus, altered smoked shrimps are valorized through this practice. In addition, in order to increase their profit, some of processors (claimed by 12.7% of surveyed) mix fraudulently the smoked shrimps with crayfish (*Astacus*) or sardine (*Sardinella*) before the grinding. These practices put in doubt the sanitary quality of smoked shrimp's powders. For grinding, the same mill is generally used to mill different types of raw materials so there is a potential cross contamination of the shrimps powder.

Other marginal preservation techniques used by the

processors are the frying of fresh shrimps and the solar drying of smoked shrimps during storage period. For frying, the fresh shrimps are washed using well water or tap water and salted using solar salt. The salted shrimps are then fried under intense fire for about 20 to 30 min using vegetal oils. The end of frying corresponds to pink colour and firm shrimps recognized by the experience of the processors. Regarding the solar drying method, the smoked shrimps are spread over a tray at ambient temperature during 2 to 3 days before been ground. During the drying, the shrimps are exposed to the environmental conditions with risk of contamination.

Packaging and storage shrimps by products

The smoked shrimps are generally packaged in baskets previously lined with cotton clothes by alternating shrimps layers with some peppers (*Capsicum annuum*) (practiced by 9% of processors interviewed) and camphor (claimed by 2% of processors). Washed without particular heat treatment these wraps could bring a share of contamination to the product. According to the majority (93%) of processor, the smoked shrimps can be stored for 2.5 to 6 months. However, when the storage duration is more than 2.5 months, the smoked shrimps are briefly smoked again once a month. The fried shrimps are also stored in basket covered with paper but without any particular precaution because they are delivered within 24 h after processing. Shrimps powder is generally conditioned in recycled bottles previously cleaned and dried (claimed by 53.9% of vendors) or in plastic bags (claimed by 27.45% of vendors). The smoked shrimps powder can be stored for 4 months (claimed by 50.9% of vendors). The main spoilage problems associated with smoked shrimps and evoked by the actors are the destruction of stocks by devastating insects of the forficula and pheidole kinds (claimed by 78% processors and wholesalers interviewed), invasion by the moulds (claimed by 30.53% processors and wholesalers interviewed), rehumidification of the products (claimed by 15%), change in colour (evoked by 11.37%) and change in smell (6.58%).

Shrimp quality assessment of fresh shrimps and by products

Regarding the quality assessment of fresh and processed shrimps a factorial correspondence analysis (FCA) was performed to reveal links between quality attributes of shrimps and categories of actors (fresh shrimp sellers, processors, hotels and restaurants) (Figure 4). The two axes obtained accounting for 98.26% of the total variation of which 92.53% was explained by the first axis (Axis 1) and 5.73% by the second (Axis 2) (Figure 4). Determinative criteria used to appreciate fresh shrimps

are grouped together with a type of actor. Thus, fresh shrimp's sellers, and restaurants and hotels prefer shrimps of grey colour and of middle or large size, while for the processors the colour and the size are not determinative criteria (Axis 1). Regarding axis 2, all the actors prefer shrimps of no putrid odour. Some previous studies showed that colour of shrimp's head and shrimps odour are used as sensory characteristics of spoilage fresh shrimps (Laghmari and El Marrakchi, 2005; Jaffrès et al., 2011).

For processed shrimps (fried and smoked), the FCA performed on quality criteria and actors resulted in two axes accounting for 100% of the total variation, of which 89.07% was explained by the first axis (Axis 1) and 10.93% by the second (Axis 2) (Figure 5). Regarding axis 1, smoked shrimps of dark-red colour, which are not beheaded and have not burnt aspect, are preferred. The best smoked shrimp's powder has reddish colour, with a special smoked shrimp's taste, and do not content foreign matters (Axis 1). This survey revealed that the quality of processed shrimps does not constitute in any way a problem of access to the local market. Each actor manages as he can so that his products interest the customers. However, in consideration of lack of hygiene (hygiene of raw material, hygiene of materials used for processing, hygiene of processing sites) and some bad processing practices observed along the local merchandising chain, fresh shrimps and its by products could be potential sources of microbial or other types of contamination. Consequently, in the artisanal shrimps sector where technology and standard are very low, fresh shrimps and its by products could be considered as potential vehicles for transmission of food borne diseases. In this respect, pathogenic microorganisms and thermostable toxins have been identified in smoked fish collected in the same conditions of distribution (Essien et al., 2005; Adu-gyamfi, 2006).

Conclusion

The current study revealed that the main preservation methods of fresh shrimps in use in the survey zones are the smoking and the cooling. Consequently, shrimps offer is mainly constituted of fresh shrimps and smoked shrimps. Quality attributes of these products are based on the colour, the odour and the size, and the importance of each of them changes according to actors of the local merchandising chain. In this sector where there is no control or regulation, every actor manages as he can, so that his products interest the customers. The survey also provides comprehensive knowledge which will facilitate the improvement of artisanal handling and processing techniques of fresh shrimps in Benin. The major problems identified during the survey included the lack of hygiene and standardization of the processing techniques, the use of inadequate packaging materials

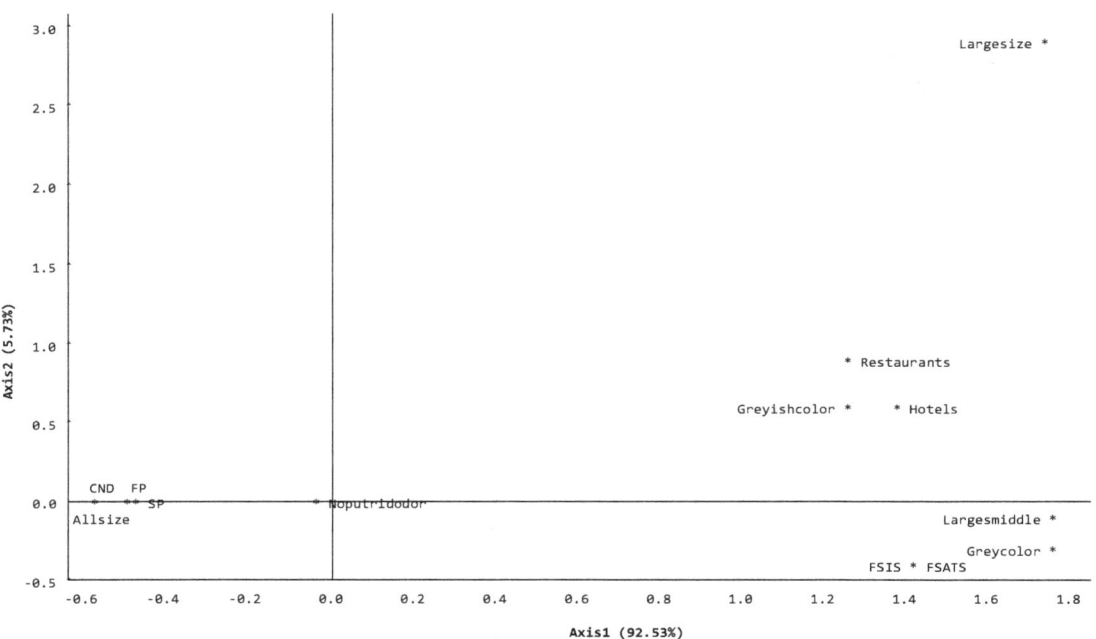

Figure 4. Factorial correspondence Analysis to reveal links between quality attributes of fresh shrimps and categories of actors. CND = Color is not determinative; Noputridodor = no putrid odor; Largesize = large size; Largesmiddle = Large size and middle size; Allsize = All size; FSATS = Fresh shrimps at ambient temperature sellers; FSIS = Fresh shrimps preserved under ice sellers; FP = Frying processsors; S P = Smoking processors.

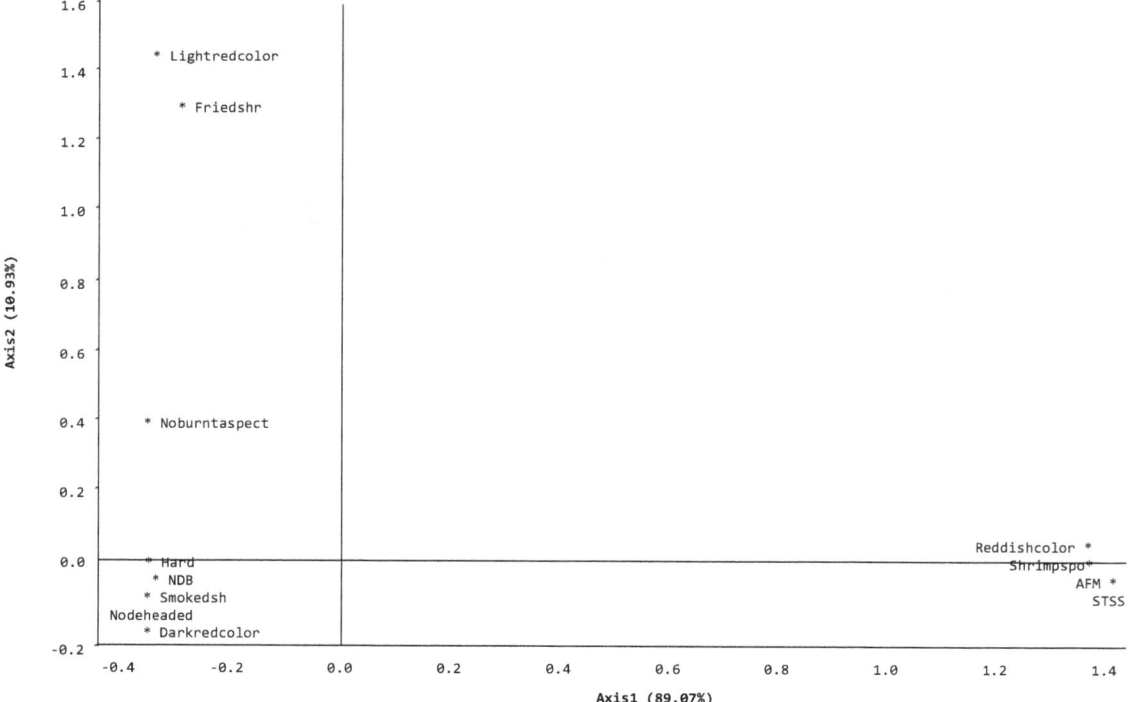

Figure 5. Factorial correspondence analysis to reveal links between determinative quality attributes and types of processed shrimps. Lightredcolor = Light red color; Darkredcolor = Dark-red color; Reddishcolor = Reddish color; Friedsh = Fried shrimps; Smokedsh = Smoked shrimps; Shrimpspo = Shrimps powder; Noburntaspect = No burt aspect; NDB = No deheaded and no burnt aspect; Nodeheaded = No deheaded; AFM = Absence of foreign matters; STSS = Spécial taste of smoked shrimps.

and the unstable nature of smoked shrimps during the storage. Future investigations need to be undertaken to characterize the fresh shrimps and its by-products commercialized in the local merchandising chain on both microbiological and physico-chemical aspects.

ACKNOWLEDGEMENTS

Financial support for this work was received from Belgium Government through UAC 01 Project. The authors are very grateful for this support. This paper is part of the PhD work for the first author.

REFERENCES

Adams MR, Cooke RD, Twiddy DR (1987). Fermentation parameters involved in the production of lactic acid preserved fish-glucose substrates. Int. J. Food Sci. Technol. 22:105-114.

Adu-gyamfi A (2006). Studies on microbiological quality of smoked fish in some markets in Accra, Ghana. Ghana J. Sci. 46:67-75.

Anacleto P, Bárbara T, Pedro M, Pedro S, Leonor Nunes M, Marques A (2011). Shelf-life of cooked edible crab (Cancer paragurus) stored under refrigerated condition. Food. Sci. Technol. 44:1376-1382.

Anonymous (2005a). La pêche au Bénin : atouts, contraintes et perspectives. Ministère de l'Agriculture de l'Elevage et de la Pêche, Direction des Pêches, Cotonou, Bénin, juillet 2005, P. 6.

Anonymous (2005b). Produit intérieur brut du Bénin: composantes et emplois. Rapport annuel INSAE/DSEE/SCN, P. 40.

Dagnelie P (1998). Statistiques théoriques et appliquées, Tome. 2:559.

Duedahl-Olesen LJH, Christensen JH, Højgardc A, Granby K, Timm-Heinrich M (2010). Influence of smoking parameters on the concentration of polycyclic aromatic hydrocarbons (PAHs) in Danish smoked fish. Food Addit. Contam. 9:1294-1305.

Jaffrès E, Lalanne V, Macé S, Cornet J, Cardinal M, Sérot T, Dousset X, Joffraud JJ (2011). Sensory characteristics of spoilage and volatile compounds associated with bacteria isolated from cooked and peeled tropical shrimps using SPME-GC-MS analysis. Int. J. Food Microbiol. 147:195-202.

Essien JP, Ekpo MA, Brooks AA (2005). Mycotoxigenic and proteolytic potential of moulds associated with smoked shark fish (Chlamydoselachus anguincus). J. Appl. Sci. Environ. 3:53-57.

Essuman KM (1992). Fermented fish in Africa: a study on processing, marketing and consumption. FAO Fisheries Technical Rome. 329:80.

FAO (2004). Mainstreaming fisheries into national development and poverty reduction strategies: current situation and opportunities, by A. Thorpe. FAO Fisheries Circular Rome. P. 997.

Gnimadi A, Gbaguidi A, Kakpo GL, Gnimadi CC, Latifou L, Salifou LL, Sohou ZL, Tossou CE (2006). Base de données sur les activités de pêche dans les lagunes du Bénin (lac Ahémé, lac Nokoué et lagune de Porto-Novo). Résultats du recensement. Programme pour des Moyens d'Existence Durable dans la Pêche (PMEDP), rapport de mission, 2:397.

Horemans B (1998). The state of artisanal fisheries in West Africa in 1997. Program for the integrated development of artisanal fisheries in West Africa, Cotonou- Benin, IDAF / WP/122:44.

Kim M, Mah J, Hwang H (2009). Biogenic amine formation and bacterial contribution in fish, squid and shellfish. Food Chem. 116:87-95.

Laghmari H, El Marrakchi A (2005). Appréciation organoleptique et physico-chimique de la crevette rose Parapenaeus longirostris (Lucas, 1846) conservée sous glace et à température ambiante. Revue Méd. Vét. 4: 221-226.

Lem A (2005). Aquaculture-world trends, opportunities for developing countries, technical and financial constraints. UNCTAD/FAO, Geneva. P. 48.

Raux J (2009). Diagnostic de la filière crevette au Bénin. Développement du secteur privé au Bénin et l'identification d'un projet de compétitivité et de croissance sous le 10eme FED, rapport de consultation, CE/FINEUROP, P. 50.

7

Fishes and smoked meat delicacies as sources of multidrug resistant bacteria and parasitic worms

Eze, E. A., Eze, Chijioke N., Amaeze, V. O. and Eze, Chibuzor N.

Department of Microbiology, University of Nigeria, Nsukka, Enugu State, Nigeria.

Frozen fishes, stock fishes, smoked pork and smoked beef (suya) that have become popular delicacies in eateries, road side stalls and similar places were analysed for bacteriological and parasitological burden. Samples were collected from Nsukka and Obollo-Afor districts of Enugu State, Nigeria and were screened using basic microbiological procedures. One hundred and eighty four persons from the study areas were also screened for parasitic worms and consumption of test delicacies. Results revealed the presence of some opportunistic and overt pathogenic bacteria, some of which exhibited resistance to a multiple of antibacterial agents. In addition, *Taenia solium* and *Taenia saginata* were detected in some smoked pork and beef samples respectively. Among the human respondents who acceded to suya and pork consumption (78.26% of total), 64.2% were positive for *T. saginata* while 28.0% harboured *T. solium*. Also detected among human respondents were members of the giant roundworms *Ascaris lumbricoides*. Occurrence of these isolated organisms in the test meat products raises hygiene and safety questions and the need for public health awareness and consciousness in this regard.

Key words: Fishes, meat delicacies, multidrug-resistant bacteria, parasitic worms.

INTRODUCTION

Emergence of antibiotic resistant pathogen is one of the most serious threats to public health in the 21st century (Willey et al., 2007). Multidrug resistance is emerging worldwide at an alarming rate among a variety of bacterial species, causing both community-acquired and nosocomial infections (Nordman et al., 2012). This may have arisen from extensive use of antibiotics in agriculture such as for treatment of infections, growth enhancement and prophylaxis in food animals at low concentration (Burgos et al., 2005). The prevalence of these drug and multi-drug resistant bacteria in food animals and their products may serve as a potential transfer route of antibiotic resistant bacteria and resistant genes into human food-chain and environment. This has the potentials to pose a health threat to lives (Tansuphasiri et al., 2006).

Some food products may be wholesome prior to processing. For instance, chicken is wholesome while the chicken is alive. However, the sterility can be compromised by some contaminants from the environment and in the process of cutting, packaging as well as during distribution (Ingham, 2001). Animals are known to constitute a vast reservoir of enteric bacteria with the general problem of drug resistance and environmental contamination through organic wastes and vectors (Bahrndorff et al., 2013; Adeleke and Omafuvbe, 2011; Olaitan et al., 2011; Samuel et al., 2011). Most of the contaminants of food animals and fishes originate from the alimentary tracts, respiratory tracts and or external surfaces of either life animal or the handlers. In Suya (a spicy, barbecued, smoked or roasted meat product) for instance, the possible contaminants come from the carcass itself, the handlers and or even from the spices used in the preparation. According to Edema et al.

(2008) all suya processors prepared the suya at their wooden stalls located by the roadsides. The surroundings are considered unhygienic given that garbage and dirty wastes litter the food processing environment with open gutters nearby, all of which attracted houseflies. None of the processors in their report was observed to wash their raw meat before suya preparation. Slabs and trays used for cutting and smoking were noted to be inadequately cleaned. Utensils made of plastic, metal or enamel were washed only once and the water used repeatedly until it becomes obviously oily, cloudy and dirty. The water used contained a significantly high coliform count in the order of 10^5 cfu/ml while the processing slab and raw meat both had counts of aerobic mesophiles above 5×10^5 cfu/g.

From the standpoint of microbiology, fishes and related products are risk foodstuff group. Particularly *Clostridium botulinum* type E and *Vibrio parahaemolyticus* rank among pathogenic bacteria associated with fishes. Freezing fish and related products in the sea-water, intensive handling, long-time transport or cooking in fishing containers straight on the deck contribute to their contamination with microorganisms (Novotny et al., 2004). Suya as a source of helminthic infection is worrisome considering that cattle and sheep are known (Adams and Moss, 1999) to be intermediate hosts of these worms from where man gets infected. The concern here is that suya is prepared (Edema et al., 2008) in such a way that intramuscular survival of these worms is not only possible but likely and man gets infected through injestion of food materials containing the encysted larval stage of these parasites.

Admittedly, the sources of contamination of suya and fishes as well as frozen meat are well known. However, little is known about the incidences of the bacteria and parasites in these food samples and the sensitivity pattern of these bacterial isolates. Edema et al. (2008) had earlier reported on the bacterial and fungal contaminant of suya in a different study area but no study has hitherto reported on this in the study area covered by this work and none has evaluated the heliminthic contamination and antimicrobial sensitivity profile of bacteria isolated there from. Importantly, there is an observable increase in the incidence of worm infection in Nsukka metropolis. The source of this infection is still obscure and the possibility of suya as a contributor is worth evaluating. This work was therefore carried out to determine the possibility of suya acting as a source of helminthic infections since these have become endemic in the study area. It is also aimed at assessing frozen and dried fishes for multidrug resistant bacteria considering the obvious public health implications.

MATERIALS AND METHODS

Sampling procedure

The samples of frozen fishes and stock fishes were randomly obtained from different grocery stores and markets in Obollo-Afor and Nsukka, Enugu State, Nigeria. Suya and smoked beef samples were randomly obtained from various suya spots within these areas. These samples were taken in sterile containers to the laboratory for analysis within six hours of collection. For each specimen (pork, frozen fishes, stock fishes and suya) a total of two samples per week were collected for 8 weeks (n = 16) and analyzed.

Parasitological analysis

One gram each of the samples (suya) and smoked pork was weighed into test tubes containing 4 ml of normal-saline for emulsification and dissolution. The suspension was made up to 7 ml by addition of 3 ml of normal-saline. The emulsified samples were sieved and 3 ml of diethyl ether was added to the resulting suspension. The suspension was shaken vigorously and centrifuged at 3,000 rpm for 5 min. The supernatant was decanted and a drop of the sediments was placed on a clean glass slide containing drops of iodine solution to stain sample. The identification of cysts in the sample was made according to the methods of worm isolation and identification (Ochei and Kolhatkar, 2007; Eggleston et al., 2008).

Examination of individuals for intestinal helminths

A total of 184 adult individuals (100 women and 84 men) from Nsukka and Obollo-Afor, both in Enugu state were examined for intestinal helminths using the methods of Adeoye et al. (2007) and Eggleston et al. (2008). Informed consent was obtained from participants who were also given questionnaires. Clean and sterilized plastic containers, appropriately labelled, were given to them for collection of fecal samples. Using an applicator, each stool sample was examined for its consistency, colour and presence of blood, mucous, adult worms and proglottides of tapeworms. Subsequently, further examinations were done on stool specimens using saline (0.85%) preparation, iodine stain, formol (10% formol saline) ether concentration for ova and cysts and the cellophane (Kato-katz) thick faecal smear techniques (Adeoye et al., 2007; Manson-Bahr and Bell, 1987). Eggs and cysts were recognized by their perculiar characteristics.

Determination of suya and smoked pork consumption pattern of participants

Questionnaires aimed at obtaining personal data and information on the period and frequency of suya consumption vis-a-vis noticeable signs and symptoms of worm infection were administered to the participants. Emphasis was placed on individuals who have consumed suya within 5-12 weeks (the usual incubation period of most cestodes) before examination while not neglecting others.

Microbiological analysis

All the samples (meat and fish) were prepared for bacteriological analysis following the methods of Stiles and Ng (1980) by weighing 10 g of each sample, blending and dispersing in 90 ml of sterile 0.1% peptone water. Appropriate dilutions were made, also with 0.1% peptone water blanks and subsequently 0.1ml of each was separately inoculated onto nutrient and MacConkey agar plates. Two sets of plates were prepared for each sample and incubated one set at 35°C and the other at 45°C for 24 h. Bacteria growths were isolated and identified by assessing colony characteristics and Gram reaction and by conduction coagulase and catalase tests,

Table 1. Genera and number of bacteria isolated from test samples.

S/N	Sample	Number of samples tested	Bacteria isolated	Number of samples harbouring bacteria species	Percentage occurrences
1	Frozen fishes	16	*Listerria* spp.	09	56.3
			Staph. aureus	12	75.0
			Aeromonas spp.	07	43.8
			Bacillus spp.	15	93.8
2	Stock fishes	16	*Bacillus* spp.	13	81.3
			Pseudomonas spp.	05	31.3
			Staph. aureus	11	68.8
			Proteus spp.	03	18.8
			Fecal *E. coli*	11	68.8
			Salmonella spp.	06	37.5
3	Suya	56	*Bacillus* spp.	37	66.1
			Fecal *E. coli*	18	32.1
			Pseudomonas spp.	22	39.3
			Staph. aureus	43	76.8
			Salmonella spp.	18	32.1
			Streptomyces spp.	12	21.4
			Enterobacter spp.	10	17.9
			Proteus spp.	06	10.7
4	Smoked pork	56	*Bacillus* spp.	26	46.4
			Fecal *E. coli*	06	10.7
			Pseudomonas spp.	13	23.2
			Staph. aureus	17	30.4
			Streptomyces spp.	09	16.1

sugar (including xylose) fermentation and other biochemical tests such as indole production, citrate utilization, urase activity, triple sugar iron (TSI) agar tests, gas and hydrogen sulphide production tests, and oxidase tests. All these were performed in accordance with the methods of Harley and Prescott (2002) and Brown (2007). Some specific tests such as low temperature (4°C) and high temperature (44.5°C) incubation, tumbling motility, colony counts, salt tolerance and Henry illumination (Adams and Moss, 1999) were performed for specific isolates showing some tell tale characteristics.

Antimicrobial sensitivity testing

This was done as earlier described (Eze et al., 2009). Briefly, overnight Mueller-Hinton (MH) broth cultures of each isolate were standardized to match 0.5 McFarland turbidity standards. Using sterile colon swabs (Evepon, Ind. Nig.) isolates were spread on dry MH agar surfaces and allowed to dry. Sensitivity discs were subsequently and carefully placed on the agar surfaces. Plates were incubated for 24 h at 37°C. After incubation and measurement of inhibition zone diameters, susceptibility ranges were scored following CLSI (NCCLS) (2006). Isolates of the same genus were numbered to reflect sources of isolation.

Statistical analysis

Analysis of variance was used to determine the level of differences

among some parameters evaluated in this study.

RESULTS

As anticipated in view of the condition under which these fish and meat products are prepared and sold, the samples had high bacterial burden as shown on Table 1. The level of worm cysts contents was relatively low with a maximum of 39.3% of the tested samples containing *Taenia saginata* (Table 2). Twenty five percent of the male respondents harboured *T. saginata* while *T. solium* were detected in 20.2%. Of the female respondents screened, 31.0% carried *T. saginata* while 24.0% had *T. solium* (Table 2). Table 3 shows that 38.0% of the female and 26.2% of the male respondents that acceded to suya consumption had worms. The percentages were lower (9.0 and 19.0% respectively) for smoked pork consumers as shown on Table 4. Among the male respondents, 26.2% harboured *Ascaris lumbricoides* while the female respondents had them up to 38.0% (Table 2).

Antibiotic resistance tests showed that up to 57.1% of species of *Aeromonas* isolated from frozen fishes were resistant to trimethoprim/sulfamethoxazole and ampicillin while 53.3% of *Bacillus* spp. isolated from the same

Table 2. Genera and number of helminths detected in samples.

S/N	Sample	Number of samples tested	Helminth detection	Number of samples harbouring helminths	Percentage occurrence
1	Suya	56	Taenia saginata	22	39.3
2	Smoked Pork	56	Taenia solium	19	33.9
3	Human (stool)	184			
3a	Men	84	Ascaris lumbricoides	22	26.2
			T. solium	17	20.2
			T. saginata	21	25.0
3b	Women	100	A. lumbricoides	38	38.0
			T. solium	24	24.0
			T. saginata	31	31.0

Table 3. Suya consumption among respondents within study period.

Sex	Number tested	Number positive	Number(percentage) of positive respondents that harboured helminths
Men	84	68	22 (26.2)
Women	100	76	38 (38.0)

Table 4. Smoked pork consumption among respondents within study period.

Sex	Number tested	Number positive	Number(percentage) of positive respondents that harboured helminths
Men	84	37	16 (19.0)
Women	100	25	09 (9.0)

source were resistant to norfloxacin, erythromycin and flucloxacillin (Table 5). Strains of fecal *Escherichia coli* isolated from stock fishes were 45.5% resistant to pefloxacin, cefalexin and nalidixic acid while more than 60.0% of *Pseudomonas* spp. from the same source were resistant to ofloxacin, pefloxacin, augumentin, gentamicin, and trimethoprim/sulfamethoxazole (Table 6). Smoked pork contained in addition to other bacteria species, strains of *Staphylococcus aureus* 29.4% or more of which were resistant to 9 antibacterial agents (Table 7). The antibiograms of bacteria isolated from suya are shown on Table 8 with species of *Pseudomonas* expectedly showing the highest resistance rates of up to 77.3% against trimethoprim/sulfamethoxazole and streptomycin. In the same vein, faecal *E. coli* from the same source were at least 22.2% resistant to all the antibacterial agents with which they were challenged. Statistical analysis showed that there were no significant differences among the resistance patterns of *E. coli*, *Pseudomonas* spp, and *Salmonella* spp. isolated from stock fishes and suya. Also *S. aureus* isolated from smoked pork, stock fishes and suya showed relatively high percentage resistance values of between 46.5 and 18.2% that did not differ significantly between source

groups. There was a significant difference between the occurrences of *S. aureus*, for example, in smoked pork on one hand (23.2%) and frozen fishes (75%), stock fishes (68.8%) and suya (76.8%) on the other hand.

DISCUSSION

Smoked beef (suya) and pork are increasingly becoming popular delicacies in our society especially in recreational facilities such as parks, restaurants and beer parlours. This trend cuts across social, ethnic and even religious divisions. The isolation of bacterial and parasitic agents from these delicacies and from frozen and dried fishes should therefore raise public health concern. Results of this investigation suggest that these ready-to-eat foods may be vectors in the transmission of overt or opportunistic pathogenic microorganisms as well as in the spread of multidrug resistant bacteria strains. For example, Table 6 shows that stock fishes harbour bacteria such as *S. aureus*, fecal *E. coli*, *Salmenella* spp. and *Pseudomonas* spp. which are known to have (opportunistic) pathogenic and public health importance (Cheesbrough, 2000). In addition, this work has shown

Table 5. Resistance pattern of bacteria isolated from frozen fishes.

Bacteria isolates	No. tested	Percentage (Number) resistant to:																
		OFX	PEF	CPX	AU	CN	S	CEP	ND	SXT	PN	NB	LC	E	APX	CH	RD	FLX
S. aureus	12	-	-	(5)41.7	-	(4)33.3	(4)33.3	-	-	-	-	(5)41.7	(4)33.3	(5)41.7	(4)33.3	(6)50.0	(5)41.7	(5)41.7
Aeromonas spp	7	(2)28.6	(1)14.3	(3)42.9	(1)14.3	(1)14.3	(1)14.3	(2)28.6	(3)42.9	(4)57.1	(4)57.1	-	-	-	-	-	-	-
Bacillus spp	15	-	-	(8)53.3	-	(7)46.7	(6)40.0	-	-	-	-	(8)53.3	(5)33.3	(8)53.3	(6)40.0	(7)46.7	(7)46.7	(8)53.3

OFX = Ofloxacin; PEF = Pefloxacin; CPX = Ciprofloxacin; AU = Augumentin; CN = Gentamicin; S = Straptomycin; CEP = Cefalexin; NA = Nalidixic acid; SXT = Trimethoprim/sulfamethoxazole; PN = Ampicillin; NB = Norfloxacin; LC = Lincomycin; E = Erythrmycin; APX = Ampicillin/cloxacillin; CH = Chloramphenicol; RD = Rifampicin; FLX = Flucloxacillin. A = All susceptible. - = Not tested (some discs were labelled as either Gram positive or negative and were used as such); Numbers in brackets are those positive for the respective tests vis-a-vis the total number tested.

Table 6. Resistance pattern of bacteria isolated from stock fishes.

Bacteria isolates	No. tested	Percentage (Number) resistant to:																
		OFX	PEF	CPX	AU	CN	S	CEP	NA	SXT	PN	NB	LC	E	APX	CH	RD	FLX
S. aureus	11	-	-	(5)45.5	-	(5)45.5	(5)45.5	-	-	-	-	(4)36.4	(2)18.2	(3)27.3	(2)18.2	(4)36.4	(3)27.3	(4)36.4
Bacillus spp.	13	-	-	(7)53.8	-	(6)46.2	(6)46.2	-	-	-	-	(3)23.1	(3)23.1	(5)38.5	(4)30.8	(6)46.2	(7)53.8	(7)53.8
Fecal E. coli	11	(4)36.4	(5)45.5	(2)18.2	(4)36.4	(2)18.2	(3)27.3	(5)45.5	(5)45.5	(6)54.5	(6)54.5	-	-	-	-	-	-	-
Salmonella spp.	6	(2)33.3	(3)50.0	(1)16.7	(3)50.0	(2)33.3	(3)50.0	(2)33.3	(2)33.3	(2)33.3	(3)50.0	-	-	-	-	-	-	-
Pseudomonas spp.	5	(4)80.0	(4)80.0	(2)40.0	(4)80.0	(4)80.0	(3)60.0	(3)60.0	(2)40.0	(3)60.0	(4)80.0	-	-	-	-	-	-	-
Proteus spp.	3	A	A	(1)33.3	(1)33.3	(1)33.3	(2)66.7	(1)33.3	(1)33.3	(1)33.3	(2)66.7	-	-	-	-	-	-	-

OFX = Ofloxacin; PEF = Pefloxacin; CPX = Ciprofloxacin; AU = Augumentin; CN = Gentamicin; S = Straptomycin; CEP = Cefalexin; NA = Nalidixic acid; SXT = Trimethoprim/sulfamethoxazole; PN = Ampicillin; NB = Norfloxacin; LC = Lincomycin; E = Erythrmycin; APX = Ampicillin/cloxacillin; CH = Chloramphenicol; RD = Rifampicin; FLX = Flucloxacillin. A = All susceptible. - = Not tested (some discs were labelled as either Gram positive or negative and were used as such); Numbers in brackets are those positive for the respective tests vis-a-vis the total number tested.

Table 7. Resistance pattern of bacteria isolated from smoked pork.

Bacteria isolates	No. tested	Percentage (Number) resistant to:																
		OFX	PEF	CPX	AU	CN	S	CEP	ND	SXT	PN	NB	LC	E	APX	CH	RD	FLX
S. aureus	17	-	-	(5)29.4	-	(6)35.3	(4)23.5	-	-	-	-	(6)35.3	(5)29.4	(6)35.3	(4)23.5	(6)35.3	(6)35.3	(5)29.4
Bacillus spp.	26	-	-	(3)11.5	-	(5)19.2	(5)19.2	-	-	-	-	(5)19.2	(4)15.4	(6)23.1	(4)15.4	(6)23.1	(5)19.2	(6)23.1
Fecal E. coli	06	(2)33.3	(1)16.7	A	(1)16.7	A	(1)16.7	(1)16.7	(1)16.7	(2)33.3	(2)33.3	-	-	-	-	-	-	-
Pseudomonas spp.	13	(7)53.8	(7)53.8	(5)38.5	(8)61.5	(6)46.2	(8)61.5	(7)53.8	(7)53.8	(8)61.5	(9)69.2	-	-	-	-	-	-	-

OFX = Ofloxacin; PEF = Pefloxacin; CPX = Ciprofloxacin; AU = Augumentin; CN = Gentamicin; S = Straptomycin; CEP = Cefalexin; NA = Nalidixic acid; SXT = Trimethoprim/sulfamethoxazole; PN = Ampicillin; NB = Norfloxacin; LC = Lincomycin; E = Erythrmycin; APX = Ampicillin/cloxacillin; CH = Chloramphenicol; RD = Rifampicin; FLX = Flucloxacillin. A = All susceptible. - = Not tested (some discs were labelled as either Gram positive or negative and were used as such); Numbers in brackets are those positive for the respective tests vis-a-vis the total number tested.

Table 8. Resistance pattern of bacteria isolated from suya.

Bacteria isolates	No. tested	Percentage (Number) resistant to:																
		OFX	PEF	CPX	AU	CN	S	CEP	NA	SXT	PN	NB	LC	E	APX	CH	RD	FLX
S. aureus	43	-	-	(18)41.9	-	(20)46.5	(18)41.9	-	-	-	-	(17)39.5	(15)34.9	(16)37.2	(20)46.5	(18)41.9	(20)46.5	(19)44.2
Bacillus spp.	37	-	-	(20)54.1	-	(21)56.8	(21)56.8	-	-	-	-	(18)48.6	(13)35.1	(17)45.9	(19)51.4	(15)40.5	(21)56.8	(21)56.8
Fecal *E. coli*	18	(6)33.3	(6)33.3	(4)22.2	(7)38.9	(4)22.2	(6)33.3	(6)33.3	(7)38.9	(7)38.9	(7)38.9	-	-	-	-	-	-	-
Enterobacter spp.	10	(4)40.0	(4)40.0	(3)30.0	(5)50.0	(3)30.0	(5)50.0	(4)40.0	(4)40.0	(5)50.0	(5)50.0	-	-	-	-	-	-	-
Salmonella spp.	18	(7)38.9	(8)44.4	(5)27.8	(9)50.0	(4)22.2	(7)38.9	(8)44.4	(8)44.4	(8)44.4	(9)50.0	-	-	-	-	-	-	-
Pseudomonas spp.	22	(11)50.0	(15)68.2	(9)40.9	(16)72.7	(10)45.5	(17)77.3	(15)68.2	(11)50.0	(17)77.3	(17)77.3	-	-	-	-	-	-	-
Proteus spp.	06	(1)16.7	(1)16.7	A	(2)33.3	A	(1)16.7	(1)16.7	(1)16.7	(1)16.7	(2)33.3	-	-	-	-	-	-	-

OFX = Ofloxacin; PEF = Pefloxacin; CPX = Ciprofloxacin; AU = Augumentin; CN = Gentamicin; S = Streptomycin; CEP = Cefalexin; NA = Nalidixic acid; SXT = Trimethoprim/sulfamethoxazole; PN = Ampicillin; NB = Norfloxacin; LC = Lincomycin; E = Erythrmycin; APX = Ampicillin/cloxacillin; CH = Chloramphenicol; RD = Rifampicin; FLX = Flucloxacillin. A = All susceptible. - = Not tested (some discs were labelled as either Gram positive or negative and were used as such); Numbers in brackets are those positive for the respective tests vis-a-vis the total number tested.

these organisms to be multidrug resistant with strains of *Salmonella* isolates for example, exhibiting more than 33% resistance to nine antibacterial agents. In a similar vein, strains of *E. coli* showed more than 36% resistance to 7 antibacterial agents while more than 60% of the *Pseudomonas* isolates were resistant to 8 antibiotics. While the medical implication of these is obvious, the public healthimport is more glaring in the light of the transferability of these traits among both pathogenic and potentially pathogenic bacteria (Eze et al., 2010). Bacteria isolated from smoked pork (Table 7) and those isolated from suya (Table 8) have shown similar attributes, thus raising the same serious concern. Bacteria isolated from frozen fishes are the least worrisome except for the drug resistant patterns and presence of *Aeromenas* spp. which have been implicated (Adams and Moss, 1999) in gastroenteritis and extraintestinal infections associated with immunocompromised hosts.

This study has also shown that the highly cherished smoked beef (suya) and pork often contain cysts of *T. saginata* and *T. solium* (Table 2). While taeniasis caused *T. saginata* can result

in allergic reactions, chronic indigestion, constipation and inflammation of the appendix, *T. solium* is acknowledged as the most harmful tapeworm in humans (because of its ability to cause cysticercosis especially neurocysticercosis). Their presence in the study samples is therefore undesirable. The risks this may portend is corroborated by the more than 26% presence of helminths among suya consumers (Table 3) and 19% presence of the same worms among smoked pork male consumers (Table 4). Although a statistical correlation has not been established, these results suggest that suya and smoked pork may be "vectors" (at least in part) of the increasing cases of worm including *Ascaris lumbricodies* (Table 1) infestation in the study areas. Overall, this study has portrayed the analyte delicacies as potential vehicles for the transmission of parasitic worms and multidrug resistant bacteria. Further (correlation) research and public awareness programmes are recommended.

REFERENCES

Adams MR, Moss MO (1999). Food Microbiology. The Royal Society of Chemistry Cambridge. pp. 156-190, 220-223.

Adeleke EO, Omafuvbe BO (2011). Antibiotic Resistance of Aerobic Mesophilic Bacteria Isolated from Poultry Faeces. Res. J. Microbiol. 6(4):356-365.

Adeoye GO, Osayemi CO, Oteniya O, Onyemekeihia SO (2007). Epidemiological Studies of Intestinal Helminthes and Malaria among Children in Lagos, Nigeria. Pak. J. Biol. Sci. 10(13):2208-2212.

Bahrndorff S, Rangstrup-Christensen L, Nordentoft S, Hald B (2013). Foodborne Disease Prevention and Broiler Chickens with Reduced *Campylobacter* Infection. Emer. Infect. Dis. 19(3):425-430.

Brown AE (2007). Benson's Microbiological Applications: Laboratory Manual in General Microbiology. McGraw Hill, Boston. pp. 109:253-293.

Burgos JM, Ellington BA, Varela MF (2005). Prevalence of Multi-Drug Resistant Enteric bacteria in Dairy farm Topsoil. J. Dairy Sci. 88:1391-1398.

Cheesbrough M (2000). District Laboratory Practises In Tropical Countries (Part 2). Cambridge University Press K. pp. 23-143.

CLSI (2006). National Committee for Clinical Laboratory Standards- Performance Standards for Antimicrobial Disc Susceptibility Tests. CLSI 26(1). Wayne Pa.

Edema MO, Osh AT, Diala CI (2008). Evaluation of Microbial Harzards Associated with the Processing of Suya (a grilled meat product). Sci. Res. Essay 3:621-626.

Eggleston TL, Fitzpatrick E, Hager KM (2008). Parasitology as a Teaching Tool: Isolation of Apicomplexan Cysts from Store-bought Meat. CBE-Life Sci. Educ. 7:184-192.

Eze E, Eze U, Eze C, Ugwu K (2009). Association of Metal

Tolerance with Multidrug Resistance Among Bacteria Isolated from Sewage. J. Rural Trop. Publ. Health 8:25-29.

Eze EA, Ngananga BC, Ugwu KO, Nwuche CO (2010). Transfer of Multidrug Resistance among Bacteria Isolated from Industrial Wastes. Bio-Res. 8(2):689-693.

Harley JP, Prescott LM (2002). Laboratory Exercises in Microbiology (5th ed.). The Mc-Graw-Hill Companies Beston. pp. 292-293.

Ingham J (2001). Food Facts for You Http//:Uwex.edu/Ces/flp/specialist/ingham.htlm.

Manson-Bahr PEC, Bell DR (1987). Manson's Tropical Diseases (19th ed.) Bailliere Tindall London. pp. 1489-1493.

Nordman P, Poirel L, Dortet L (2012). Rapid Detection of Carbapenemase-producing Enterobacteriaceae. Emerg. Infect. Dis. 18(9):1503-1507.

Novotny L, Dvorska L, Lorencova A, Beran V, Pavlik I (2004). Fish: A potential source of bacteria pathogens for humans. Vert. Med.-Czech. 49:343-358.

Ochei J, Kolttatkon A (2007). Medical Laboratory Science Theory and Practice. Tata McGraw-Hill Publishing Co. New Delhi. pp. 1029-1033.

Olaitan JO, Shittu OB, Akinliba AA (2011). Antibiotic resistance of enteric bacteria isolated from duck droppings. J. Appl. Biosci. 45:3008-3018.

Samuel L, Marian MM, Apun K, Lesley MB, Son R (2011). Characterization of Escherichia coli isolated from cultured catfish by antibiotic resistance and RAPD analyses. Int. Food Res. J. 18(3):971-976.

Stiles ME, Ng L (1980). Estimation of Escherichia coli in Raw Ground Beef. Appl. Environ. Microbiol. 40(2):346-351.

Tansuphasiri U, KhaminthakulD, Pandii W (2006). Antibiotic resistance of enterococci isolated from frozen foods and environmental water. Southeast Asian J. Trop. Med. Public Health 37(1):162-170.

Willey JM, Sherwood LM, Woolverton CJ (2007). Prescott, Harley, and Klein's Microbiology (7th Ed.) McGraw-Hill Companies Inc. New York. pp. 53-854.

8

Clinical changes observed in *Clarias gariepinus* (Burchell 1822) fed varying levels of ascorbic acid supplementation

Gbadamosi O. K., Fasakin E. A. and Adebayo O. T.

Department of Fisheries and Aquaculture Technology, P. M. B 704, Akure, Ondo State, Nigeria.

This study was undertaken to observe the clinical changes associated with the dietary ascorbic acid supplementation in the diet of African catfish, *Clarias gariepinus*, using the presence or absence of scorbutic (ascorbic acid deficiency) symptoms as indices. *C. gariepinus* fingerlings weighing 6.02±0.4 g were randomly distributed into glass tanks of 60 × 45 × 45 cm^3 dimension at ten fish per tank in a triplicate treatment. Five isonitrogenous and isocalorific diets containing 40% crude protein was formulated. Ascorbic acid (AA) was supplemented in the diets as ascorbyl-2-polyphosphate (a mixture of phosphate esters of ascorbate). Each treatment had varying levels of ascorbic acid (AA) supplementation, at 0 (Control) 50, 100, 150 and 200 mg AA/kg, in Treatment 1, 2, 3, 4 and 5 respectively. Fish were fed practical diets twice daily at 900 and 1600 h GMT. Weekly weighing of fish was done and the data collected were subjected to statistical analysis. At the end of week 4, fish fed scorbutic diets (diets without ascorbic acid supplementation) had significantly lower weight than fish fed AA supplemented diets (P<0.05). After week 6 fish fed scorbutic diet began to develop clinical symptoms including lordosis (lateral curvature), broken skull, pigmentation and scoliosis (vertical curvature). Radiographs confirmed defects in the vertebral columns of fish in this treatment. Highest mortality was recorded in this treatment as 30%, followed by treatment two fed 50 mg AA/kg which had 10% mortality, fish in this treatment also showed some clinical signs like eroded fins and pigmentation. In all parameters considered, treatment 4 fed 150 mg AA/kg gave the best AA supplementation that prevented growth reduction and clinical signs of AA deficiency in this study.

Key words: Clinical, ascorbic acid, *Clarias gariepinus*, scorbutic, supplementation.

INTRODUCTION

The inability of many fish species to synthesize ascorbic acid (AA), or vitamin C, which is essential for fish growth, reproduction and health is well documented (Dabrowski, 1990; Soliman et al., 1986; Yamamoto et al., 1985). Ascorbic acid must therefore be supplied via feed. Major clinical sign of ascorbate deficiency include reduced growth, scoliosis, lordosis, internal and fin heamorrhage, distorted gill filaments, fin erosion, abnormal pigmen-tation, increased capillary fragility, reproductive performance (Halver, (1989);Sadnes et al., 1992). Bone rare faction spinal deformity and crackhead diseases have recently been traced to Vitamin C deficiency syndrome in intensive catfish production systems (Halver, 1990). The recommended dietary level of ascorbic acid (Vitamin C) in the 1993 NRC bulletin is 100 mg/kg for fish (NRC, 1993).

Table 1a. Composition of the experimental diet in g/100 g dry matter containing various inclusion level of ascorbic acid supplementation for *Clarias gariepinus*.

Ingredient	Treatments				
	1 (control)	2	3	4	5
Fish meal	22.00	22.00	22.00	22.00	22.00
Groundnut cake	28.00	28.00	28.00	28.00	28.00
Soy bean meal	24.00	24.00	24.00	24.00	24.00
Yellow maize	11.00	11.00	11.00	11.00	11.00
Vegetable oil	5.00	5.00	5.00	5.00	5.00
Oyster shell	2.00	2.00	2.00	2.00	2.00
Rice bran	4.00	4.00	4.00	4.00	4.00
*Vit/Min premix	2.00	2.00	2.00	2.00	2.00
Salt	1.00	1.00	1.00	1.00	1.00
Starch	1.00	1.00	1.00	1.00	1.00
Ascorbic acid (mg/kg)	0.00	50	100	150	200

*Premix as supplied by Animal Care, Limited, Lagos, Nigeria. Vitamins supplied mg/100 g diet: thiamine (B_1) 2.5 mg: riboflavin (B_2), 2.5 mg pyridoxine 2.0 mg: pantothenic acid, 5.0 mg: inositol, 3 mg: biotin, 0.3 mg: folic acid, 0.75 mg para-amino benzoic, 2.5 mg: chlorine, 200 mg; niacin, 10.0 mg, cynobalamin (B_{12}), 10.0 mg; menadione (k), 2.0 mg. Minerals: $CaHPO_4$, 727.8 mg: Mg SO_4, 1275 mg , 60.0 mg; kCl 50.0 mg; $FeSO_4$, 250 mg, $ZnSO_4$, 5.5 mg; $MnSO_4$, 2.5 mg $CuSO_4$, 0.79 mg; $CoSO_4$, 0.48mg: $CaClO_3$, 0.3 mg; Cr Cl_3.

The clariid catfish, *Clarias gariepinus*, (Burchell, 1822) is the most important fish species cultured in Nigeria. This species has shown considerable potential as a fish suitable for use in intensive aquaculture. This fish grows rapidly, it is disease and stress resistant, sturdy and highly productive in polyculture with many other fish species. However, with the exception of α-tocopherol (Baker and Davies, 1997), there is dearth of work in both qualitative and quantitative ascorbic acid requirement for the African catfish *C. gariepinus*. The dearth of information on the importance of vitamin C as an immunomodulator and a key nutritional element in the African catfish *C. gariepinus* and the need to establish the mechanisms through which ascorbic acid as a nutritional element influences the immunological and haemolytical systems in modern intensive fish farming prompted this study.

MATERIALS AND METHODS

Experimental diets

Five isonitrogenous and isocalorific diets containing 40% crude protein and 12% lipids were formulated for fingerlings catfish, *C. gariepinus* in a ten-week trial experiment (Table 1a). Ascorbic acid, commercially available as ROVIMIX STAY C- (Roche, Istanbul, Turkey) was used. Scorbutic diets (without ascorbic acid supplementation) served as the control. Ascorbic acid supplementation in diets 2 to 5 were 50.0, 100.0, 150.0 and 200.0 mg/kg respectively. All dietary ingredients were weighed with a weighing top balance (Metler Toledo, PB8001 London). The ingredients were then ground to a small particle size. Ingredients including vitamin/premix and ascorbic acid were thoroughly mixed in a Hobbart A-200 T pelleting and mixing machine (Hobart Ltd London, England) to obtain a homogenous mass, cassava starch was added as a binder. The resultant mash was then pressed

without steam through a mincer with 0.9 mm die attached to the Hobart pelleting machine. Diets were immediately sun-dried. After drying the diets were broken up, sieved and stored at (-8°C) in refrigerator until immediately prior to feeding.

Experimental fish management

Clarias gariepinus fingerlings with average weight of 6.0g ± 0.4 g were obtained from the hatchery of Ondo State Agriculture Development Programme, Akure, Ondo State, Nigeria and were randomly distributed into glass tanks (60 × 45 and 45 cm) at ten fish per tank (replicate) in a flow through system, with three replications per treatment. Tanks were supplied with water from a borehole powered by 1.5 HP Pumping machine. Water temperature was maintained at 24 ± 0.5°C. Dissolved oxygen was kept at a saturation level of 6.00 ± 0.1 mg/L. The fish were fed with their respective diets at 5% body weight per day in two equal portions at 9.00 - 10.00 and 1600 - 17.00 h GMT for 70 days. Fish responded to feeding well on the diets immediately. Fish weights were determined at the 7th day each week by weighing all the fish in the tank. The quantity of feed was adjusted based on the changes in the body weight of fish for subsequent feeding.

Proximate composition

Proximate composition of diets and fish carcasses before and after the experiment, were performed according AOAC (1990) for moisture content, fat, and ash. The result of the proximate analysis of experimental diet is as shown in Table 2.

Ascorbic acid determination

Ascorbic acid concentration in diets and target organs, that is, liver and kidney was determined by semi-automated flourometric method as described by AOAC (1990). Samples were weighed into a plastic cup containing 100 ml solvent extract, 4% metaphoporic acid was added and the mixture filtered. The filtrate was passed through an

Table 1b. Proximate composition of experimental diets (% DM).

Ingredient	T_1	T_2	T_3	T_4	T_5
Crude protein	40.28	40.19	40.21	40.13	40.09
Lipid	12.39	12.21	12.33	12.17	12.03
Crude fibre	5.09	5.28	5.11	5.19	5.42
Ash	8.35	8.36	8.48	8.33	8.44
Moisture content	13.41	13.54	13.61	13.48	13.37
Nitrogen-free extract (NFE)[1]	20.48	20.42	20.26	20.70	20.65
Added ascorbic acid mg/kg	0.0	50.00	100.00	150.00	200.00
Measured ascorbic acid mg/kg	6.40	56.20	109.70	165.90	204.83
Gross energy[2] kcal/100 g	431.30	429.00	429.40	429.70	427.50

[1]Nitrogen free extract: calculated as 100- (crude protein + ash + crude fibre + ether. [2]Gross energy (kcal/100 g) based on 5.7 kcal protein; 9.5 kcal/g lipid; 4.1 kcal/g carbohydrate.

Table 2. Proximate composition (% wet weight) of the carcass of *C. gariepinus* fed experimental diets containing varying inclusion levels of ascorbic acid.

Parameter	Sample initial (%)	Final sample of Fish					S.E. ±
		T_1	T_2	T_3	T_4	T_5	
Moisture	73.05^c	73.01^c	72.18^b	71.48^a	71.49^a	72.01^b	0.16
Protein	15.15^a	15.23^b	16.15^c	17.11^d	17.16^d	17.01^d	0.28
Fat	6.30^b	7.03^a	7.10^a	7.20^a	7.19^a	6.91^b	0.08
Ash	4.70^a	4.33^b	4.56^{ab}	4.21^{bc}	4.16^{cd}	4.07^d	0.93

Figures in each row having the same superscripts are not significantly different (P > 0.05).

Autoanalyzer (Technicon corporation Ubana, Illnois). Ascorbic acid concentration was recorded in mg/100 g.

Performance evaluation

Fish performances during the experiment were based on productivity indices, growth performance described by Fasakin et al. (2003) as follows:

Total weight gain = final weight - initial weight

Mean daily weight gain/weeks

This was determined by dividing the total weight gain by the number of weeks of the experiment:

Final mean weight - initial mean weight / number of weeks

Total percentage weight gain (%)

This was calculated using the formula:

TPWG = Total weight gained × 100 / Initial weight

Specific growth rate (SGR) = $(\log_e Wt - \log_e Wi) / T \times 100\%$

Where Wt = final weight (g), Wi = initial weight (g), and T = rearing periods.

Feed conversion ratio (FCR) = dry weight of fish (g)/ fish weight gain (g).

X-ray photographs

At the end of the experimental period, three fish in each group were sampled for vertebral column integrity by X-ray Radiograph (Fracalossi et al., 1998).

Data analysis

Biological data resulting from the experiments was subjected to one-way analysis of variance (ANOVA) using the SPSS (Statistical Package Computer Software). Duncan's multiple range was used to compare differences among individual means (Zar, 1984). Differences were considered significant at P-levels < 0.05.

RESULTS

Proximate composition of feed

The proximate composition of the experimental feeds is shown in Table 1b. The initial and the final carcass proximate analysis (% wet weight) of the experimental fish, African catfish are presented in Table 2.

The results of growth performance of *C. gariepinus* fingerlings fed varying inclusion levels of ascorbic acid are shown in Table 3. In the feeding trials fish fed diet without ascorbic acid (AA) (Control) had a significantly lower weight gain than fish fed AA supplemented diets (P<0.05). Fish fed 150 mg/kg ascorbic acid T_4 showed

Table 3. Cumulative growth performance and nutrient utilization of *C. gariepinus* fed varying levels of ascorbic acid.

Parameter	Treatment 1	Treatment 2	Treatment 3	Treatment 4	Treatment 5
Final weight	18.81 ± 0.10^a	32.12 ± 1.40^b	40.61 ± 1.10^c	70.01 ± 0.10^e	50.76 ± 1.20^d
Initial weight (g)	6.06 ± 0.30^a	6.02 ± 0.50^a	6.06 ± 0.40^a	6.05 ± 0.30^a	6.02 ± 0.20^a
Weight gain (g)	12.75 ± 0.50^a	26.10 ± 0.50^b	34.55 ± 0.15^c	63.96 ± 0.17^e	44.74 ± 0.22^d
Daily feed int. (g)	0.32 ± 0.18^a	0.55 ± 0.24^b	0.70 ± 0.19^c	1.20 ± 0.38^e	0.87 ± 0.15^d
FCR	1.77 ± 0.02^a	1.48 ± 0.01^b	1.28 ± 0.03^c	0.94 ± 0.08^e	1.13 ± 0.04^d
SGR (%)	1.61 ± 0.06^a	2.38 ± 0.04^b	2.72 ± 0.05^c	3.50 ± 0.02^e	3.03 ± 0.15^d
% Weight gain	310.40 ± 0.71^a	433.60 ± 0.65^b	570.13 ± 0.9^c	1057.2 ± 1.5^e	743.19 ± 0.89^d
Mortality %	30^a	10^b	0.00^c	0.00^c	0.00^c

Figures in each row having the same superscripts are not significantly different (P < 0.05).

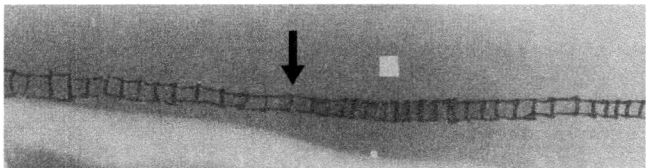

Plate 1. Radiograph of the backbone of *Clarias gariepinus* that were not fed ascorbic acid supplemented diets in Treatment 1, showing deformed backbone.

the highest percentage weight gain. Fish without AA supplementation (Control) showed reduction in growth.

Fish without dietary AA supplementation in Control, developed deformities in the vertebral column as shown in Plate 1, while fish fed dietary ascorbic acid showed no deformities in their vertebral column. Radiographs of the vertebral column further confirmed the deformity of the vertebral column in T_1 (Control) as shown in Plate 1 (arrow). Other clinical signs such as skin pigmentation, heamorrhage in the eye, fins and broken skull were detected in Treatment 1 (T_1) (Control) and scorbutic related signs, skin pigmentation, heamorhage of the eye were also noticed in Treatment 2 (T_2), whereas fish in Treatment (T_3), Treatment (T_4) and Treatment 5 (T_5) did not show any clinical signs .

DISCUSSION

The results of this study showed the clinical and deficiency symptoms in *C. gariepinus* fed ascorbic acid supplemented diets. The temperature (24.6 ± 0.1°C) and dissolved oxygen (6.68 ± 0.1 mg/L) values were within the range recommended for African Catfish (Boyd, 1986). The proximate analysis of experimental the diets revealed that they contain a mean of 40% crude protein. This range is suitable for *C. gariepinus* growth and agrees with 8 to 14% fat content recommended for *C. gariepinus* fingerlings (Jauncey and Ross, 1982). The highest growth rate in this study was recorded for fish fed

diet 4 while the lowest was found in fish fed the control diet.

From the result of the weight gain, it was obvious that fish fed diet 4 had the highest weight gain. This was in agreement with the result recorded by Li and Lovell (1985) for channel catfish fed different levels of AA after bacterial ingestion, where the fish in the control stopped growing from the 8th week of the experiment (Fracalossi et al., 1998) also recorded that Oscar fish *Astronotus ocellatus* without AA supplementation stopped growing at the 14th week.

Fish fed 150 mg/kg AA supplemented diets showed the best growth profile during the 10th week feeding period as shown in Table 3. In Treatment 2, growth rate was poor and inconsistent from week 2 to week 9, with fish fed 50 mg/kg showing some form of AA deficiency signs like pigmentation, and lesions. Dalta and Kaviraj (2003) also recorded deficiency symptoms in African Catfish, *C. gariepinus* fed less than 100 mg/kg ascorbic acid when exposed to deltamethrin induced stress.

Radiograph of the *C. gariepinus* backbone in the study was shown in Plate 1. The vertebral column of fish in treatment 1 (control) showed deformity, in the backbone. This further confirmed the vertebral column deformity in *C. gariepinus* fed scorbutic diets. No deformity was noticed in *C. gariepinus* that received ascorbic acid supplementation this also agreed with the result of Kumari et al. (2005) which recorded that *C. gariepinus* fed 150 mg/kg AA and above did not show any deformity in their backbone. *C. gariepinus* in control without ascorbic supplementation presented typical clinical signs reported for vitamin C deficiency such as reduced growth, impaired collagen formation and lordosis, crackhead as shown in Plates 2 and 3. Some clinical signs like depigmentations were also noticed in fish fed 50 mg/kg, as shown in Plate 4, while fish in T_3, T_4 and T_5 did not manifest any clinical deficiency signs of ascorbic acid.

Mortality was highest in the fish fed control diet, where there was no AA supplementation. Death in the control tank did not occur suddenly. There was usually a gradual withdrawal from feeding, inactivity, emaciated, serious underunning necrosis of the skull, grayish skin lesions

Plate 2. Deformity in the vertebral column of *C. gariepinus* after feeding on ascorbic acid-deficient diet in treatment one.

Plate 3. Crack head disease noticed in *C. gariepinus* after feeding on ascorbic acid deficient diet in treatment one.

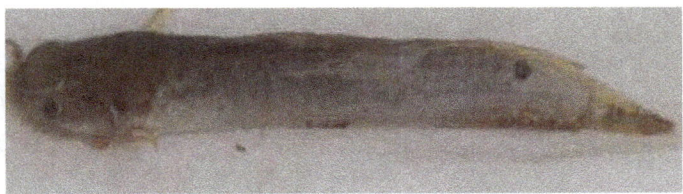

Plate 4. Dermopathy induced in *C. gariepinus* after feeding on 50 mg AA/kg ascorbic acid diet in treatment two.

occurs leading to "crack head" Plate 4. No mortality occurred in T_3, T_4 and T_5. This was in concordance with the work of Fracalossi et al. (1998) on rainbow trout and Oscar, respectively where no mortality was recorded apart from accidents for fish fed with AA supplementation above 100 mg/kg for rainbow trout and 50 mg/kg for Oscar, *Astronotus ocellatus*. The great improvement of *C. gariepinus* survival by adding ascorbate 2 polyphosphate in diets indicates that ascorbate 2 polyphosphate is an effective dietary AA source and can be utilized by *C. gariepinus* to prevent AA related mortality. This improvement is in accordance with the percentage mortality recorded by Kumari et al. (2005) in Asian Catfish, *Clarias batrachus* fed 100 mg/g fed dietary vitamin C.

Conclusion

The result of this study confirmed the efficacy of ascorbic

acid (AA) as an essential nutritional element in the diet of African catfish *C. gariepinus* in promoting optimal survival and normal growth performance. When this nutrient was absent from the diet, *C. gariepinus* developed clinical signs of ascorbic acid deficiency. *C. gariepinus* without ascorbic acid presented clinical signs such as reduced growth, impaired vertebral column, lordosis, and vitamin C deficiency syndrome which includes crackhead disease (Halver, 1989). Apart from spinal deformities and clinical ascorbic acid deficiencies signs, ascorbic acid deprivation in *C. gariepinus* as noted in the fish fed scorbutic (lacking vitamin C) challenged the physical integrity and survival of fish. Ascorbic acid has been reported to be a dietary facilitator (Verlhac et al., 1996). In all the treatments when compared *C. gariepinus* in treatment 4 with 150 mg/kg acid supplementation had the best performance. Based on this result, it can be concluded that the dietary ascorbic acid supplementation is necessary in *C. gariepinus* farming.

RECOMMENDATION

A dose of 150 mg/kg ascorbic acid supplementation per kilogram diet which was sufficient to prevent reduction in growth or development of the clinical signs of scurvy/ascorbic acid deficiencies and at the same time gave the best performance in this study is therefore recommended *prima facie* for catfish *C. gariepinus* farmers.

REFERENCES

AOAC (1990). Association of Official Analytical Chemists. Official method of analysis. (15 ed) P. 1. (K. Heltich ed.), Arlington, Virginia.

Baker RTM, Davies SJ (1997).The quantitative requirement for α-tocopherol by juvenile African catfish, Clarias gariepinus, Burchell. Ani. Sci. 65:135-142.

Boyd CE (1986). Water quality in warm water fish ponds. Agricultural experiment J. Aubum Univess. P. 35.

Dabrowski K (1990). Ascorbic acid status in the early life of white fish (Coreyonus lavaretus L.). Aquaculture 84:61-70.

Dalta M, Kaviraj A (2003). ascorbic acid supplementation of diets for reduction of deltamethrin induced stress in freshwater Catfish, Clarias gariepinus. Aquatic science Abstracts (ASFA). 53(8):883-888.

Fasakin EA, Balogun AM, Daramola AG (2003): A linear programming approach to low-cost fish feed formulation using water fern and duckweed as protein source .14th Annual National Conference of Fisheries Society of Nigeria. 19-23 January, 1998, Ibadan, Nigeria.

Fracalossi DM, Allen ME, Nichols DK, Oftedal OT (1998). Oscars Astronotus Ocellatus, have a dietary requirement for Vitamin C. Department of Zoological Research Paper. Smithsonoan Inst. Washington, DC 20008.

Halver JE (1989). The Vitamins. In: Fish Nutrition, 2nd edn (ed. By J. E. Halver), Academic Press New York. NY. pp. 32-102.

Halver JE (1990). Fish Nutrition 3rd Edition. Academic Press. London.

Jauncey K, Ross B (1984). A guide to tilapia feed and feeding. Institute of Aquaculture. University of Stirling, Scotland. P. 111.

Kumari T, Jaya O, Sahoo PK (2005). High dietary Vitamin C affects growth, non-specific immune responses and disease resistance in Asian Catfish.

Li Y, Lovell RT (1985). Elevated levels of dietary ascorbic acid increases immune response in channel catfish. J. Nutri. 115:23-131.

NRC (1993). (National Research Council). Nutrient Requirement of Fish. National Academy Press. Washington D.C. P. 114.

Sadnes K, Torrisen O, Waagbo R (1992). The minimum dietary requirement of Vitamin C in Atlantic salmon (Salmon salar) fry using Ca ascorbate –2- monophosphate as dietary source. Fish phiscol. Biochem. 10:315-319.Soliman AK, Jauncey K, Roberts RJ (1986). The effect of varying forms of dietary ascorbic acid on the nutrition of juvenile tilapias (Oreochromis niloticus) Aquaculture. 52:1-10.

Verlhac V, Gabaudan J, Obach A, Schuep W, Hole A (1996). Influence of dietary glucan and vitamin C on non-specific and specific immune responses of rainbow trout (Oncorhynchus mykiss) Aquaculture 143:123-133.

Yamamoto Y, Sato M, Ikeda S (1985). Existence of L-gulonolactone oxidase in some telecosts. Bull. Jpn. Soc. Sci. Fish 44:775-779.

Zar JH (1984). Biostatistical analysis. Prentice-Hall, Englewood Cliffs, New Jersey. USA.

9

Effects of aqueous extract of *Ricinus communis* on radial growth of *Alternaria solani*

Bayaso I.[1], Nahunnaro H.[2], and D. M. Gwary[3]

[1]Ministry of Agriculture headquarters P. M. B. 2079, Yola, Adamawa State, Nigeria.
[2]Department of Crop Production and Horticulture, Federal University of Technology P. M. B. 2076, Yola, Adamawa State, Nigeria.
[3]Department of Crop Protection, Faculty of Agriculture, University of Maiduguri, P. M. B. 1069, Maiduguri, Borno State, Nigeria.

The study evaluated the effect of *Ricinus communis* aqueous extract on radial growth of *Alternaria solani* and its most effective concentration. The experiment was laid out using a Completely Randomized Design (CRD) with *R. communis* extract tested at 3 concentrations of 25, 50 and 100% plus control on Potato Dextrose Agar (PDA) amended medium in 3 replications for 2 separate experiments. Isolation and identification of the early blight pathogen was made by symptoms on tomato plants, macro and microscopic observations in pure culture. Data on radial growth were collected, statistically analysed and percent inhibition (I%) was calculated. Results revealed that, *R. communis* at 100% concentration recorded the lowest radial growth either at 24, 48 and 72 h post inoculation (hpi) in the first, second, and combined results. The combined results further revealed that, the lowest radial growth 1.43, 2.00, and 2.72 cm were recorded in *R. communis* treatment at 24, 48 and 72 hpi, respectively. I% varied from 26 to 59%, according to the experimental conditions, and it was then concluded that, *R. communis* extract used at different concentrations had inhibitory effect on *A. solani*. It is was then suggested that, *R. communis* at 100% concentration could be put to field trial to evaluate its effectiveness in the control of the early blight pathogen *A. solani*.

Key words: *Alternaria solani, Ricinus communis*, radial growth, concentration.

INTRODUCTION

Tomato is attacked by many diseases which constitute a serious setback to its production (Peet, 2003). Some of these diseases include: early blight (*Alternaria solani*), late blight (*Phytophthora infestans*), damping off (*Phythium solanacerum*), bacterial wilt (*Burkholderia solanacearum*), fungal wilt (*Fusarium oxysporum*) and nematode (*Meloidogyne javanica*). The early blight of tomato induced by *A. solani* (Ell. and Mart.) has in no small measure contributed to yield losses of up to 79%, as reported from Canada, India, USA, and Nigeria (Sherf and MacNab, 1986; Gwary and Nahunnaro, 1998).

In the past, growers reported increasing incidence of this disease and decline in its control and have to deal with development of resistance of *Alternaria* species toward over-use of fungicides (Iacomi-Vasilescu et al., 2004). Hence, systematic screening of plant extracts may result in the discovery of novel effective compounds (Tomoko et al., 2002). Therefore scientific research is important in order to determine the presence of antifungal activity in the crude extracts of some common plants, known or not for their biological activity. Consequently, this study was aimed at evaluating *Ricinus communis* extract in control of the early blight pathogen *A. solani in vitro* and at determining the most effective concentration.

MATERIALS AND METHODS

Experimental sites

This study was carried out in the Laboratory of the Crop Production and Horticulture Department, Federal University of Technology, Yola, Adamawa State, Nigeria which lies between latitude 8°N and 11°N and longitude 11.5°E and 13.5°E.

Experimental design and layout

The experiment was conducted in the laboratory using a Completely Randomized Design (CRD) and aqueous plant extracts of Castorbean plant *(R. communis)* used at 3 concentrations (25, 50 and 100%) with a control to give a total of 4 treatments.

Collection of plant materials and preparation of extracts

R. communis plant was collected from around the Federal University of Technology, Yola. The collected plant materials were rinsed, washed with 10% sodium hypochlorite (NaOCl), air dried and later packed in brown envelops and oven-dried at 70°C for 20 mins according to Akinbode and Ikotun (2008). Thereafter, the plant materials were grinded using mortar and pestle, sieved using 40 mm sieve. About 200 g of the plant powder was added into 500 ml of distilled water and stirred to get uniform suspension. The suspension was allowed to stand for 24 h and the content was filtered using a muslin cloth and kept in glass bottles until needed. The different concentrations were then prepared by taking 25, 50, and 100 ml of the stock preparation and dissolved in distilled water to give 25, 50, and 100% concentrations.

Isolation and identification of *A. solani*

Diseased tomato leaves with the symptoms of *A. solani* were collected from farms around the University. A small piece from the advancing margin of a lesion on diseased leaf was cut with a sterile pair of scissors sterilized with 10% NaOCl (Larone, 1995). The tissues were then washed thoroughly in several changes of sterile distilled water and placed aseptically into 9 cm diameter Petri dishes containing 15 ml of sterile PDA. The plates were incubated for 7 days at room temperature (28 to 30°C). Distinct colonies on the plates were selected, purified by repeated culturing and maintained on PDA slants. The fungus isolated was identified as *A. solani* using the macroscopic and microscopic identification guide according to Larone (1995).

Preparation of growth medium and inoculation

About 39 g of PDA powder (Sigma GMBH) was dissolved in 1000 ml of distilled water and the content was stirred and autoclaved for 25 mins at 115°C (Awale, 2001). The medium was allowed to cool down and was then aseptically poured into 25 ml flavour bottles. Thereafter, 5 ml of the plants extract prepared was poured into the Petri dishes. About 15 ml of molten PDA at 45 to 50°C was poured aseptically onto the plant extract in the Petri dish and swirled round 5 times for even dispersion of the extract into the agar and allowed to solidify, before the pathogen was inoculated (introduced) into the middle of the 'poisoned agar'. A mycelial plug of 5 mm diameter from 3 days old fungus was cut using a 5 mm sterile cork borer and transferred to the PDA plate in the center of the Petri dish and was kept in a sterilized fume cupboard kept at room temperature of 28 to 30°C.

Data collected

Radial growth

Measurement of the radial growth in centimeters (cm) was done and the radial growth was determined by using the formula K_r according to Reeslev and Kjoller (1995):

$$\text{Radial growth } (K_r) = \frac{(R_1 - R_0)}{(t_1 - t_0)}$$

Where, R_0 and R_1 are the colony radii at time t_0 and t_1 respectively, determined after 24, 48 and 72 h from inoculums.

Inhibition percentage

The inhibition percentage was calculated measuring the radial growth of the fungus grown on control and amended plates after 24, 48, and 72 hpi, using the following formula (Harlapur et al., 2007):

$$I\% = \frac{100 \times (C - T)}{C}$$

Where, I% = inhibition percentage of pathogen growth, C = average radial growth in control plates and T = average radial growth in plates amended with *R. communis* extract suspension.

Data analysis

Data collected were subjected to analysis of variance (ANOVA) for a completely randomized design using SAS (1999) statistical package. The means were separated using the Least Significant Test (LSD) at $P = 0.05$.

RESULTS

Radial growth of *A. solani* at 24, 46, and 72 h

Results of the first laboratory experiment at 24, 48 and 72 h shows that, the lowest radial growth was recorded in *R. communis* 100% concentration with means of 1.57, 2.16, and 2.88 cm (Plate 1) with the control recording the highest radial growth of 3.53, 4.09 and 4.59 cm (Plate 2). Also in the second experiment the treatment means showed that, *R. communis* at 100% concentration recorded the lowest radial growth with 1.30, 1.83, and 2.55 cm at 24, 48 and 72 h. In addition the control recorded the highest radial growth with mean of 3.40, 3.98 and 4.43 cm at 24, 48 and 72 h (Table 1).

The combined results in Table 2 revealed that, at 24, 48, and 72 h *R. communis* at 100% concentration had the lowest means of 1.43, 2.00, and 2.72 cm, followed by 50% concentration with 1.83, 2.32, and 2.93 cm. The control recorded the highest radial growth of 3.47, 4.04 and 4.51 cm at 24, 48, and 72 h.

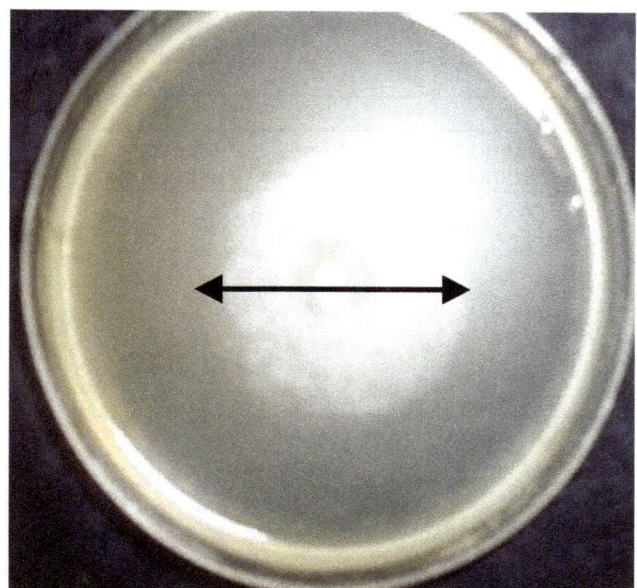

Plate 1. The lowest radial growth in *R. communis* 100% (72 h).

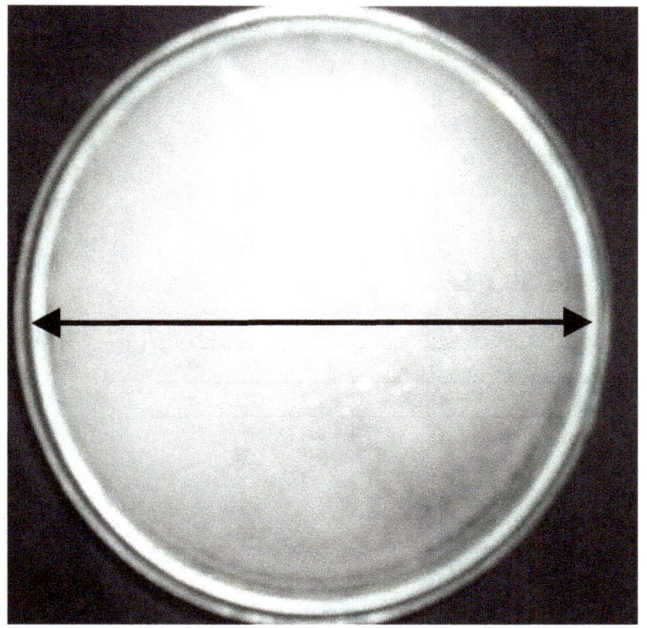

Plate 2. The highest radial growth in control (48 h).

Inhibition percentage of radial growth of *A. solani* at 24, 48 and 72 h

The I% for the combined results (Table 3) revealed that, *R. communis* extracts inhibited radial growth in the range of 41 to 59% at 24 h in comparison to the control. At 48 h, the result indicated that, *R. communis* extracts inhibited fungal radial growth in the range of 33 to 51%. At 72 h, the result of the calculated I% showed that, *R. communis*

extracts showed inhibitory effect on the radial growth in the range of 26 to 40% as shown in Table 3.

DISCUSSION

In this study, screening of the *in vitro* effect of *R. communis* aqueous extract for antifungal activity against *A. solani*, responsible for early blight of tomato, was carried out at 3 concentrations supplied to PDA medium. Results of the effects of plant extract on radial growth of *A. solani in vitro* revealed that, *R. communis* extract was able to reduce the radial growth and at the end of the experiment (72 h) the percentage at 100% concentration was still 40%. This may be attributed to the presence of toxic compounds, like ricin and ricinine in the extracts, as reported by Ukpabe (2002) who stated that, leaf extracts of *R. communis* inhibited the growth of *F. oxysporum*. Comparing the rate of radial growth in medium amended with *R. communis* aqueous extract with that of the control; it could be deduced that, the pathogen grew freely on the control medium, establishing itself and using up the food, while on the "poisoned food" of PDA containing the plant extract, growth was significantly reduced. The inhibitory effects of the extracts might be due to the chemicals present in the plant. Ricin is a proteinaceous toxin as reported by Lowery et al. (2007). This therefore suggests that, the plant extracts of *R. communis,* posses inhibitory effects on the growth of the fungus. Similarly the seeds, leaves, and stems of the plant contain ricin and ricinine, dihydroxystearic, linoleic, oleic, and stearic acids, β-sitosterol (Oplinger et al., 1997).

Tariq (2009) also reported that, growth in fungi is affected by the availability of substrate. He further stated that, the growth of most fungi is rapid at the exponential phase until one or more nutrients become limiting or depleted and/or metabolic products accumulate to low level. This could explain why the percentage was higher at 48 h than at 72 h in comparison with the control, since the fungus already had occupied all the surface of PDA plates and ceased to grow.

Similar results on extracts from five Chinese medicinal herbs were reported by Tongle et al. (2002) in their study: *Galla chinensis, Rheum palmytum* (root), *Sophora flavescens* (root), *Terminalia chebula* (fruit), and *Magnolia officinalis* (bark) showed inhibitory effects against *Phytophthora infestans in vitro* by inhibiting sporangia germination, mycelial growth and/or infection on potato leaves.

In this study, results on the plant extracts showed that, *R. communis* at 100% concentration inhibited radial growth by 40 to 59% in the combined results. This finding concurs with that of Baldrian and Gabriel (2002) who reported that, *Piptoporus betulinus* growth was found to be concentration-dependent as fungal growth was inhibited at higher concentration. This could be inferred that *R. communis* could inhibit the growth of *A. solani*. In

Table 1. Effects of plant extracts on radial growth (cm) of *A. solani* first and second experiments.

Treatment	Hours					
	24		48		72	
	First experiment	Second experiment	First experiment	Second experiment	First experiment	Second experiment
[†]Ricom25	2.27	1.85	2.66	2.76	3.35	3.33
Ricom50	1.75	1.91	2.43	2.20	2.95	2.90
Ricom100	1.57	1.30	2.16	1.83	2.88	2.55
Control	3.53	3.40	4.09	3.98	4.59	4.43
Mean	2.28	2.12	2.84	2.70	3.44	3.30
LSD	0.53	0.11	1.04	0.24	0.84	0.18
Prob. F	**	**	**	**	**	**

[†]Ricom = *Ricinus communis*; **, highly significant ($P = 0.01$)

Table 2. Combined means of the effect of *Ricinus communis* extract on radial growth (cm) of *Alternaria solani*.

Treatment	Hours		
	24	48	72
Ricom[†] 25	2.06	2.71	3.34
Ricom50	1.83	2.32	2.93
Ricom100	1.43	2.00	2.72
Control	3.47	4.04	4.51
Mean	2.20	2.77	3.38
LSD	0.27	0.47	0.50
Prob. F	**	**	**

**, Highly significant ($P = 0.01$).

Table 3. Combined results of the effect of *A. solani* on inhibition percentage (%) on radial growth of *A. solani*.

Treatment	Concentration (%)	% I (hpi)		
		24	48	72
Control	0	0	0	0
Ricom	25	41	33	26
Ricom	50	47	43	35
Ricom	100	59	51	40

addition, the fungus is much more sensitive to *R. communis* at higher concentration in affecting radial growth at lower concentrations.

Conclusion

The plant extracts at different concentrations tested on *A. solani* showed promising prospects for its utilization in plant disease control. The results showed that, *R. communis* at 100% concentration could reduce radial growth and inhibit growth Therefore, plant extracts are a potential source of botanical fungicides, after successful completion of wide range of field trials.

RECOMMENDATIONS

Based on the findings of this study, it was recommended that, *R. communis* at 100% concentration should be put to field trail to confirm its effectiveness.

REFERENCES

Akinbode OA, Ikotun T (2008). Evaluation of some bioagents and botanicals in *in vitro* control of *Colletotrichum destructivum*. Afr. J. Biotechnol. 7(7):868-872. http://www.academicjournals.org/AJB.

Baldrian P, Gabriel J (2002). Intraspecific variability in growth response to cadmium of the wood-rotting fungus *Piptoporus betulinus*. *Mycologia*, 94(3):428-436. www.mycologia.org/misc/terms.shtml.

Gwary DM, Nahunnaro H (1998). Epiphytotics of Early Blight of Tomatoes in Northeastern Nigeria. Crop Prot. 17(8):619-624.

Harlapur SI, Kulkarni MS, Wali MC, Srikantkulkarni H (2007). Evaluation of Plant Extracts, Bio-agents and Fungicides against *Exserohilum turcicum* (Pass.) Leonard and Suggs. Causing Turcicum Leaf Blight of Maize. Karnataka, India Karnataka. J. Agric. Sci. 20(3):541-544.

Iacomi-Vasilescu B, Avenot H, Bataillé-Simoneau N, Laurent E, Guénard M, Simoneau P (2004). *In vitro* fungicide sensitivity of *Alternaria* species pathogenic to crucifers and identification of *Alternaria brassicicola* field isolates highly resistant to both dicarboximides and phenylpyrroles. Crop Prot. 23(6):481-488.

Larone DH (1995). Medically Important Fungi - A Guide to Identification, 3rd ed. ASM Press, Washington, D.C.

Lowery C, Auld D, Rolfe R, McKeon T, Goodrum J (2007). Barriers to commercialization of a Castor Cultivar with Reduced Concentration of Ricin. Reprinted from: Issues in new crops and new uses. 2007. J. Janick and A. Whipkey (eds.). ASHS Press, Alexandria, VA.

Peet M (2003). Sustainable Practice for Vegetable Production in the South. North Carolina State University.

Reeslev M, Kjøller A (1995). Comparison of Biomass Dry Weights and Radial Growth Rates of Fungal Colonies on Media Solidified with Different Gelling Compounds. Appl. Environ. Microbiol. 61(12):4236-4239.

Sherf AF, MacNab AA (1986). Vegetable diseases and their control. John Wiley and Sons, New York. pp. 634-640.

Tariq V (2009). Fungi online: Growth kinetics. British Mycological Society. Accessed from http://www.britmycolsoc.org.uk.

Tomoko N, Takashi A, Hiromu T, Yuka I, Hiroko M, Munekazu I, Kazuhito W (2002). Antibacterial activity of extracts prepared from tropical and subtropical plants on methicillin-resistant *Staphylococcus aureus*. J. Health Sci. 48:273-276.

Tongle H, Wang S, Cao K, Forrer HR (2002). Inhibitory effects of several Chinese medical herbs against *Phytophthora infestans*. *ISHS*. Acta Horticulturae 834. 3rd International Late Blight Conference Proceedings. http://www.actahort.org/members/showpdf?booknrarnr=834_23.

Ukpabe R (2002). Effect of four plant crude extracts on the growth of *Fusarium oxysporum f.sp lycopersici*. Student Project Report, University of Ibadan, Ibadan, Nigeria (unpublished).

10

Design and implementation of a fishery science management information system based on client/server and browser/server models

Hui Fang, Ying Jing, Gang Han and Yingren Li

Chinese Academy of Fishery Sciences, Beijing 100141, China.

A scientific management information system is the way forward for modern scientific management. The management status and layout of the Chinese Academy of Fishery Sciences is introduced. By comparing the Client/Server and Browser/Server models, a structure combining both models is described together with its network topology. Having analyzed the functional requirements of the management information system, we present a suitable composition of the system. In consideration of the development environment of the system, key technologies are investigated.

Key words: Client/Server, Browser/Server, scientific management, information system.

INTRODUCTION

In recent years, thanks to the rapid development of computer and network technology, the information age has dawned. However, manual management using pen paper and a single type of management, that is, the traditional scientific management model using Word/Excel files, is still widely used. To establish a scientific, systematic, standardized management information system using information technology to reduce the workload, improve efficiency, and promote sharing of research information is the current goal of scientific management (Zhu, 2008; Luo, 2004).

The Chinese Academy of Fishery Sciences (CAFS) is a national fisheries research institution in China, shouldering the responsibility for key basic and applied research and high-tech industrial development. CAFS includes 3 institutes along the coast, 4 institutes situated on rivers, 2 specialty research institutes, as well as 4 fishery resource enhancement stations and the Beijing headquarters. In addition, there are 5 co-built institutes with local governments, located in the 12 provinces (or cities) nationwide. The two-level management structure and geographically wide distribution forces the information in the Chinese Academy of Fishery Sciences to be centralized, making it difficult to integrate and sustain. Therefore, re-planning and organizing scientific management, establishing a networked information system for scientific management to realize "whole-process management", improving the quality and efficiency of management, and providing a better service for the Academy, is now an urgent task facing the administrative department of CAFS.

SYSTEM STRUCTURE AND DESIGN

Client/Server and Browser/Server models

At present, there are two popular scientific management models: the Client/Server (C/S) and Browser/Server (B/S) models. The C/S model consists of two-tiers, connected

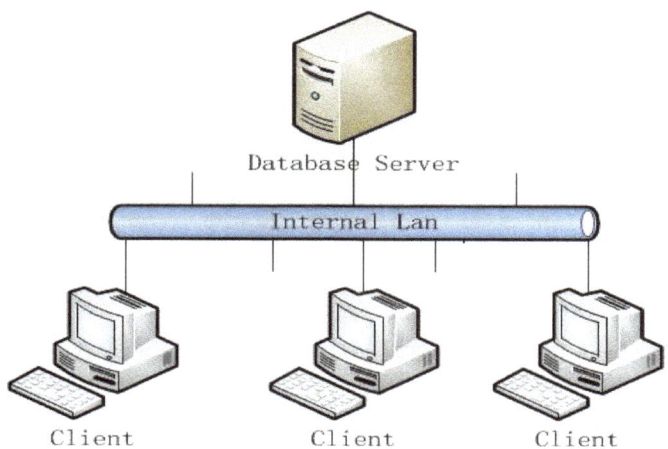

Figure 1. C/S mode.

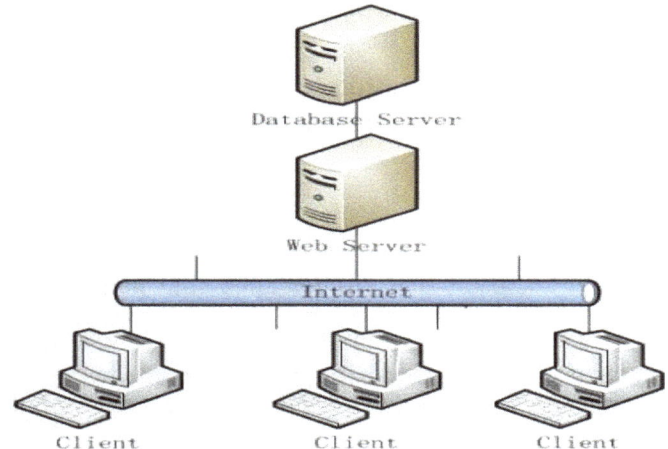

Figure 2. B/S mode.

1. A scientific management information system should be open-source, scalable, and easy to maintain and upgrade. Since the B/S model is based on general standards accredited by ISO, it is open-source, a general browser is applicable as the Client, and maintenance and upgrades occur at the server side. In addition, since the 3 tiers are independent of each other, they are unaffected by maintenance and expansion in the other tiers (Zhang and Zhao, 2000).

2. The Academy has a 2 tier management structure, but the geographically wide distribution of affiliated institutes makes an internal LAN-based C/S model impracticable. Using the Web based B/S model makes realizing interoperability and data sharing between institutes comparatively easy.

3. However, since the main bodies conducting research work are independent institutes, this facilitates the establishment of separate internal LANs. Therefore, to improve the processing speed of the system, ensure data access security, and reduce the network communication load, the C/S model is adopted in each institute.

For the above reasons, a combined B/S and C/S model is adopted for the Academy's scientific management information system. With this combination, the Web based B/S model used in the Academy and institutes offers the advantages of the Browser/Server model, that is, being light, efficient, and stable, while the C/S model used within the institutes (the research division) contributes the characteristic powerful and high security features. The topology of the information system is depicted in Figure 3.

DESIGN OF SYSTEM FUNCTIONALITY

Needs of the system

The underlying principle of the management system is that the process of scientific management should be integrated and statistical analysis should be comprehensive. The overall objective is "the whole process of management" for all research in the Academy, implemented by establishing a network for the scientific management information system. To achieve this objective, the scientific management information system, based on the principle set, should enable unified management of research projects' implementation and cultivation of achievements/awards, while also being able to generate various reports required by clients automatically (Lang et al., 2005).

Modules in the system

In accordance with the needs of the system, the scientific management information system is divided into 4 modules as depicted in Figure 4: Scientific management,

by a local area network, as shown in Figure 1. It has the advantages of quick processing speed, safe access to data, and small network communication load. However, it is flawed by poor cross-platform service, lack of openness, inconvenient upgrades, redundancy of resources, and so on. The B/S model is Web-based, comprising three tiers, as shown in Figure 2. It offers convenient maintenance and upgrades, consistency of the user interface, as well as openness and good scalability (Wang and Ge, 2006; Liu et al., 2010).

Design of models

Considering the advantages and disadvantages of the C/S and B/S models, together with the nature of the scientific management of the CAFS, we adopted a combined C/S and B/S model to compensate for each model's deficiencies. The choice of using a combined model was made for the following reasons:

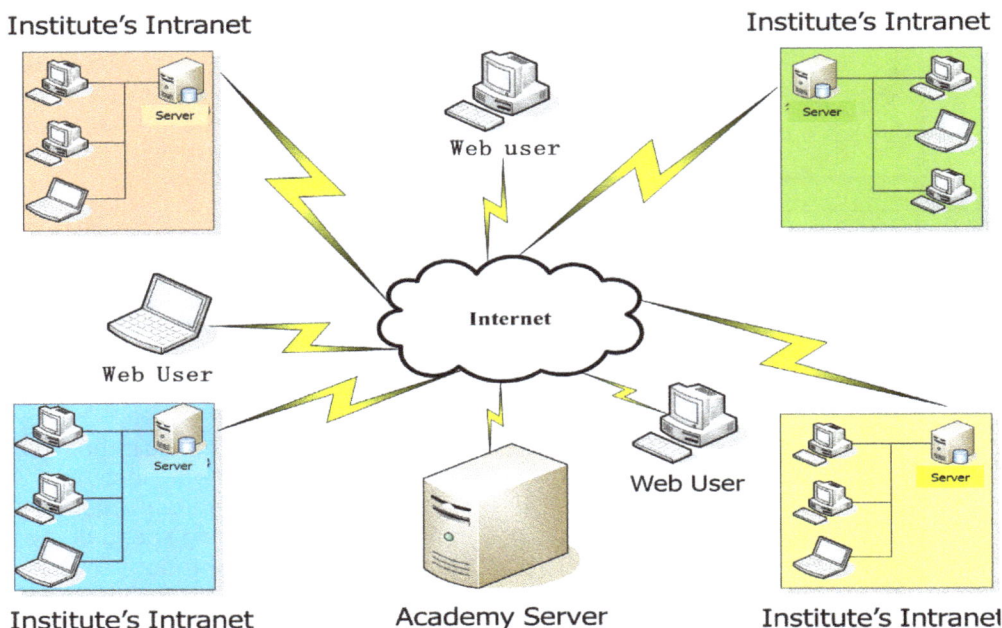

Figure 3. The topology of the information system.

office management, statistics analysis, and maintenance of the system (Ye and Zhang, 2005; Zhu, 2009). The functionality of each module is described below:

(1) Scientific management: to establish the Academy's project library, sci-tech achievements library, equipment library, intellectual property library, publications and papers library, talent databank, and standards databank, and to collect and store the Academy's scientific management related data and information.

(2) Office management: to achieve the daily office management functions, including document registration, leaders' instructions, archiving of records, as well as management of research expenditure, search engine service, printing, and so on.

(3). Statistical analysis: To search and release information by carrying out simple and combined queries on the data within the system, and according to the data selected, to generate various statistical reports, including charts, histograms, and pie charts.

(4) Maintenance of the system: to manage user passwords and access permissions and to import/ export data in a particular format of text or for Excel.

SYSTEM DEVELOPMENT AND KEY TECHNOLOGIES

Environment for system development

The environment for developing the scientific management information system is configured as follows:

Microsoft Windows 2000 Professional as the operating system, Microsoft SQL Server 2000 with powerful data processing capabilities as the database management system, Microsoft IIS 5.0 or above as the Web server, Borland Together for VS.NET as a modeling tool, Microsoft Visual SourceSafe for source control management, and Nunit for unit text.

Key technologies for system development

Design of software architecture

Due to the fact that the institutes in the Academy are geographically widely distributed, a hierarchical structure for the scientific management information system, distributed top-down in three tiers, has been adopted. The 3 tiers include a user presentation tier, business logic tier, and data tier.

(1) The User Presentation tier provides users with a graphic interface, which helps them understand and position targets efficiently.

(2) The Business Logic Tier implements the application strategies, packages the application modules associated with the system, and presents the packaged module to the user to apply.

(3) The Data Persistence Tier converts the relational database to an object database, allows access to maintain and update data, and realizes communication with the same database server.

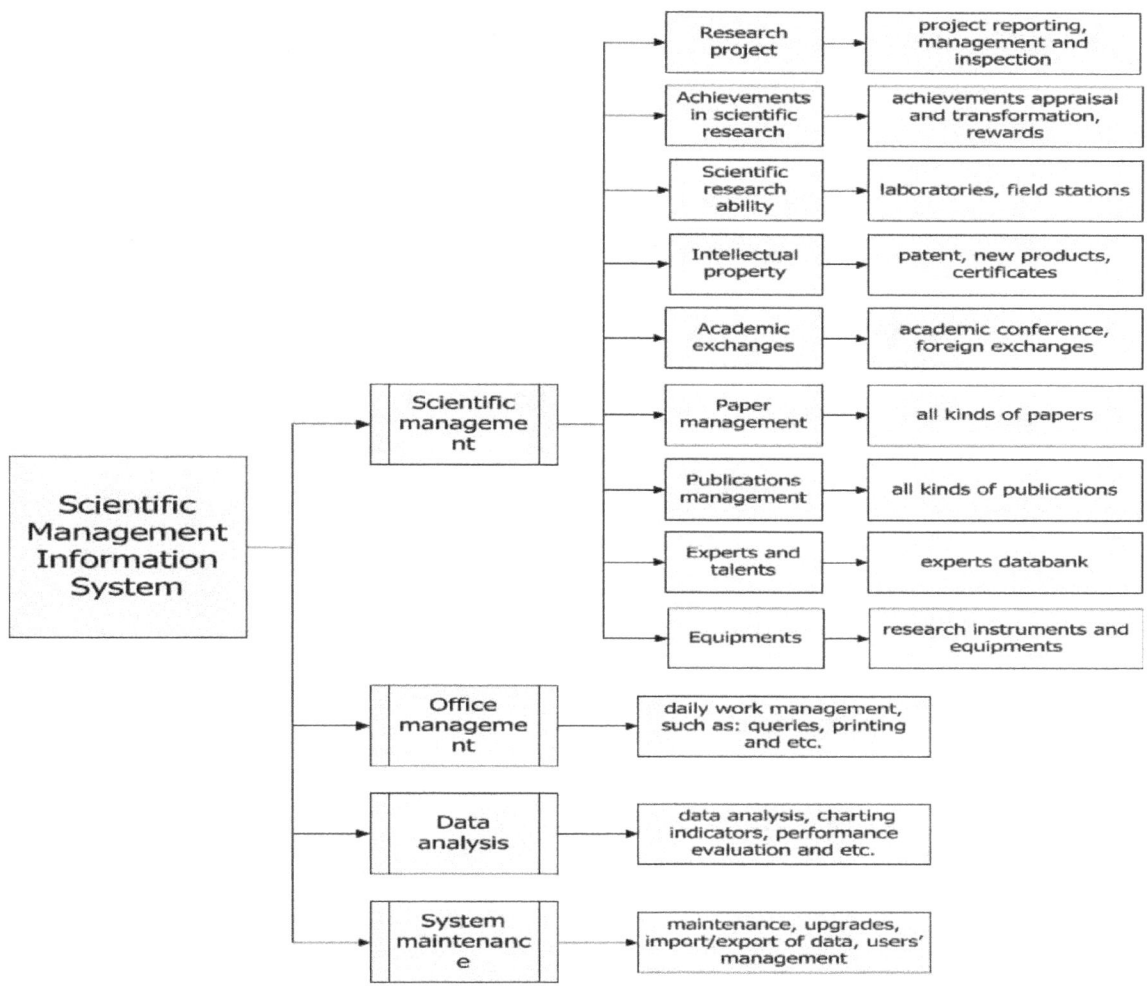

Figure 4. Modules of scientific management information system.

The hierarchical structure above should abate the dependence between tiers. In terms of coding standards, this implies improved reusability of each tier's logic, alleviation of concerns, a loosely coupled system, ability to define standards, and so on. The software architecture is depicted in Figure 5.

Database design and data synchronization

Ensuring that data within the institutes' databases under the C/S model is consistent with that within the Academy's database under the B/S model, is key to the success of the system. To achieve consistency of data at the institute level, ID and InsID are used jointly as the primary key in the data tables under the C/S model, of which, the ID is the unique identifier for data in the C/S model, while InsID is used to distinguish the source of the data. Meanwhile, the management information system randomly generates a GUID (Globally Unique Identifier) when data is created. This is a character sequence generated automatically, which means that within the same time and space, it is the unique identifier for each computer. In the process of synchronizing the data in the C/S and B/S models, the system identifies data based on the GUID, and performs the required operations.

Automatic backups and backups by administrators are combined with data synchronization. For important data that needs timely updates, the backup is operated by administrators and synchronized to the Academy's server. All other data is backed-up automatically in accordance with the backup cycle set by the system and synchronized to the Academy's server (Yang and Lin, 2004; Guo et al., 2003).

CONCLUSION

The aquatic research and management information system presented in this paper is currently operational and running stably. Application of the system has greatly reduced the workload of research managers, significantly

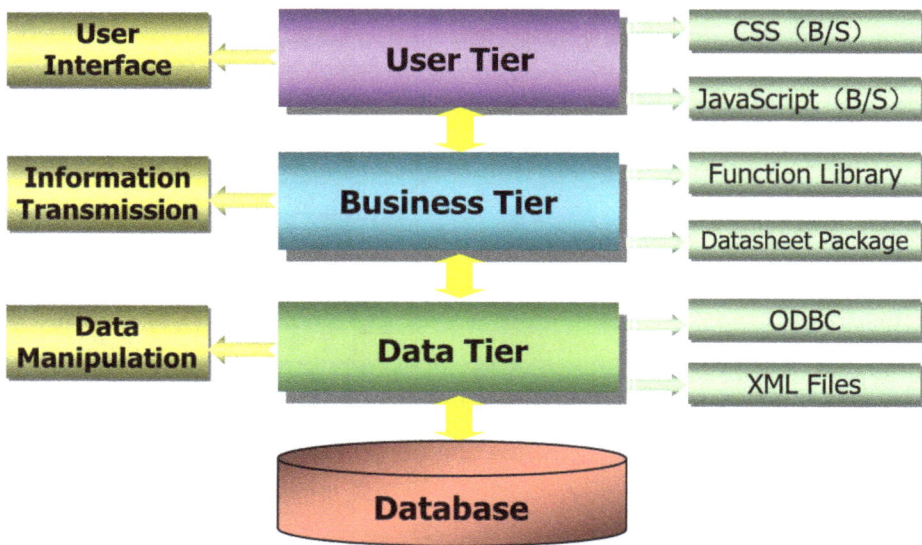

Figure 5. Architecture of the system software.

improved the efficiency and qualifications of scientific management, realized scientification, systematization, standardization, and informationization of the scientific management in the Academy, and provided an important mechanism for scientific management and decision-making. It is, therefore, of great value to all users.

REFERENCES

Guo JY, Shen LZ, Ma GF (2003). Design and Implementation of Collegiate Science Research Management Information System Based on C/S and B/S. Comput. Eng. Application 1:212-214.

Lang Q, Feng L, Xia X (2005). Scientific Management Information System Based on Three- tilter Structure. Comput. Era 5:24-25.

Liu P, Song W, Wan J (2010). Scientific Research Project Management System Based on the Structure of C/S Model and B/S Model. Software Guide 9:110-111.

Luo HX (2004). Analysis and Design of the Scientific Research Management Information System Based on Web in Colleges and Universities. Sci. Technol. Manage. Res. 5:122-123.

Wang CD, Ge T (2006). Construction of the Information System for the Management of University's Scientific Researches Based on Internet. J. Anhui Univ. Sci. Technol. Nat. Sci. 1:27-32.

Yang XM, Lin DY (2004). The current status of Management Information Systems. Chinese J. Med. Sci. Res. Manage. 17:219-222.

Ye ZW, Zhang XX (2005). The Design and Implement of University Scientific Research Management Information System Based On . NET. J. Jiaying Univer. Nat. Sci. 3:71-73.

Zhang XY, Zhao WY (2000). Research and Design of Management Information System of Scientific Research Item. J. Xi'an Univ. Sci. Technol. 20:246-249.

Zhu ZY (2008). The Design and Implementation of Management Information System of University Scientific Research. J. Yancheng Institute. Technol. Natural Sci. Edition. 4:35-37.

Zhu XD (2009). The Construction of Scientific Management Information System. J. Harbin Institute. Technol. Nat. Sci. 11:110-114.

11

Fishing in oil spillages zone: A case study of Burutu Local Government Area of Delta State, Nigeria

Gbigbi, Theophilus Miebi

Department of Agricultural Services, Ministry of Agriculture and Natural Resources, Asaba, Delta State, Nigeria.

The study considered fishing in oil spillages zone with special reference to Burutu local Government area of Delta State, Nigeria. The study area was particularly chosen for this study because of its prime place in artisanal fishing and oil exploitation activities. Data collection was through well structured questionnaire administered to 120 artisanal fishermen selected through random sampling technique. The method of analysis used were descriptive statistics, costs and return analysis and production function analysis using ordinary least square (OLS) criterion to estimate the parameters of the production function. The software used was SPSS 9.0. Results showed that majority of the fishermen were ageing. The result also showed that we have more male (61%) artisanal fishermen than females (39%) artisanal fishermen in the study area. Again, the results further revealed that there was high level of illiteracy as about 63% of total respondents did not attend primary school, while 28, 8 and 1% had attended primary, secondary and tertiary institution, respectively. The result of the analysis showed that the mean household size was 6 persons and mean annual farm income was about 32,883.33. The results of the regression analysis showed that the independent variables, labor, fishing experience, income level of fishermen, cost of fishing nets/bait were positive and significant at 5% level of probability except age of fishermen that was negative and significance at 5% probability level and all the factors were related to output. Artisanal fishing was not profitable in the study area with gross margin and net returns of 45,550 and 34,350 per annum respectively. The result showed that the surveyed fishermen were producing at a diminishing return to scale.

Key words: Fishing, fish production function, costs and returns analysis, Delta State, Nigeria.

INTRODUCTION

Delta State which is one of the nine (9) States in the Niger Delta region of Nigeria is blessed with abundant natural water resources. According to Ita and Sado (1987), Nigeria has an estimated inland water mass of about 12.5 million hectares capable of producing about 512,000 metric tons of fish annually. Nigeria is blessed with a long coastline, extensive brackish and mangrove swamps supporting a wide range of fish species, such as tilapia, thread fins, moon fish, sea cat fish, snappers, cray fish, sea turtle, lobsters, sardines and razor fish West African Croakers, Bonga fish, shark, shrimps, bivalves, periwinkle and many others. Available statistics showed that Nigeria's inland water bodies are producing less than 13% of their estimated fishery potential (Sule et al., 2002). The effect of oil resource extraction on the environment of the Niger Delta has been very glaring in terms of its negative effect on the region. Eteng (1997) stated that oil exploration and exploitation has over the last four decades impacted disastrously on the socio-physical environment of the Niger bearing communities, massively threatening the subsistent peasant economy and the environment and hence, the entire livelihood and basic survival of the people. Suffice it to say that, while oil exploitation has caused negative socio-economic and environmental problems in the Niger Delta, it has contributed enormously to the country's economic growth since it was discovered in commercial quantities in 1956 at Oloibiri located in Bayelsa State. NEST (1991) reported that oil spillages in the country's oil producing riverine areas were increasingly reducingsome of the Nigerian water bodies to veritable sewage depots for toxic chemicals which either kill or contaminate fish and other aquatic life. It was further reported that at 1981, about six million tones of petroleum was discharged into off-shore water annually. Out of this amount, about 600,000 tones resulted from sudden accidental spillages while the constant dripping of petroleum products from activities in all sphere of the oil industry accounted for the remainder. Following this development, fisheries occurrence in Nigeria's wetlands could be limited by activities which pollute water and the greatest threat in this regard is oil pollution resulting from crude oil exploitation (NEST, 1991). The exploration and exploitation has impact on the environment through frequent spills, pipe explosions, pollution, sabotage, gas flaring and effluent emission. Other sources of oil include transportation and marketing, effluent water from oil refineries, lubrication oils and other wastes in the form of sludge, bitumen, slops and oil sand/sediment present in large amount within oil flow stations, storage terminals and tanks (Nwilo et al., 2000; Ogri, 2001).

Nwilo and Badejo (2001) posited that where there is oil spill, it covers the surface of the water. This reduces oxygen exchange thereby, causing death of fishes because the oil coats the gills of the fishes preventing them from inhaling oxygen. In addition, oil spills endangers fish hatcheries in coastal waters, contaminates the flesh of commercially valuable fish and oil slicks prevent sunlight from reaching deeper levels of water where coral thrive, thus, limiting food production of plants (photosynthesis).

Further reports on the effect of oil exploration and exploitation activities on aquatic lives showed that an oil spill can directly damage the boats and gear used for catching or cultivating marine species. Floating equipment and fixed traps extending above the sea surface are more likely to become contaminated by floating oil. In a study, on the effect of hydrocarbon pollution on water in the Niger Delta, Ukpong and Akpabio (2003) reported that hydrocarbon pollution causes great damage to spawning grounds; aquatic vegetation having economic values had degenerated in productivity while fish and crustaceans become carriers

of toxic hydrocarbon substances along the food chain and fishing as an economic activity is lost or threatened, exacerbating hunger and poverty in fishing communities. Anderson et al. (1974) reported cases of oil spillage in Sangana, Koluama, Akassa and Brass communities in Bayelsa State, in which tremendous damage was done to fisheries in the wild. Oil spillage has a list of deleterious effect on the biota particularly, the fish which includes fin erosion, respiratory difficulties and mortality (Ziskowki and Murchelano, 1995). Keke (1989) observed massive fish deaths at Bayelsa State because of the incessant oil pollution in the coastal waters.

In the Gulf region, TED case studies (1996) asserted that the fishing industry was deleteriously affected by the oil spillage into the Gulf, which was important due to the fact that it is one of the most vibrant productive activities in the region after the production of oil. As an example of the vibrancy of this industry, prior to the Iraqi Invasion of Kuwait the Gulf yielded harvests of marine life up to 120,000 tons of fish a year; after the oil spillage, these numbers significantly dropped. As a matter of fact, the incidence of oil spillage often results in total extermination of fish, leading to reduced fish output. Such losses adversely affected fishermen active economic livelihood. This has a backward integration in the national economic development. Although, the impacts of oil pollution in the Niger Delta are enormous, the objective of this study is to examine fishing in oil spillage zone and its effects on total fish catch and profitability in the study area.

MATERIALS AND METHODS

The study area

Burutu Local Government Area (LGA) which is a home to several oil producing communities is the area of study; and it is a leading source of on-shore crude oil production in Delta State. Delta State is located between Latitudes 4°N and 6°N and Longitudes 5°E and 7°E. Bayelsa, Anambra, Edo and Bight of Benin bound the State on the Southeast, Northeast, North, Northwest and South respectively. It has a land area of 17,011 km^2 and a population of 4, 0981,391 (Federal Republic of Nigeria Official Gazette, 2007).The topography of the area is low, lying with a Coastline of about 160 km on the River Niger with rivulets and streams, criss-crossed with creeks through which the River Niger empties into the Atlantic Ocean, thus, forming the larger part of the Niger Delta area.

The State has a tropical climate marked by two distinct seasons, the dry season lasting November to March and rainy season lasting April to October. Average rainfall ranges from about 267 in Coastal areas to 191cm in the North of the State. It has a minimum temperature of 28°C and a maximum of 34°C. Inhabitants of communities are mainly fishermen.

The area is endowed with mangrove swamps, rivers, creeks and flood plains which offer great opportunities for fishing. The communities or ethnic groups are Ekeremo, Sokebolou, Yeye, Ogunlagha, Forcados, Yonkri, Odimodi, Ezon-Burutu and Ojobo. The fishing season spans seven months from the end of one rainy season (usually in October) to the beginning of another rainy season (most commonly in April).

The bulk of oil exploration and exploitation activities both on-shore and off-shore is concentrated in the area, and the fact that fishing is the major occupation besides farming of the people. Cases of incessant oil spillages have been reported there.

Sampling and data collection

Ten (10) major fishing communities were purposely selected from the Local Government areas which were affected by oil spills in the study area. Twelve fishermen were randomly selected from them to make a total sample size of 120 fishermen used for the study. Primary data were collected using a set of pre-tested structured questionnaire administered through the interview schedule method during the last fishing season. Information collected included age, years of fishing experience of the fishermen, their household size, and level of education and level of income. Other data include number of canoes, fishing nets together with their acquisition costs, quantities of fish caught, and quantities sold together with their sales revenues as well as, man days of labor utilized during the season.

Analytical framework

Koutsoyiannis (2003) observed that regression analysis is one most commonly used techniques in analyzing dependence among variables. According to Eboh (1998) and Kedison (2003), the aim of regression analysis is to establish and prove how one variable is related to another variable. It is based on the functional relationship between the variables. They noted that the key relationship in a regression is the regression equation which contains the regression parameters whose values are to be estimated using the data. The parameters measure the relationship between dependant and each of the explanatory (independent) variables. The X (s) are fixed or predetermined outside the model and are called independent variables. The Y (s) are to be determined within the model by the X (s) hence, they are called dependent variables because their value depend on the values of the X (s).

The way the Xs are transformed to Y is the functional relationship, the error term 'e' is introduced into the function to capture the effects of omitted variables, erratic nature of human behavior, errors of measurement, and the effects of aggregation (Awoke, 2001). Mbanasor and Obioha (2003) pointed out that, there are many functional forms that could be used to describe production relationship, but in practice the commonly used forms include linear, quadratic, semi-log and Cobb- Douglas functional forms. Depending on the number of independent variables used to estimate the dependent variable, regression analysis is divided into simple and multiple regression analysis. Multiple regression analysis is an econometric method used to study relationship involving more than two variables. The variation in the dependent variable is explained by more than one independent variable. Gujarati (2003) and Gbigbi (2008) opined that most regression models are multiple regression models because few economic phenomena can be explained by only one variable. The ordinary least squares (OLS) is one of the most commonly employed techniques in the multiple regression analysis, especially, in constructing the production models as it gives the best fit (Subair, 2009). The ordinary least squares (OLS) are based on certain assumptions, namely, the distribution of the random variable, and the relationship between the explanatory variables themselves (Koutsoyiannis, 2003).

Model specification and estimation

In order to estimate fish production in oil spillages zone, the following econometric models relating to quantity of fish caught (Kg), with the under listed explanatory variables were specified and subsequently estimated. The implicit form of equation for this study is specified as follows:

$$Q = f(X_1, X_2, X_3, X_4 \ldots \ldots X_{n, e}) \qquad (1)$$

Where Q = Quantity of fish caught (kg), X_1 = Labor input in man days, X_2 = Fishing experience (years), X_3 = Income level of fishermen (₦), X_4 = Age of fishermen (years), X_5 = Cost of fishing nets / baits incurred by fishermen (₦), e = Error term (which was assumed to be normally distributed with zero mean and constant variance).

Three functional forms of the equation, linear, Semi-log and double log were tried and the one with the best fit was chosen as the lead equation. The criteria for selecting the lead equation were based on the value of "Coefficient of determination" (R^2), the significance of the Coefficients as well as the a priori expectation (Olayemi, 1998). The Cobb-Douglas functional form (Linearized in Logarithm) fitted the observed data very well. Thus, the explicit format of the specified fishermen production function is presented:

$$Log\ Q = b_0 + b_1\ logX_1 + b_2logX_2 + b_3logX_3 + b_4logX_4 + b_5logX_5 + e \qquad (2)$$

Where b_0 = Intercept, b_1 - b_6 = regression coefficients and X_1 = X_5 = explanatory variables.

Cost and returns analysis

This analytical technique is otherwise referred to as enterprise budget. It provides information on the financial and physical transaction or plan for the farm enterprise for a given production period. Costs and returns analysis is often composed of two components, the costs or expenditure component and the returns, revenue or income components. According to Olukosi and Erhabor (1988), the costs component is sub-divided further into variable costs and fixed costs.

Costs

Variable cost (VC): This cost changes with the variation in the output level. All costs are associated with the variable inputs such as labor, fuel and lubricant boat repair, net repair fish processing, transportation as well as service and maintenance charges.

Fixed cost (FC): This involves depreciation of the set of tools and equipments used. Straight line method was used and the useful life was estimated using the average life span (year) obtained during the study. Therefore, the fixed cost involves cost of fishing canoe, paddle, fishing net and other implements which do not change at least in the short run. The total cost of production, however, is the total sum of the variable costs and fixed costs.

Income

Gross revenue (GR): This component shows the outputs or returns both in physical terms and the corresponding monetary values or fish revenue.

Gross margin (GM): This is the difference between the gross revenue and variable costs. The formula used for this calculation is given as:

Table 1. Distribution of socio-economic characteristics of respondents (n = 120).

Parameter	Frequency	Mean
Age of fishermen		
27-35	4 (3.3)	
36-44	13 (10.8)	
45-53	62 (51.7)	51 years
54-62	33 (27.5)	
63-71	8 (6.7)	
Gender of fishermen		
Male	73 (60.8)	
Female	47 (39.2)	
Educational level		
No primary education	75 (62.5)	
Primary education	34 (28.3)	
Secondary education	10 (8.3)	1.0
Tertiary education	1 (0.8)	
Household size		
2-5	68 (56.7)	
6-9	41 (34.2)	
10-13	10 (8.3)	6 persons
14-17	1 (0.8)	
Annual farm income		
19,000 – 34,000	81 (67.5)	
35,000 – 50, 000	29 (24. 2)	
51,000 – 56, 000	8 (6.7)	32,883.33
57,000 – 82,000	2 (1.7)	

Figure in parenthesis are percentages (Field survey, 2009).

$$G.M = \sum_{i=1}^{n} Piqi - \sum_{i=1}^{m} CjXj$$

Where; GM = Gross margin, Pi = Market unit price of output I, qi = Quantity of output I, Cj = Unit cost of the variable input j, Xj = Quantity of variable input j, m = Number of input used and n = Number of output produced

Net returns / profit (NR): This is the difference between gross revenue and total cost. The formula used for this calculation is given as:

NR = TR-TC

Where; NR= Net return (in Naira), TR= Total revenue (Naira) and TC= Total cost (Naira)

The use of costs and returns technique as an analytical tool is often criticized on the ground that it does not provide satisfactory information on the relative importance of the various inputs in contributing to outputs. Besides, the use of data obtained can only be applied in the area from which the data were generated since it uses only money as a unit of measurement. However, its ease of computation and simplicity appropriate once the data have been generated. This method was used by Olagunju et al. (2007) to analyze the economic viability of cat fish production in Oyo State, Nigeria and it was found to be suitable. Other researchers stressed that the method is used extensively to measure profitability of farm enterprise (Gbigbi, 2008; Ogundari and Ojo, 2009; Ojo and Ehinmowo, 2010).

RESULTS AND DISCUSSION

Socio-economic analysis

The socio-economic characteristics of the artisanal fishermen surveyed are presented in Table 1. The results indicate that 86% of the fishermen had ages ranging between 45 to 71 years with an average age of 51 years. With such an aged agricultural work force, agricultural productivity is bound to be low. As the fishermen grow older, their performance drops and so does the general fish catch levels. These results are, however, in agreement with the findings of Akwiwu (2002), Olomola (1991) and Mabawonku et al. (1984). This development could be attributed to youth's abandonment of fishing

Table 2. Estimates of input elasticities and returns to scale for artisanal fishermen.

Production inputs	Elasticity estimated	t –values
Labor input in Man days	0.124	5.90
Fishing experience in years	0.101	2.53
Income level of fishermen	0.137	4.28
Age of fishermen	-0.142	2.73
Fishing nets / baits	0.214	6.69
Return to scale	0.576	-

Field survey (2009).

work for highly paying oil company seismic jobs and white-collar jobs. Table 1 also indicates that 60.8% of the fishermen in the study area were males, while the females were 39.2% indicating that more males are involved in artisanal fishing than female. The result might be attributed to high energy, labor intensive demand of fishing which female fisher folks could not give. This result is also in agreement with the traditional gender pattern of fishing (Williams and Awoyomi, 1998).

A major proportion of the artisanal fishermen did not attend primary school (62.5%), while 28.3% of them had primary education. On the whole, only 9.1% of the respondents have secondary school and above. This is an indication that the illiteracy rate is high in rural communities.

This finding is consistent with the result by Forde (1994) that the low levels of education of the artisanal fishermen were some of the constraints to their fishing catching levels and indeed their development. However, a higher level of education attainment many discourage some people from participating actively in artisanal fishing operators.

A relative household size was found in the study with a mean size of 6 persons per household. About 34% of thehouseholds have a family size that range between 6 to 9 persons despite the relative family size of 6 persons per household; it is plaque with rural-urban migration of able-bodies young men and women.

This could be attributed to the fact that a reduced family size that is prevalent in the oil producing areas is an indication of non-availability of enough family labor for the fishing operations. In fact, the intensity of agricultural production has been found to have a direct relation to household size.

The level of income realized from fishing by the respondents reveals that fishing income is very low. This is not unexpected given the rate of incessant oil spills that have destroyed aquatic life and traditional fishing grounds. Annual farm income ranged between ₦19, 000 to N82000, though, about 92% of the fishermen actually earned income of between ₦19, 000 and N50, 000 from fishing activities. The average fishing income was ₦32, 883.33. This findings supports that of Inoni and Oyaide (2007); NEST (1991) and Awobajo (1993) that the low

level of fish income may be due to oil spillage.

Regression result

Results of the fish production function are given as:

Log Q = 2.306 (0.047) + 0.124*logX$_1$ (0.021) + 0.101*logX$_2$ (0.040) + 0.137*logX$_3$ (0.032) - 0.142*logX$_4$ (0.52) + 0.214*logX$_5$ (0.032) (5)

The figures in parenthesis are the standard errors; the coefficient asterisk (* are significant at 5% probability level; R- square = 0.839 and F- ratio = 111.68 (significant at 5% level).

The regression results as indicated by the co-efficient of multiple determination (R^2 = 0.839) showed that the combined effects of the identified production resources explained 84% of the total variation on fish caught. Other factors not reflected in the model might have combined with the stochastic term to account for the remaining 16% variation in output of fish caught not explained by the combined effects of the stated resources. The F statistics indicated that the model was highly significant at the 5% level. The t-test of significance for the independent variables were positive except the age of the fishermen that was negative and significant at 5% level of probability indicating that the production factors were related to output for artisanal fishermen in a manner consistent with a prior expectation. This implies that each of these variables are increased. While the negative coefficient of age shows that as fishermen become ageing, fish output decreases, reflecting the mean age of 51 years obtained from the analysis. This implies that the fishermen are relatively old; hence, they lack vigor to accomplish the task associated with fishing that depends heavily on human labor. This confirmed with the reports by Ojo (2000). The coefficients or b-values; 0.124 for b$_1$; 0.101 for b$_2$; 0.137 for b$_3$; 0.142 for b$_4$ and 0.214 for b$_5$ indicated that one percent increase in labor, fishing experience, income level of fishermen, age of fishermen and fishing nets/baits would bring about 12, 10, 14, 14 and 21% increase in fish caught respectively. The results of the estimated model indicate that the b-valves for b$_1$, b$_2$, b$_3$, b$_4$ and b$_5$ added together resulted to 0.576 (Table 2).

Table 3. Average costs and returns per Artisanal fisherman in Burutu L.G.A.

Items /operations	Total value (₦)
Variable cost (₦)	
Labor	30,750
Fuel and Lubricant	3000
Boat repair	2600
Net repair	1800
Fish processing	4500
Transportation	3800
Miscellaneous	4000
Total variable cost	50,450
Fixed cost	
Fishing canoe	8,000
Fishing paddle	700
Fish Net	2500
Total fixed cost	11,200
Total cost	61,650
Output revenue	
Fish catch (kg)	800
Price (N/kg)	120
Gross return	96,000.00
Gross margin	45,550.00
Net fishing income	34,350.00

Field survey (2009).

This is less than unity implying that the survey fishermen were producing at a decreasing or diminishing return to scale. This means that a unit increase in all the production resources put together would bring about less than unit increase in output of fish caught. Hence, it is advisable that the production units should maintain the level of input utilization at this stage as this will ensure maximum fish output from a given level of input ceteris paribus.

Analysis of cost and returns

Analysis of cost: The costs concept can be viewed from many perspectives. The incurred cost items were grouped as either variable or fixed costs. The variable cost items considered included expenses on labor, fuel lubricant, boat repair, net repair, fish processing and transportation. The fixed cost items were depreciation on equipment used such as fishing canoe fishing paddle and fishing net. Straight-line depreciation method was used. The average cost composition per harvest for fish caught is presented in Table 3. It could be noticed that the variable cost made up the bulk of the total cost of production.

This high level of the variable cost shows the flexibility of the business. According to Table 3, the labor cost accounted for about 61% of the variable costs for the artisanal fishermen. This is followed by expenditure on fish processing.

Gross return: The gross return that accrued to individual fish farmer during the survey year was calculated by multiplying their respective fish output with the market price. On the average, the selling price was N120 per kg. Table 3 shows the average fish caught and revenue per harvest. The revenue from the sales of fish caught was ₦96,000.00. The study reveals that average gross revenue of ₦96, 000 per harvest was realized by the artisanal fishermen.

Gross margin and net returns: The gross margin for each artisanal fisherman was calculated as the difference between the gross revenue and variable costs. The average gross margin per harvest by artisanal fishermen was ₦45, 550. The net return is the difference between the gross revenue and total costs. The average net returns on artisanal fisherman per harvest was N34, 350. The result of the study revealed that artisanal fishermen in the oil producing areas incurred higher costs of production and poor fish harvest presumably as a result of oil exploitation activities leading to lower profit for fishing activities in the study area.

RECOMMENDATION AND CONCLUSION

Based on the findings of the study, it can be concluded that majority of the artisanal fishermen in the study area were ageing. Majority of the people are males with household size of 6 to 9 persons. The result shows that the average income earned by artisanal fishermen was N34, 350.00. The study revealed that oil spillage has presumably had a negative impact on fishing activities leading to reduced agricultural output, poor harvest and low income level among the artisanal fishermen. These have led to calls for resource control by oil producing areas in the country. The result further revealed that artisanal fishermen in the study area incurred higher cost of production leading to lower profit and poverty.

Based on the findings of the study, the following recommendations are made:

1. That the impact assessment of oil exploration and exploitation on fishing be carried out,
2. The petroleum industry should work closely with government agencies, Universities and research centres so as to reduce the frequency and impact of oil spills.

REFERENCES

Akwiwu CD (2002). Determinants of adoption of Agroforestry Technologies and the Market of the produce in crude oil producing Areas of Imo State. Unpublished M.Sc. Thesis, Agricultural Economics and Extension Department, Imo State University, Owerri.

Anderson JW, Nelt JM, Cox BA, Taken HE, Hightower GM (1974). Characteristics of Dispersions and water soluble Extracts of crude oil and Refined Oils and their Toxicity to Esturine crustaceas and fish. Mar. Biol. P. 27.

Awobajo SA (1993). An Analysis of oil spill incidents: 1980-1987: In Thomopulose, A.A. (ed). The Petroleum Industry and the Nigerian Environment: Proceeding of the Warri 1981, International Seminar, Nov. 9-12. Lagos.

Awoke MU (2001). Econometrics Theory and Application. Will Rose and Apple seed Publishing Company. Abakaliki, Nigeria.

Eboh EC (1998). Social and Economic Research: Principles and Methods; Academic publication and development Resource Ltd, Lagos.

Eteng IA (1997). "The Nigerian State, oil Exploration and community Interest: Issues and perspectives" University of Port Harcourt, Nigeria.

Federal Republic of Nigeria Official Gazette (2007). N0. 24, Government notice No. 21, Lagos. P. 94.

Forde AC (1994). Update on Sierra Leone Fisheries. IDAF Technical Report N0: 69. FAO, Rome. P. 9.

Gbigbi MT (2008). Socio-Economic Implicaiton of oil Exploitation on farming communities in oil Producing Areas of Delta State of Nigeria. Unpublished M.Sc Dissertation, University of Nigeria, Nsukka.

Gujarati DN (2003). Basic Econometrics New York. McGraw-Hill Companies Inc 4th edition.

Inoni OE, Oyaide NJ (2007). Socio-Economic Analysis of artisanal fishing in the South Agro-Ecological Zone of Delta State, Nigeria. Agric. Trop. Subtrop. 40(4):135-149.

Kedison A (2003). An Essay in Production Function and empirical study in total factor productivity growth. Htt://www.geocities.com/jeabcu/paper/product.htmaccesseddated 16/03/09.

Keke LA (1989). Fin Erosion in Phrastolaemus ansorgii (Bolenger) Exposed to crude oil in Water Dispension. J. Aquat. Sci. P. 4.

Koutsoyiannis A (2003). Theory of Econometrics. Macmillan Publishers Ltd, London, 2nd Edition.

Mabawonku AF, Olayemi JK, Ogunfowora O, Akinyemo O (1984). Evaluation of the Artisanal and Inshore Fisheries Development project. A study commissioned by the Federal Department of fisheries, Lagos. pp. 34-41.

Mbanasor JA, Obioha LO (2003). Resource Productivity Under Fadamas Cropping System in Umuahia North Local Government Area of Abia State, Nigeria. Trop. Subtrop. Agric. 2:81-86.

Nigerian Environmental Study/Action Team (NEST) (1991). Nigeria's Threatened Environment, A National Profile. NEST, Ibadan.

Nwilo PC, Badejo OT (2001). Impact of Oil spills Along the Nigerian coast the Association for Environmental Health. AEHS magazine 2001. www.aehsmag.com. Accessed 12/7/2005.

Nwilo PC, Peter KO, Badejo OT (2000). Sustainable Management of Oil Spill Incidents Along the Nigerian Coastal Areas. Electronic conference on sustainable Development Information System CEDARE.

Ogri OR (2001). A Review of the Nigerian Petroleum Industry and the Associated Environmental Problem. Environmentalist 1:11-21.

Ogundari K, Ojo SO (2009). An Examination of Income Creation Potential of Agriculture farms in Alleviating Household Poverty: Estimation and policy Implications from Nigeria. Turk. J. Fish. Aquat. Sci. 9:39-45.

Ojo SO (2000). Factor Productivity in Maize Production in Ondo State, Nigeria. School of Agriculture and Agricultural Technology, FUTA, Akure, Ondo State, Nigeria. Appl. Trop. Agric. 5(1):57-63.

Ojo SO, Ehinmowo OO (2010). Economic Analysis of Kola-nut Production in Nigeria. J. Soc. Sci. 22(1):1-5.

Olagunju FI, Adesiyan IO, Ezekiel AA (2007). Economic Viability of cat fish production in Oyo State, Nigeria. J. Hum. Ecol. 21(2):121-124.

Olayemi JK (1998). Elements of Applied Econometrics, Department of Agricultural Economics, University of Ibadan, Ibadan.

Olomola AS (1991). Capture Fisheries and Aquaculture in Nigeria: A Comparative Economic Analysis. In: C. Doss and C. Olson (Eds). African Rural Social Sciences Research Networks: Issues in African Rural Development. Winrock International Institute for Agricultural Development, Arkansas. pp. 343-361.

Olukosi JO, Erhabor PO (1988). Introduction to farm Management Economics: Principles and Application. Department of Agricultural Economics and Rural Sociology, A.B.U. Zaria, Nigeria.

Subair K (2009). Environment-Productivity Relationship in the South West Nigeria's Agriculture. J. Hum. Ecol. 27(1):75-84.

Sule AM, Ogunwale SA, Atala TK (2002). Factors Affecting Adoption of fishing innovations among fishing Entrepreneurs in Jebba Lake communities". J. Agric. Ext. P. 6.

Ted case studies (1996). Oil production and Environmental Damage.

Ukpong I, Akpabio E (2003). Petroleum Polution in part of Niger Delta, Implications on Sustainable Agriculture, Fishing and Health. In Onokala, P.C, Phil-Eze, P.O. and Madu. I.A. (eds) Environment and poverty in Nigeria. Jamoe Enterprises (Nigeria). pp. 57-61.

Williams SB, Awoyomi B (1998). Fish as a prime mover in the Economic life of Women in a fishing Community. In Eide A, Vassdal, P. (eds): Proceedings of the 9th International Conference of the International Institute of Fisheries Economics and Trade. Tromso, Norway.

Ziskowki J, Murcheland F (1995). Fin erosion in Winter Flounder Pseudopleuronects americanus New York Bright. Marine Pollut. Bull. P. 6.

12

Availability and utilization of instructional materials for effective teaching of fish production to students in senior secondary schools in Benue State, Nigeria

Asogwa V. C.[1], Onu D. O.[1] and Egbo B. N.[2]

[1]Department of Agricultural Education, University Agriculture, P. M. B. 2373 Makurdi, Nigeria.
[2]Department of Agricultural Education, Enugu State College of Education (Technical), Enugu, Nigeria.

This study paper assessed the availability and utilization of instructional materials for effective teaching of fish production to students in senior secondary schools in Benue State, Nigeria. Four specific objectives and four research questions were developed to guide the study. Descriptive survey research design was adopted for the study. The study was carried out in Benue State. The population of the study was 284 teachers of Agricultural science. The sample for the study was 142 teachers of Agricultural science obtained through stratified random sampling technique. A Fish Production Instructional Material Questionnaire (FPIMQ) was used for data collection. Six field work assistants helped in administering and retrieving the questionnaire. The data collected were analyzed using descriptive statistical tools such as frequency table, percentage, and mean scores. The results of this study revealed that out of all the instructional materials recommended for teaching fish production to students, 8 of them are available, 5 are accessible, and 8 are often utilized by teachers in senior secondary schools. The study also revealed that 12 challenges were encountered by teachers in accessing and utilizing available instructional materials in fish production in senior secondary schools in Benue State. It was recommended that teachers of Agricultural science should improvise some of the instructional materials lacking in the school locally, among others.

Key words: Availability, utilization, instructional materials, fish production.

INTRODUCTION

Fish is an aquatic "cold blooded" animal. It lives in water and the body temperature changes with the temperature of the surrounding environment. Fish is used primarily as human food because it is rich in protein, iron, zinc, magnesium, phosphorus, calcium, vitamin A and C while marine fish is a good source of iodine (Asogwa, 2012). Other usefulness of fish includes the following:

(i) The skin of some cat fish makes useful leather and polishing material.
(ii) The stem of fishes yield substances that when coated are used as glass beads to make beautiful artificial (Iwena, 2008).
(iii) Fish oil is used for human consumption and manufacturing of soap.
(iv) Some species of fishes are used to beautify aquarium.
(v) Fishes are used for educational and research purposes.
(vi) Whale fish especially clupeid, anchovy and other spp used for fish meal or condiment (Olaitan and Omomia, 2009).

The consumption of fish is drastically increasing because it has low calories and cholesterol levels with high content of protein. The need to produce more animal proteins for human consumption has led to fish production in specially constructed ponds. Fish production is the process of raising fish in an enclosed area for use (Benson, 2011). It is the principal form of aquaculture which involves raising fish commercially in tanks or enclosures, usually for food (Stephanie, 2011). (Oluwatomi, 2012) opined that fish production is an ancient practice which has many profitable opportunities because it involves raising fish in tanks or enclosures, usually for food or commercial purposes. Writing on the importance of fish production, Daramola et al. (2008) stated that it is a source of food for man as it is a rich source of animal protein, supplying essential amino-acids and vitamins, serves as a source of raw materials for industries, income generation to producers, provides employment opportunities for interested individuals, serves as means of foreign exchange and efficient land utilization and conservation of natural resources. The Federal Government appreciates the importance and impacts of fish production in this country and reviewed its production in the curriculum of Agricultural Science through the Nigerian Educational Research and Development Council (NERDC, 2009). The objectives of fish production in the curriculum for senior secondary school are:

(1) For students to have fishery as a trade for livelihood on completion of fish studies.
(2) To produce fish that will increase the nutritive value of man's diet.
(3) To be able to meet with the gap between the demand for fish and its supply.
(4) To bridge the gap between poverty and hunger.

The achievement of these objectives depends on several factors among which include teachers' competence and instructional materials. A teacher, according to Tarum (2009), is someone who impacts knowledge and skills to students and prepares them with the vision of being leaders of tomorrow through motivated educational system. Soni (2012) described a teacher as a leader who is always dynamic and believes in change and have the capacity to prepare future leaders and develop in them the skills that they may need to succeed. Teachers, in this context, refer to individuals who are trained technically and pedagogically in agriculture and are charged with the responsibility of impacting knowledge and skills in fish production to students in senior secondary schools. These teachers discharge their duties to students effectively with the use of instructional materials in fish production. Instructional materials are objects used by a teacher in passing across essential facts of a lesson to students to facilitate their understanding and appreciation of the objectives of the lesson. Onyejemezi (1998) explained instructional materials as resources or teaching materials which a teacher utilizes in the course of presenting a lesson in order to make the content of the lesson understandable to the learner. Nwachukwu (2008) postulated that one of the principles of vocational agriculture is that the instructional materials to be used in training the learners should be a replica of what is obtained in the field or industry. Abimbola and Udonsoro (1997) posited that instructional materials are two or three dimensional aids used by a teacher in order to save students from wondering in imagination and to help their understanding. According to Agbulu and Wever (2011), instructional materials are important because they are used for the transference of information from one individual to another, help the teacher in extending his learner's horizon of experience, stimulate learners' interest and help both teachers and students to overcome physical limitations during the presentation of subject matter, among others. The instructional materials required for effective teaching of fish production to students in senior secondary schools include nursery tanks/ponds, demonstration ponds, scoop nets, hatching troughs, aquaria tanks, compounded feeds, charts and pictures, video clips (NERDC, 2009). It is the view of the council that the recommended instructional materials of fish production in the curriculum should be made available in schools by the school authority.

Availability, in the opinion of Ibrahim (2007) refers to the condition of being obtainable or accessible at a particular point in time. It expresses how a material can easily be gotten and used for a particular purpose and time. It also states how operable or usable resources are upon demand to perform its designated or required functions. In this study, availability means the condition with which teachers have access and make use of functional instructional materials for effective teaching of fish production to students in senior secondary schools. It refers to the quality, quantity, functionality and disposability of such instructional materials to teachers at every point in time for effective utilization.

Utilization, according to Raghu (2009) is the primary method by which asset performance is measured and business determined. It is the transformation of a set of input into goods or services (Subba, 2009). It involves creation of value in things. Utilization, in this context, refers to the rate or how often an instructional material in fish production is put into use or services by teachers of agriculture in senior secondary schools. Utilization of instructional materials depends on their availability in the school.

Due to the fact that fish production is given a specific curriculum with new objectives for senior secondary schools, it becomes absolutely necessary to ask: what are the instructional materials in fish production available in schools as recommended by the ENRDC in the curriculum, how accessible are they to teachers in schools and how often are they used by teachers for effective instruction? Positive answers to the questions could be used to predict the effectiveness of students' learning in fish production. This will likely enhance the performance of the senior secondary school students in both internal and external examination, their interest and employability in fish production industry on graduation.

Statement of problem

The Agricultural science curriculum reform by NERDC (2009) made fish production a single subject for students in senior secondary schools 1 to 3 as against its initial position in the entire curriculum of agricultural science. This implies that new interest, knowledge, skills and instructional materials need to be acquired in order to achieve the stated objectives. On the contrary, the researchers observed that among all the elective subjects in agriculture, most secondary school authorities selected fish production for their student but still implement it the same way fish production was in the former curriculum. In an interaction with some agricultural science teachers in 18 secondary schools visited in a pilot study in Benue State, it was realized that many instructional materials in fish production are lacking. Even the few available in the schools are not in good condition while teachers hardly have easy access to the few available ones. This became a concern to the researchers because of what would be the interests, competence and performance of the students in fish production on graduation if they are not trained with relevant and current instructional materials in fish production.

Purpose of study

The purpose of this study was to access the availability and utilization of instructional materials for effective teaching of fish production to students in senior secondary schools in Benue State, Nigeria. Specifically, the study determined:

(1) Available instructional materials in fish production in senior secondary schools.
(2) Extent to which the available instructional materials in fish production are accessible to teachers in senior secondary schools.
(3) Frequency at which the available instructional materials in fish production are used by teachers in senior secondary schools.
(4) Challenges encountered by teachers in utilizing available instructional materials in fish production in senior secondary schools.

Research questions

(1) What instructional materials in fish production are

available in senior secondary schools?
(2) To what extent are the available instructional materials in fish production accessible to teachers in senior secondary schools?
(3) How often are the available instructional materials in fish production used by teachers in senior secondary schools?
(4) What are the challenges faced by teachers in utilizing the available instructional materials in fish production in senior secondary schools?

METHODS

A descriptive survey research design was adopted for the study. This design, according to Nworgu (2006), is a design in which group of people or items is studied by collecting and analyzing data from a few people , or items considered to be representative of the entire group. The study was carried out in Benue State, Nigeria. The population of the study was two hundred and eighty-four teachers of Agricultural science in senior secondary schools in the three education zone in Benue State. The sample for the study was one hundred and forty-two teachers obtained through stratified random sampling technique based on three education zones in the State. Fifty-seven teachers were selected from Makurdi Education Zone, forty-five from Otukpo Zone and forty from Obi Zone. The total number of 142 respondents corresponds to a 50% sample size. This is in agreement with the statement of Boll, Boll and Gall in Uzoagulu (1998) which stated that when a defined population of a study is less than one thousand, 50% could be used to reduce sampling error. A Fish Production Instructional Material Questionnaire (FPIMQ) was used for data collection. The questionnaire contained 65 items which were grouped into four sections: A, B, C and D. The items in section A were structured based on a two-point rating scale of available or not available while items in sections B, C and D were structured on a four-point rating scale on the accessibility, utilization and acceptance respectively.

Six field work assistants who were familiar with the zones were recruited and trained locally on what to do when administering and retrieving the questionnaire. One hundred and forty-two copies of the questionnaire were administered to the respondents but one hundred and forty-one were retrieved. The data collected were analyzed using descriptive statistical tools such as frequency table, percentage, and mean scores. In order to determine the degree of agreement or disagreement in each of the statements in the questionnaire, values were assigned to the scale used. The figure 50% and 2.50 were used to establish cut off points for judgment. The decision rule was that a percentage of 50% or above was considered available while a percentage below 50% was considered not available. A mean of 2.50 or above was considered accessible, utilized or accepted while a mean below 2.50 was considered not accessible, not utilized or not accepted.

RESULTS

The results of the study were obtained from the data collected and analyzed as shown in the Tables 1 to 4.

Research question 1

What instructional materials in fish production are available in senior secondary schools?

Table 1. Frequency and percentage ratings of the responses of teachers on the availability of instructional materials in fish production.

S/N	Instructional materials	Available (A)	Not available(NA)	Percentage		Decision
1	DO(Dissolved Oxygen) meter	105	145	42	58	Not available
2	pH meter	190	60	76	24	Available
3	Conducting meter	65	185	26	74	Not available
4	Thermometer	180	70	72	28	Available
5	Water test kits	63	187	25.2	74.8	Not available
6	Microscopes	192	58	76.8	23.2	Available
7	Magnifying glass	175	75	70	30	Available
8	Aquaria tanks	65	185	26	74	Not available
9	Hatching troughs	83	167	33.2	62.7	Not available
10	Nursery tanks/ponds	20	187	8.7	90.324	Not Available
11	Demonstration pond	35	215	14	86	Not available
12	Scoop nets	203	47	81.2	19.8	Available
13	Aerators and accessories	10	240	4.0	96.0	Not available
14	Plastic sieves	209	41	83.6	23.4	Available
15	Compounded feeds	150	100	60	40	Available
16	Grinding machine	43	207	17.2	82.8	Not available
17	Charts and pictures	206	44	82.4	17.6	Available
18	Video clips in fisheries	75	175	30	70	Not available
19	Pelleting machine	67	183	26.8	73.2	Not available
20	Dissecting kits	98	152	39.2	60.8	Not available
21	Water pumps	100	150	40	60	Not available
22	Sec chi disc	66	184	26.4	73.6	Not Available

Table 1 revealed that amongst the 22 instructional materials recommended by the NERDC (2009), as required by teachers for effective teaching of fish production to students, only 8 instructional materials were considered available in schools and they are: pH meter, Thermometer, Microscopes, Magnifying glass, Scoop nets, Plastic sieves, Compounded feeds, Charts and pictures. This is because the percentage availability of the instructional materials was 50% or above which were above the cut-off point. The remaining 14 instructional materials had their percentage availability below 50% which is the cut-off point. This indicates that they were not available on the average.

Research question 2

To what extent are the available instructional materials in fish production accessible to teachers in senior secondary schools?

Table 2 revealed that 5 out of 22 items had their mean values ranged from 2.64 to 3.42 which were above the cut-off point of 2.50. This indicated that the 5 items are available and accessible by the teachers in schools. The table also revealed that 17 out of 22 items had their mean values ranged from 0.06 to 2.16 which were below the cut-off point of 2.50. This indicated that the 17 items including some available in schools were not accessible

to teachers for effective teaching of fish production to students. The table also showed that the standard deviation of the responses of the respondents ranged from 0.38 – 0.94, indicating that the respondents were not too far from the mean and from the opinion of another in their responses.

Research question 3

How often are the available instructional materials in fish production utilized by teachers in senior secondary schools?

Table 3 revealed that 6 out of 8 available instructional materials in schools had their mean value ranged from 2.92 to 3.82 which were above the cut-off point of 2.50. This indicated that the 6 items are available, accessible and often utilized by the teachers for effective teaching in schools. The table also revealed that 2 out of 8 items had their mean value 2.24 and 2.43 which were below the cut-off point. This indicated that the 2 items (microscope and magnifying glass) were available and accessible in schools but not often utilized by teachers for effective teaching of fish production to students. The table also showed that the standard deviation of the responses of the respondents ranged from 0.54 – 0.93, indicating that the respondents were not too far from the mean and from the opinion of one another in their responses.

Table 2. Mean ratings and standard deviation of the responses of teachers on their accessibility of instructional materials in fish production.

S/N	Items	X	SD	Decision
1	Dissolved Oxygen meter	1.48	0.81	Not accessible
2	pH meter	2.20*	0.67	Not accessible
3	Conductivity meter	2.16	0.56	Not accessible
4	Thermometer	3.42	0.71	Accessible
5	Water test kit	1.06	0.38	Not accessible
6	Microscope	0.80*	0.89	Not accessible
7	Magnifying glass	0.64*	0.92	Not accessible
8	Aquaria tanks	0.96	0.55	Not accessible
9	Hatching troughs	2.04	0.73	Not accessible
10	Nursery tanks/troughs	1.70	0.72	Not accessible
11	Demonstration pond	1.44	0.90	Not accessible
12	Scoop nets	2.88	0.63	Accessible
13	Aerators and accessories	0.06	0.84	Not accessible
14	Plastic sieves	2.82	0.51	Accessible
15	Compounded feed	2.80	0.56	Accessible
16	Grinding machines	1.18	0.77	Not accessible
17	Charts and pictures	2.64	0.94	Accessible
18	Video clips in fisheries	2.22	0.67	Not accessible
19	Pelleting machine	0.88	0.43	Not accessible
20	Dissecting kits	1.52	0.62	Not accessible
21	Water pumps	2.12	0.74	Not accessible
22	Sec chi disc	2.07	0.92	Not accessible

X = mean, SD = standard deviation, * = available but not accessible.

Table 3. Mean ratings and standard deviation of the responses of teachers on the accessibility of instructional materials in fish production by the teachers.

S/N	Items	X	SD	Decision
1	pH meter	2.92	0.57	Utilized
2	Thermometer	3.80	0.67	Utilized
3	Microscopes	2.24*	0.78	Not utilized
4	Magnifying glass	2.43*	0.93	Not utilized
5	Scoop nets	3.82	0.77	Utilized
6	Plastic sieves	2.96	0.83	Utilized
7	Compounded feeds	3.72	0.69	Utilized
8	Charts and pictures	3.56	0.54	Utilized

SD = standard deviation, * = available but not often utilized.

Research question 4

What are the challenges faced by teachers in accessing and utilizing the available instructional materials in fish production in senior secondary schools?

Table 4 revealed that 12 out of 13 challenge items had their mean values ranged from 2.55 to 4.00 which were above the cut off point of 2.50. This shows that the teachers accepted that the 12 items were the challenges they face in accessing and utilizing instructional materials in fish production. The table also revealed that 1 out of 13 challenge items had its mean value 1.23 which was below the cut off point of 2.50. This shows that teachers rejected that the item (number 12) was a challenge they face in accessing and utilizing instructional materials in fish production. The table also shows that the standard deviation of the items ranged from 0.09 – 1.07, indicating that the respondents were not too far from the mean and

Table 4. Mean ratings and standard deviation of the responses of teachers on their challenges in accessing and utilizing available instructional materials in fish production.

S/N	Challenges	X	SD	Remark
1	The teachers' knowledge and technical know-how on the use of some of the instructional materials in poor	4.00	0.31	Accepted
2	Many of instructional materials in the schools are out of use in fish industries	2.79	0.90	Accepted
3	Most of the materials are under the custody of principal of the school which hinders utilization	2.90	0.45	Accepted
4	The materials are not maintained for proper functioning during lessons	3.61	1.02	Accepted
5	Most of the available instructional materials are not in good working condition	3.11	0.67	Accepted
6	There is no steady power supply in the school to operate some of the materials	3.02	0.87	Accepted
7	The instructional materials are not sufficient for students to practice with during lesson	3.00	1.00	Accepted
8	Students damage some of the instructional materials during instruction	3.77	0.88	Accepted
9	The number of students is large too to evenly see the materials during instruction	3.68	0.72	Accepted
10	The lesson duration on the time table does not favour the use of instructional materials.	2.55	0.41	Accepted
11	The interest and participation of students in fish production is low and does not favour the use of instructional materials	1.97	0.34	Accepted
12	The value of fish in the society does not favour the use of instructional materials	1.23	0.09	Not accepted
13	The female genders are indifferent on the use of instructional materials	3.21	1.07	Accepted

X = mean, SD = standard deviation.

from the opinion of one another in their responses.

DISCUSSION OF FINDINGS

The results of this study revealed that out of all the instructional materials recommended by the NERDC (2009) for teaching fish production to students, 8 of them are available, 5 are accessible, and 8 are often utilized by teachers in senior secondary schools. It implies that teachers probably improvise for the 3 instructional materials that were not easily accessible to them but were often utilized. The study also revealed that 14 recommended instructional materials in fish production were not available in schools while 12 challenges were encountered by teachers in accessing and utilizing available instructional materials in fish production in senior secondary schools.

The findings of the study are in line with findings of Nwosu (2010) in a study on utilization of information and communication technology (ICT) as a tool and strategies for improving teacher professional development for effective service delivery, where it was found out that teachers can, to a very low extent, utilize 10 ICT resources for their professional development to enhance service delivery in schools. It also revealed that slow access to ICT equipments, low interest connectivity, lack of sufficient computers and high cost of laptop, lack of qualified personnel, interrupted power supply among others constitute a hindrance to ICT usage. Enwereuzor (2011) in a study on utilization of antenatal care facilities

in Ikeduru Local Government Area, Imo State, found out that women of child bearing age in Ikeduru LGA do not effectively utilize antenatal care services during and after pregnancy. It also revealed that the level of educational attainment, poor staffing of the maternal health facilities, distance, cost and preference for traditional birth home go a long way to determine the extent of utilization of maternal centres for antenatal services. The study is in consonance with the findings of Daudu (2012) in a study on assessment of availability and use of information resources and services in the Institute of Education Library, Ahmadu Bello University, Zaria, who found out that human resources of the library are quite adequate. Materials resources are not very current except for newspapers. Equipment such as photocopying machine though essential is not available in the library. The findings of the authors cited above help to add credence to the findings of this study.

CONCLUSION AND RECOMMENDATION

In 2009, the NERDC reformed the curriculum of Agricultural science in senior secondary schools. The reform made some components of Agriculture single subjects for the students among which include animal husbandry, fish production. This development demands that the teaching of fish production takes a new dimension with regards to the competence of teachers, duration of lessons, ratio of theory to practical, instructional materials and so on. Even though, fish production is optional, it was observed by the researchers that many secondary schools selected this option without any adjustment to provision of instructional materials and other relevant factors as recommended by the NERDC. This suggests that the interest and competence of the students in fish production on graduation may be equivocal. It then becomes necessary to assess the availability and utilization of instructional materials for effective teaching of fish production to students in senior secondary schools in Benue State. The study found out that out of all the instructional materials recommended by the NERDC (2009) for teaching fish production to students, only 8 of them are available, 5 are accessible, and 8 are often utilized by teachers in senior secondary schools. It also revealed that 12 challenges were encountered by teachers in accessing and utilizing available instructional materials in fish production in senior secondary schools. Based on this, it was recommended that:

(1) Teachers of Agricultural science should improvise some of the instructional materials locally.
(2) The school authority should provide and make accessible to teachers those materials that are lacking in the school.
(3) The Federal and the State Ministry of Education should provide instructional materials in fish production to schools in the State especially those that cannot be afforded by the school authority.

REFERENCES

Abimbola AA, Udonsoro VN (1997). Instructional materials for senior secondary schools. Nigeria: University press.

Agbulu ON, Wever DG (2011). *Introduction to Vocational* Agricultural Education. Makurdi, Benue State: Selfers Academic Press Ltd.

Asogwa VC (2012). Introduction to wildlife and fisheries. Lecture note, Department of Agricultural Education, University of Agriculture, Makurdi.

Benson T (2011). Advancing Aquaculture: Fish welfare Slaughter". Retrieved from www.hptt:/ fishculture.edu.org on 2011-06-12.

Daramola A, Igbokwe EM; Mosuro GA, Abdullahi JA (2008). Agricultural Science for WASSCE and SSCE. Ibadan: University Press Plc.

Daudu HM (2012). Assessment of availability and use of information resources and services in the Institute of Education Library, Ahmadu Bello University, Zaria. Nig. J. Teacher. Edu. Teach. 10(1):230-241.

Enwereuzor AI (2011). Utilization of antenatal care facilities in Ikeduru Local Government Area, Imo State. In Nigeria at fifty: Issues, Challenges and Agenda. Ed by Ezeudu SA, Ezeani, EO, Onuoha JO, Nwizu SC 1(25):223-234.

Iwena OA (2008). Essential Agricultural Science for Senior Secondary Schools. Ogun State. Tonad Publishers Limited.

Nigerian Educational Research and Development Council (NERDC) (2009). Senior Secondary School Curriculum: Fishery for SSS1-3. Sheda, Abuja: University Press Plc.

NWachukwu CU (2008). Opinion of the scientific panel on animal health and welfare on a request from the commission related to welfare aspects of the main system of stunning and killing the main commercial species of animal. EfSA J. Retrieved 2011-06-12.

Nworgu BG (2006). *Educational Research Basic Issues and Methodology.* Ibadan: wisdom publishers Ltd.

Nwosu EN (2010). Utilization of information and communication technology (ICT) as a tool and strategies for improving teacher professional development for effective service delivery. Int. J. Edu. Res. 11(2):86-95.

Olaitan SO, Omomia OA (2009). Round up Agricultural science: A complete guide. Enugu; Longman Nigeria PLC.

Oluwatomi O (2012). *Fish farming.* Retrieved from www.google.com on 3/4/13.

Onyejemezi OE (1998). Practical instructional materials. Retrieved from www.google.com on 3/4/13.

Raghu U (2009). *Utilization of economic resources.* Retrieved from www.google .com on 6/5/13.

Soni MO (2012). Fish processing plants. Retrieved from www.slaughter. lt/en/ 04/05/13.

Stephanie Y (2011). The welfare of farmed fish at slaughter. Humane society of the United States. Retrieved 2011 -06-12.

Tarum FC (2009). Passion for Agriculture. Retrieved from Naijagreen. Com on 4/07/13.

Uzoagulu AE (1998). Practical Guide to Writing Research Project Reports in Tertiary Institutions. John Jacobs Classic Publishers Ltd.

13

Impact of cooperative society on fish farming comercialization in Lagos State, Nigeria

Odetola S. K.[1], Awoyemi T. T.[1] and Ajijola S.[2]

[1]Department of Agricultural Economics, University of Ibadan, Nigeria.
[2]Institute of Agricultural Research and Training, Moor Plantation, Ibadan, Nigeria.

This study was carried out to determine the impact of cooperative society among the fish farmers in Lagos State. A multi stage purposive sampling techniques was used to select five Local Government areas notable for fish farming business. 30 fish farmers were selected from each of the Local Government areas for cooperative society and 30 farmers from non cooperative society having a total of 150 respondents each. A well structured questionnaire was used to obtain information and 130 questionnaires were retrieved each from cooperative and non-cooperative members. Analytical techniques used include descriptive statistics and Tobit regression Analyses. The results show that the mean age of the farmers is 56 and 57 for cooperative and non-cooperative fish farmers, respectively. Majority (83%) and (93%) of the cooperative and non-cooperative fish farmers respectively were males. It was discovered that both farmers have an average of 8 household members. It was revealed that larger percentage of the cooperative fish farmers (50%) used amount ₦100,000 to ₦500,000 as the initial investment while (56%) of the non cooperative used the same amount as capital investment. The result of the Tobit regression analysis indicates that gender of farmers is significant at 5%, years of formal education; membership of cooperative and the cost of inputs were significant at 1%. Since majority were producing for profit making, it is suggested in the paper that government should increase the supply of credit to cooperative farmers and embark on enlightenment campaign to increase the participation of rural farmers in cooperative activities.

Key words: Impact, cooperative society, fish farming, commercialisation, Lagos State.

INTRODUCTION

In developing countries in which Nigeria is one, agriculture dominates the economy of the nation. It has been established that about 70% of Nigeria population is engaged in agriculture while 90% of Nigeria total food production comes from small farms and 60% of the country population earn their living from these small farms. The fall in agricultural production could be attributed to inadequate infrastructure, under mechanization and inadequate finance (Oluwatayo et al., 2008). One of the major problems of agricultural development in Nigeria is that of developing appropriate organization and institution to mobilize and induce members of the rural sector to a greater productive effort (ICA, 2010). As such rural farmers who are characterized by low income, low resource utilization, small farm holdings and scattered nature of farmland, finds it difficult to pool their resources together in order to raise their farm income and substantially improve their living conditions (Ibitoye, 2012).

Inadequate finance has remained the most limiting problem of agricultural production. This is because capital is the most important input in agricultural production and its availability has remain a major problem to small scale farmers who account for the bulk of agricultural produce of the nation. In Nigeria, credit has long been identified as a major factor in the development of agricultural sector (Ndifon et al., 2012). Cooperative societies in Nigeria perform multipurpose functions. They are engaged in the production, processing, marketing, distribution and financing of agricultural products. It is an established fact that many household in the country today, live below the poverty line, in fact, investigation has shown that the highest percentage of Nigeria's workforce work in the public sector and earn their monthly salary of below one dollar per day (Awotide et al., 2012). The rural community, whose main occupation is agriculture, produces the food consumed in the country, but which is hardly sufficient to feed the people, because farmers still use crude farming implements to till the land. The federal government, in a bid to fight the menace of poverty therefore, has set up some agencies essentially to provide financial assistance particularly to youths and women involved in small scale businesses. So recently, Cooperate Societies, a concept that was given birth from the traditional thrift collection, began to spread like wild fire in virtually every part of Nigeria. There is hardly any workplace in Nigeria today particularly government establishments, where a cooperative society is not operational. It is quite effective because transactions of money are carried out in conjunction with employers of labour on behalf of their staff (Godwin, 2011).

Agricultural commercialization is the share of agricultural produce that is marketed. Commercialization is the process through which increased amount of small farm resources (land, labour e.t.c) is transferred from self consumption production to market oriented production. As such commercialization can be measured along a continuum from zero (total subsistence oriented production) to unity (100% production is sold). Commercialization of agriculture involves a transition from subsistence oriented to increasingly market – oriented patterns of production and input use (Nweze, 2003).

In spite of the importance of loan in agricultural production, its acquisition is fraught with a number of problems. The small scale farmers are forced to source for capital from relations, money-lenders and contribution clubs. All of these are known to be ineffective in providing capital for substantial increase in agricultural production. The last hope for the small scale farmers then lies with the cooperative societies, the cooperative has been identified to be better channel of credit delivery to farmers the NGO's in term of its ability to sustain the loan delivery function (Alufohai, 2006).

Adekunle and Henson (2007) studied the effect of cooperative thrift and credit societies on personal agency belief: A study of entrepreneurs in Osun State, Nigeria. He opined that little or no attention has been paid to the role of entrepreneurship and the capacity of institutions like Cooperative Thrift and Credit societies to promote entrepreneurship. Cooperatives are defined as "an autonomous association of persons who unite voluntarily to meet their common economy and social needs and aspiration through a jointly owned and democratically controlled enterprise. Cooperatives are established by like-minded persons to pursue mutually beneficial economic interest. Researchers are of the opinion that under normal circumstance cooperative play significant role in the provision of services that enhance agricultural development (Ndifon et al., 2012).

Regular and optimal performance of these roles will accelerate the transformation and sustainability of not only the cooperatives but the enhancement of agricultural and rural economic development. Cooperative embraces all type of farmers and a well organized and supportive cooperative is a pillar of strength for agriculture in Nigeria. Previous studies have shown that cooperative carryout the function of credit delivery to farmers but there is ample evidence that farmers face difficulties in obtaining credit and the problem of sourcing for capital still lingers on. Therefore, any cooperative society to be effective and successful, it must continuously achieve two inter-related goals: enhance viability and improve ability to service its members; and remain an economically viable, innovative and competitive enterprise (Dogarawa, 2005).

Fish farmers in Lagos State are generally involves in one form of self help group or cooperative organization to carry out their production activities such as improvement on fish farming practices (that is, adoption of new technology) income growth and stability, business growth, purchase of inputs like fingerlings, feed and other basic needs such as clothing, food and shelter. One of the ways to improve the lots of these fish farmers' welfare and productivities is cooperative society membership and participation. Without an iota of doubt, the cooperative society will help the farmers a lot to improve their productivities as well as their welfare. Through cooperative, fish farmers will be able to access more fund for their fish production hence engage in fish farming commercialization.

Nigeria being a coastal country has about 1,280 km marine areas and about 124,878 km of inland waterways. Lagos State with a general area of 3,577 km representing 0.4% of Nigeria territorial land mass is one of the maritime states of Nigeria and as such share a potion of the Atlantic Coast of the Gulf of Guinea which is rich in fisheries resources. In spite of this potential, domestic fish production is grossly inadequate to meet even domestic demand (FAO, 1990). Fish is the cheapest sources of protein and because of its low cholesterol level which makes it medically acceptable to young and old people. The demand for fish protein according to Federal Department of Fisheries (FDF) was 2.6 million tonnes in 2007 while domestic production was 634,370 tonnes. The deficit was partly augmented by massive importation of frozen fish of about 740,000 tonnes valued at 94.- a big draw – down on scarce foreign exchange. This leaves a huge deficit of 1.3 million tonnes and hence the concerted efforts to ensure self sufficiency in fish production through fish farming (aquaculture). Aquaculture has been estimated to have a potential of producing 2.5 milloin ones annually which is fully harnessed can almost satisfy the demand for fish in Nigeria alone. The estimated total law available for aquaculture production is 1.7 million hectares excluding marine brackish water bodies. Unfortunately, aquaculture production was only 85,087 tonnes in 2007 despite its potential and its enormous water resources in contrast with the state fish production capacity of about 157,000 tonnes (Kareem et al., 2012).

In view of the above, this study therefore deals with the effect of cooperative society on fish farming commercialization, determined the problems faced by the artisan and identified the factors that affect participation in fish farming in Lagos State. This study is significant in the sense that the assessment of co-operative development will further serve as framework for formulating new and better policies for agricultural co-operative development in Nigeria.

MATERIALS AND METHODS

Area of study

The area of study is Lagos State which was created in 1967. Lagos State is located on the coast in the most South Western corner of Nigeria. It is the smallest but most densely populated state in the federation with land of 3,586 km^2 which is about 0.39% of the Nations 923,768 km^2 area.

Sampling procedure and sample frame

The sampling method adopted for the study was the multistage purposive random sampling; Lagos State comprises of twenty local government areas which was divided into five geographical zones namely, Ikeja, Ikorodu, Epe, Lagos Island and Badagry.

The research was carried out in five local government areas of Lagos State which represent geographical zones of the state and notable for fish farming in large production. The list of cooperative fish farmers in each local government were obtained from the Lagos State agricultural development project, Oko – Oba, Lagos since they coordinate the activities of the cooperative society.

A total of 150 cooperative fish farmers and 150 non – cooperative fish farmers were interviewed. That is, 30 cooperative fish farmers and 30 non – cooperative fish farmers from each local government. However, 130 questionnaires were retrieved each from cooperative farmers and non cooperative farmers for analyses making a total of 260 farmers.

Data collection and analytical procedures

The data used was obtained mainly from primary source through the use of structured questionnaires that was administered to fish farmers. The questionnaires contain both open and close ended questions covering the social and personal characteristics of the respondents and other related variables such as awareness and participation in cooperative activities, income and expenditure, pond size. The instrument for data collection is subjected to expert validation.

Data collected during the study was analysed using descriptive statistics and Tobit regression analysis. Descriptive Statistics – Tables was used to present frequency distribution, percentages and averages on demographic and non-demographic characteristics of the cooperative fish farmers. Tobit regression analysis – Tobit regression analysis was employed to examine the functional relationship among the variables.

The Tobit model is expressed as – $\Upsilon^* = \beta\chi + \mu$; $\Upsilon^* = Y = $ Income; β = Vector of parameter estimated; X = Set of explanatory Variables; μ = The disturbance term; X_1 = Age (years); X_2 = Gender; X_3 = Fish farming Experience (years); X_4 = Education; X_5 = Size of pond (m^2); X_6 = Marital Status; X_7 = Cooperative membership (Members = 1, Non member = 0); X_8 = Cost of input in naira and X_9 = Household Size.

RESULTS AND DISCUSSION

Table 1 shows the socio economic characteristics of the fish farmers. It reveals that 43% of the cooperative fish farmers were within the age of 56 years while 46% of the non-cooperative fish farmers were in the same age range. There is no significant difference between the mean ages of the cooperative and non-cooperative farmers. About 46% of the cooperative fish farmers and 47% of the non-cooperative fish farmer have secondary school education. Majority (83%) and 93% of the cooperative and non-cooperative fish farmers were males, respectively. It was also discovered that both cooperative fish farmers and non-cooperative has an average of 8 household members.

Table 2 showed the initial capital outlay and sources of fund for both cooperative and non-cooperative fish farmers in the study areas. The result shows that higher percentage (45%) sourced their fund through personal savings, 20% sourced fund through friends while about 36% sourced fund through cooperative society. It was revealed that larger percentage of the cooperative fish farmers (50%) used amount #100,000 to #500,000 as the initial investment while (56%) of the non cooperative used the same amount as capital investment. About 53% cooperative fish farmers and 14% non cooperative fish farmers were operating with over half a million (above #500,000.00) as initial capital investment in fish commercialisation. The results revealed that the involvement in cooperative society had made great impact in fish commercialisation and the fish farmers have been able to increase their initial capital investment

Table 1. Socio-economic characteristics of respondents (Co-operatives & Non cooperative Fish farmers).

Variables	Cooperative farmers	Frequency	Percentage	Variables non cooperatives farmers	Frequency	Percentage
Mean age (Yrs)	56	56	43.08	57	60	46.15
Sex	Male	108	83.08	Male	122	93.85
	Female	22	16.92	Female	8	6.15
Marital status	Single	6	4.62	Single	4	3.08
	Married	98	75.38	Married	104	80.00
	widowed	26	20.00	widowed	22	16.92
Religion	Christianity	70	53	Christianity	62	47.69
	Islam	48	36.92	Islam	54	41.54
	Tradition	12	09.23	Tradition	14	10.77
Education	No formal edu.	8	6.15	No formal education	12	9.23
	Primary	20	15.38	Primary	24	18.46
	Secondary	60	46.15	Secondary	62	47.69
	Tertiary	42	32.31	Tertiary	32	24.62
Years of experience	1 - 5	72	55.38	1 - 5	64	49.23
	6 - 10	30	23.08	6 - 10	38	29.23
	11 - 15	18	13.85	11 - 15	22	16.92
	Above 15	4	3.08	Above 15	6	4.62
H/H size	1 - 5	46	35.38	1 - 5	44	33.85
	6 - 10	62	47.69	6 - 10	58	44.62
	11 - 15	18	13.85	11 - 15	22	16.92
	Above 15	4	3.08	Above 15	6	4.62

Source: Field survey, 2014.

in the enterprise. The larger number of the side of the cooperative fish farmers might not be unconnected to the financial assistance obtained from the cooperative society for fish farming.

Table 3 shows the purpose for engaging in fish farming in the study area. The results show that about 88% of the cooperative farmers and 98% of the non-cooperative farmers were running the business for profit making; that is, they were fully commercialised while only 12% engaged in the fish farming for sustaining the family.

The problems encountered in the fish farming are inadequate capital, marketing problem and high cost of input (Figures 1 and 2). Tax from government was not posturing too much problem for both cooperative and non-cooperative fish farmers in the study areas.

Table 4 shows the factors that affect farmers' participation in fish farming commercialization using Tobit regression model. Nine explanatory variables were considered in the model. However,

Table 2. Sources of fund and Initial Capital Outlay.

Variable	cooperative fish farmers	Frequency	Percentage	Variables non-cooperative farmers	Frequency	Percentage
Sources of Fund	Own Savings	58	44.62	Own Savings	114	87.69
	Friend	26	20.00	Friend	16	12.31
	Co-operatives	46	35.38	Cooperatives	-	
Initial Capital investment (#)	Less than 100,000	27	20.77	Less than 100,000	32	24.62
	100,001 - 500,000	50	38.46	100,001 - 500,000	56	43.08
	500,001 - 1,000,000	48	36.92	500,001 - 1,000,000	12	9.23
	Above 1,000,000	5	3.85	Above 1,000,000	2	1.54
Income Group	Less than 100,00	12	9.23	Less than 100,00	25	19.23
	100,001 - 500,000	48	36.92	100,001 - 500,000	64	49.23
	500,001 - 1,000,000	66	50.77	500,001 - 1000,000	40	30.77
	Above 1,000,000	4	3.08	Above 1000,000	1	0.77

Source: Field survey, 2014.

Table 3. Purpose for engaging in fish farming in the study area.

Variable	cooperative fish farmers	Frequency	Percentage	Variables non-cooperative farmers	Frequency	Percentage
Purpose of engaging in fish farming	Profit	114	87.69	Profit	98	75.38
	To maintain family	16	12.31	To maintain family	32	24.62

Source: Field survey, 2014.

only four were significant. They are sex of farmers, years of formal education, membership of cooperative and the cost of inputs. The log likelihood ratio of - 2006 and the P - Value of 0.0001 reveals that the model as a whole is statistically significant.

Education is significant (P < 0.029) and positively related to fish farming commercialization. This shows that at higher level of education, fish farming commercialization is high. This is due to the fact that formal education can improve technical know-how in fish production and marketing. Gender is significant (P < 0.0449) and negatively related to fish farming, this shows that female fish farmers tend to be involved more in fish farming commercialization. This may be as a result of the fact that women are producing mainly to sell and not to feed their household. Membership of cooperative is significant (P < 0.0001) and is positively related to fish farming commercialization. This may be as a result of the cooperative assistance obtained from the cooperative societies to promote fish farming commercialization.

The cost of input is significant (P < 0.0001) and positively related to fish farming commercialization because as the input cost increases more fish will be produced and fish farming commercialization will be promoted. This will also motivate the farmers to seek for assistance when the cost of

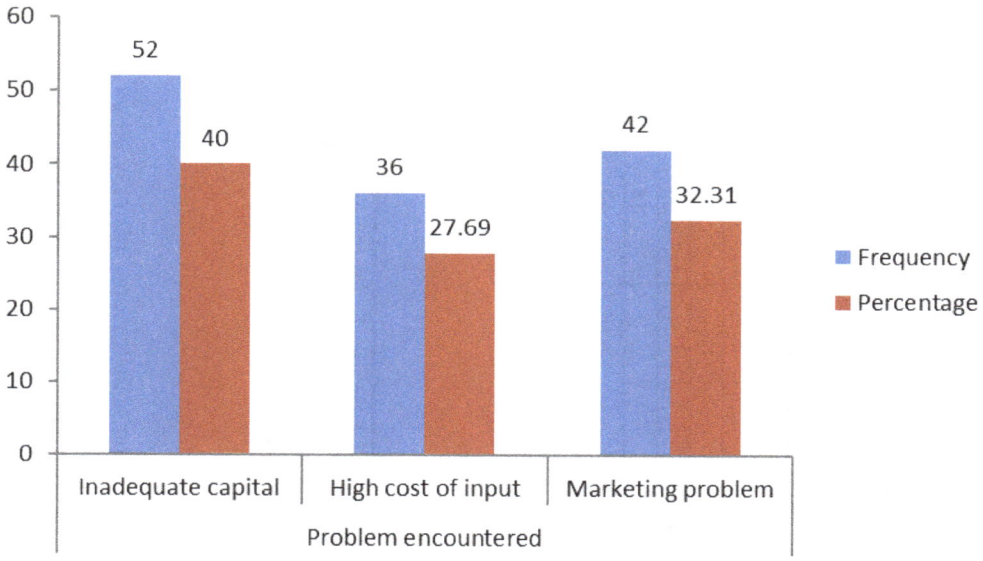

Figure 1. Problem encountered by cooperative farmers.

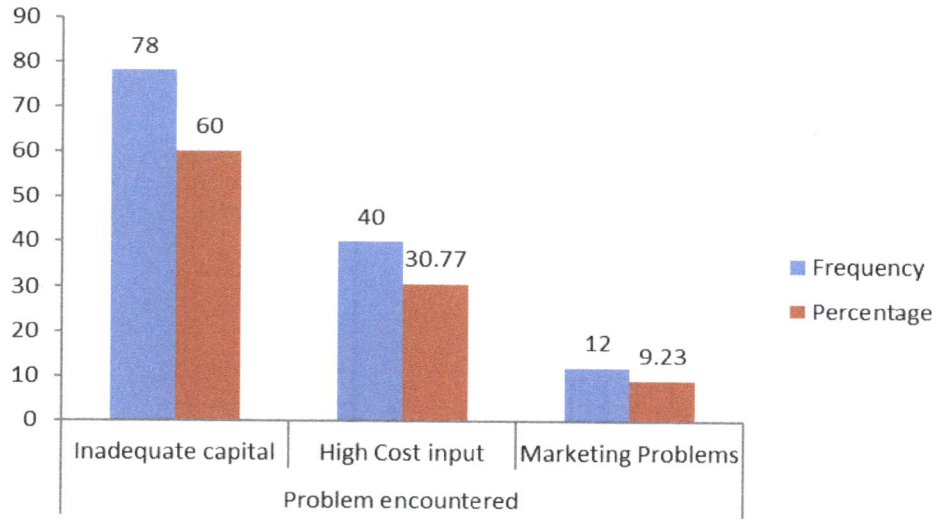

Figure 2. Problem encountered by non cooperative farmers.

production is high in which the cooperative society will be of advantage to them. The size of pond is negatively related to fish farming commercialization. This may be due to the fact that the size of pond does not determine pond stocking density and fish output. Fish output was determined by quantity of fish stocked and proper management practices.

CONCLUSION AND RECOMMENDATION

This study had shown that there is great prospect for fish farmers in Lagos State since fish farming commercialisation is profitable. Since the respondents confirmed that their income is always higher than the capital outlay in fish farming. it was concluded that co-operative societies have effect on member's welfare and the role of co-operative society in poverty reduction and capital formation cannot be overlooked in the development process of any country particularly the less developed countries like Nigeria.

However, the findings revealed the importance of cooperative societies arises from the fact that the rural poor (farmers) are not properly served by formal

Table 4. Tobit Regression Analysis for the identification of factors that affect participation in fish farming commercialization.

Variable	Coefficient	Std error	T	9 > (t)
Age	9.24	6.96	1.33	0.184
Gender	- 151.40*	75.49	- 2.01	0.044
Fishing experience	36.97	19.90	1.86	0.063
Formal education	22.66**	7.61	2.98	0.0021
Size of pond	-56.03	87.08	-0.64	0.52
Marital status	147.86	80.05	1.85	0.064
Cooperative membership	439.68***	88.08	4.99	0.0001
Cost of input	1.40***	0.08	17.21	0.0001
Household size	4.86	15.90	0.31	0. 760

Source: Field survey, 2014. Log likelihood - 2006; No of Observation 260; Schwarz Criterion 4073. *, **, *** significant at 10, 5, and 1% level respectively.

institution agencies (viz, commercial banks and other government owned financial institution). These institutions refrain advancing loan to the rural poor because of the bureaucratic procedures and high cost service involved in lending. Therefore, this study gives credence to the use of cooperative as machinery for rural transformation and agricultural development in Nigeria. The continued existence and operation of cooperative societies have to be encouraged by both individuals and government. They have been able to make impart in the area of membership enrolment, farm input procurement through loan disbursement and training of members.

Based on the findings, the following recommendations were made:

i. Fish farmers should be encouraged to join cooperative societies as this promotes fish farming commercialization.
ii. Women should be encouraged to go into fish farming.
iii. Fish farmer should be supported financially by the government and financial organization through provision of loans.
iv. Government should increase the supply of credit to cooperative farmers and embark on enlightenment campaign to increase the participation of rural farmers in cooperative activities vis a vis improve fish commercialization.

Conflict of Interest

The authors have not declared any conflict of interest.

REFERENCES

Adekunle B, Henson SJ (2007). The effect of cooperative thrift and credit societies on personal agency belief: as study of entrepreneurs in Osun State Nigeria. Afr. J. Agric. Res. 2(12):678-686.

Alufohai GO (2006). Sustainability of Farm Credit delivery by Cooperatives and NGO's in Edo and Delta State, Nigeria. Edu. Res. Rev. 1(8):262-266.

Awotide DO, Aihonsu JOY, Adekoya AH (2012). Cooperative societies' effectiveness in credit delivery for agricultural enterprises in Ogun State, Southwest Nigeria. Asian J. Bus. Manage. Sci. 2(3):74–79.

Dogarawa B (2005). The Role of Cooperative Societies in Economic Development. MPRA Paper No. 23161.

FAO (1990). Food and Agriculture Organisation.

Godwin S (2011). Poverty Reduction Through the Use of Cooperative Societies. Kaduna: Rev. Int. Cooperatives. 4:85–86.

Ibitoye SJ (2012). Survey of the Performance of Agricultural Cooperative Societies in Kogi State, Nigeria. Eur. Sci. J. 8(28):98–114.

ICA (2010). International Cooperative Alliance. Retrieved 1, October, 2011 from http://www.ica.coop/.ss

Kareem RO, Arigbabu YD, Akintaro JA, Badmus MA (2012). The Impact of Co- Operative Society on Capital Formation (A Case Study of Temidere Cooperative and Thrift- Society, Ijebu- Ode, Ogun State, Nigeria). Global J. Sci. Frontier Res. Agric. Vet. Sci. 12(11):1.0.

Ndifon HM, Agube EI, Odok GN (2012). Sustainability of Agricultural Cooperative Societies in Nigeria: The Case of South-South Zone, Nigeria. Mediterranean J. Soc. Sci. 3(2):19–25.

Nweze NJ (2003). "Cooperative promotion in rural communities: The project approach". Nig. J. Cooperatives 2(2):76- 89.

Oluwatayo AB, Sekumade O, Adesoji SA (2008). Resource Use Efficiency of Maize farmers in Rural Nigeria: Evidence from Ekiti State, Nigeria. World J. Agric. Sci. 4(1):91-99.

14

Utilization of common carp fish surimi in baby food products

Emad M. El-Kholie[1,3], Mohammed A. T. Abdelreheem[2] and Seham A. Khader[3]

[1]Research Center, College of Science, King Saud University, P. O. Box 2455, Riyadh11451, Kingdom of Saudi Arabia.
[2]Department of Biochemistry, Faculty of Agriculture, Ain Shams University, Egypt.
[3]Department Nutrition and Food Science, Faculty of Home Economics, Menufiya University, Egypt.

This work was conducted to increase the nutritional value of some baby food namely Cerelac (rice-base) widely distributed in the Egyptian markets. Raising protein and minerals of baby food (Cerelac) increases its nutritional value. Dried surimi from common carp fish (*Cyprinus carpio*) fortified at 10, 20 and 30% levels. Chemical composition, microbiological, quality aspects and sensory evaluation were determined. Results indicated that moisture; protein, fat, ash, carbohydrates and energy values of surimi were 12.71, 60.77, 0.35, 4.29 and 21.88%, 333.75 kcal/100 g w/w, respectively. Aerobic bacterial counts detected *Staph.* spp., Coliform group, *Salmonella* and *Shigella,* while anaerobic bacteria and mold and yeasts did not detect. Increasing levels of surimi result an increment of moisture, protein, and ash contents. Fat, fiber and carbohydrates contents were reduced. Grams daily requirement (GDR) and percent satisfaction (PS/150) for protein decreased, while GDR and PS/150 for energy value increased. Fortification with 30% dried surimi leads to a maximum improvement of all tested sensory evaluation by different rates. This work strongly recommend that the fortification of Cerelac with 30% dried common carp fish surimi due to a maximum improvement of all tested sensory evaluation and nutritional value by different rates.

Key words: Surimi, fortification, Cerelac, nutritional value.

INTRODUCTION

The important sections of less developed countries of the world are believed to be facing problem of malnutrition due to deficiencies in protein. Fish protein has been proposed as a possible solution to this problem. Meat and fish protein provide approximately one-third of the dietary protein requirements (Buffa, 1971). With the decrease in the availability of traditional caught fisheries products, the demand for other sources of protein appears to be growing.

Carp fish is considered the main type of fish widely produced in fish farms all over the world. Their advantages are likely due to faster of their growth and easy to breed. Common carp fish is unacceptable to the consumer in the fresh form due to numerous species penetrating the flesh. However, ready to eat products (processed fish) were developed in an attempt to increase the acceptability and utilization of carp fish (Ganesh et al., 2006).

Surimi is a Japanese term for mechanically deboned fish flesh that has been washed with water and mixed with cryoprotectants for good frozen shelf-life (Toyoda et al., 1992). It is used as an intermediate product for a variety of seafood derivatives (Park et al., 2005), such as the crab legs and flakes. Minced fish, on the other hand, is a mechanically separated flesh that has not been washed and does not have good freeze storability. Washing not only removes fat and undesirable materials, such as blood, pigments and odor substances, but more importantly, increases the concentration of myofibrillar protein (actomyosin), thereby improving gel strength and elasticity, being the essential functional properties for surimi-based products (Guenneugues and Morrissey, 2005).

Surimi is made from minced meat, providing opportunities to use different sources of protein in its production, such as underutilized species with little or no commercial value, including non-fish species. The surimi process, parts of it or modified versions, could be a way of exploiting resources that otherwise would be neglected by the food industry and consumers. Fortunately, the use of novel species for the production of surimi is increasing. Besides fish, the potential for other resources, such as cephalopods (Cortés-Ruiz et al., 2008) and giant squid (Campo-Deaño et al., 2009). In addition, crabs are being studied for surimi production with the incorporation of new methods and technologies according to Luo et al. (2004).

Surimi has a high protein (14%) and low fat content (0.21%) as determined by proximate chemical analysis. The addition of surimi to food systems has been suggested as a way of improving of myofibrillar proteins to polymerize and entrap water in a network structure (Lanier, 1984).

Abd El–Aal and Ibrahim (2001) reported that minced fillet of silver carp fish had 78.97% moisture, 18.39% protein, 0.42% fat and 1.15% ash, while surimi had 79.39% moisture, 11.68% protein, 0.28% fat and 0.55% ash. Garcia et al. (1993) found that boar surimi-like material tended to have lower microbial counts and less lipid oxidation than unwashed counterparts.

As for fish meat, it is characterized by its high nutritional value due to its content of protein, minerals and vitamins. It is also easy to be digested and its aroma is attracting for consumer. Yet, the catfish and carp fish have fewer acceptances because of its low eating qualities and toughness, in addition the catfish was dark colored muscles and carp fish was more spines (fishbone) between flesh. Accordingly, in present study, it is suggested that minced meat of catfish and carp fish may be used to produce an intermediate raw material (surimi) which is characterized by its high nutritional value. This raw material should have good functional properties and may be used to improve the nutritional value of variety of food products. Surimi has a white color and no fish odor.

It can also be added to baby foods as a supplement to raise the nutritional value according to Luo et al. (2004).

Hence, the objective of this work is to increase the nutritional value of some baby food namely Cerelac (rice-base) using dried surimi from common carp fish (*Cyprinus carpio*) fortified with 10, 20 and 30% levels.

MATERIALS AND METHODS

Source of materials

Male and female common carp fish (*C. carpio*) was purchased from the local wholesale market at Cairo City, Cairo Governorate. Samples were put into an ice box and transported to the laboratory for Faculty of Home Economics, Minufiya University.

Baby food (Cerelac) was purchased from a pharmacy at Sheben El-Kom City, Minufiya Governorate. Sugar (sucrose) was obtained from the local market at Sheben El-Kom City, Minufiya Governorate. Sorbitol and sodium tripolyphosphate (STPP) were purchased from El-Gomhoria Co., Cairo, Egypt. This work has been carried out in Minufiya University, Egypt starting at April, 2010.

Preparation of fish samples

Common carp fishes (about 5 kg of each) were washed, packed in polyethylene bags and stored at -20°C for 2 weeks until used. After thawing (overnight at 4°C.) the fish were eviscerated, headed, skinned, cleaned, washed and filleted by hand. The fillets without skin were ground using a meat grinder (Moulinex, HV2, Model A14, Moulinex, France) with 4 mm whole plate.

Preparation of surimi from common carp fish

Surimi was prepared from fillets using the methods described by Park et al. (1990) with some modification. The minced fish meats were immediately washed three times in stainless steel container (25 L) with water 1 part and crushed ice. 2 parts (iced water) at a ratio of 1 part minced fish to 3 parts iced water (W:W). Hand whipped was used to stir slurry 5 min, and excess water was removed between washing using cheese cloth. After the final washing cycle the minced fish was put in a cheese cloth bag and water removed by compression. Raw surimi was either directly packaged or mixed with cryoprotectants (4% sucrose, 4% sorbitol and 0.25% STPP) and chopped for 2 min using a meat blender (Moulinex, HV2, Model A14, Moulinex, France) and then packaged in polyethylene bags and stored at -4°C until analysis.

Dried surimi production

Frozen surimi from carp fish was thawed over night at 5°C in a refrigerator. Samples distributed in pan and put in vacuum oven to dry for 35 min at 55°C to obtain dried surimi, samples were weighted (dried weight). Preliminary dehydration trails were conducted to determine the most suitable time and temperature for the treatment.

Supplementation of baby foods (Cerelac)

Dried carp fish meat surimi was crushed to obtain homogenized dried surimi (surimi supplement) and added to Cerelac supplemented at 10, 20 and 30% to obtain supplemented formulas (W.W.). These supplemented formulas with dry surimi were

Table 1. Chemical composition of dried surimi from common carp fish.

Chemical composition (%)	Fish	
	Dried common carp fish surimi	
	WW (g)	DW (g)
Moisture (%)	12.71 ± 0.00115	-
Protein (%)	60.77 ± 0.00107	69.62 ± 0.0022
Fat (%)	0.35 ± 0.15744	0.40 ± 0.0206
Ash (%)	4.29 ± 0.00056	4.91 ± 0.0021
Carbohydrates (glycogen) (%)	21.88 ± 0.00156	25.07 ± 0.0017
Energy value (K.cal/100 g)	333.75 ± 0.00012	-

WW, Wet weight; **DW,** dry weight.

analyzed for chemical and microbiological characteristics. Preliminary supplement baby foods (Cerelac-rice base) with dried surimi were evaluated to determine the suitable treatment required for these baby foods.

Analytical methods

Moisture, protein (N × 6.25 Keldahl method), fat (hexane solvent, Soxhiet apparatus), fiber and ash were determined according to the method recommended by AOAC (2003).

Carbohydrates (as glycogen) and energy value

Carbohydrate was calculated by differences as follows:

% Carbohydrates = 100 - (% moisture + % protein + % fat + % fiber + % ash).

Energy value was estimated by multiplying protein and carbohydrates by 4.0 and fat by 9.0 (AOAC, 1995).

Microbiological methods

Preparation of fish samples and supplemented baby food samples for microbiological investigation

Total aerobic plate count (TAPC) determined on nutrient agar media according to the method described by Oxide Manual (1979), Staphylococcus aureus determined on Paird parker agar base media (ICMSF, 1996), while molds and yeast, enumerated in potato dextrose agar (ICMSF, 1996), Coliform bacterial (Oxoid) enumerated on Endo agar media (WHO, 1988), salmonella sp. and Shigella SS agar modified Oxoid according to Bryan (1991) and anaerobic bacteria was examined using nutrient agar media (Difco Manual, 1970).

Organoleptic evaluation

Baby foods (Cerelac) supplemented were subjected to organoleptic tests (by 10 judges) according to Watts et al. (1989). Judging scale for color, aroma, taste, texture and overall acceptability was as follows: Very good 8 - 9, Good 6 - 7, Fair 4 - 5, Poor 2 - 3 and Very poor 0 - 1.

Statistical analysis

Statistical analysis were performed by using computer program statistical package for social science (SPSS), and compared with each other using the suitable tests. All obtained results were tabulated. Significant differences between treatments means were determined using Duncan's multiple test (1955).

RESULTS AND DISCUSSION

Chemical composition of dried surimi from common carp fish

The chemical composition of dried surimi from common carp fish is shown in Table 1. On the other hand, the energy value of dried surimi from common carp fish was 333.75 kcal/100 g on wet weight basis. These results are in agreement with Abd El-Aal and Latif (2002) and Ibrahim et al. (2005).

Microbiological aspects of dried surimi from catfish and common carp fish

Data presented in Table 2 show the microbiological aspects of dried surimi from common carp fish. The results showed that the TAPC was the only detected microorganisms in dried surimi from common carp fish. The Value was 1.7×10^1 cfu/g. On the other hand, Staphylococcus spp., Coliform group, Salmonella and Shigella, anaerobic bacteria and mold and yeasts were not detected in dried surimi of common carp fish. Results from the same table revealed that drying process had tremendous effect on the number of microorganisms. This observation may be due to the flow chart of processing of dried surimi on the microorganisms by destroying their tissues. Niki et al. (1982) published the results microbial tests of spray dried surimi made from Alaska Pollock which were as follows: bacterial count 1×10^4 cfu/g, Coli-aerogenes group not detected, Psychrotrophic bacteria 8×10^3 cfu /g, Salmonella not detected, Yeasts 30 cfu /g, Mold 20 cfu/g and Vbrio

Table 2. Microbiological aspects of dried surimi from common carp fish (cfu/g).

Test of microorganisms	Dried common carp fish surimi
Total aerobic plate count (TAPC)	1.7×10^1
Staphylococcus spp.	N.D.
Coliform group	N.D.
Salmonella and *Shigella*	N.D.
Anaerobic bacteria	N.D.
Mold and yeast	N.D.

Table 3. Chemical composition of baby food (Cerelac).

Chemical composition (%)		Cerelac of baby food
Moisture	WW	2.92
	DW	-
Protein	WW	7.10
	DW	7.32
Fat	WW	1.70
	DW	1.75
Ash	WW	1.66
	DW	1.71
Fiber	WW	2.10
	DW	2.16
Carbohydrates (glycogen)	WW	84.49
	DW	87.06
Energy value (k.cal/100 g)	WW	381.66
	DW	-

WW, Wet weight; **DW,** dry weight.

paratheempolyticus not detected.

Chemical composition of baby food (Cerelac)

Data given in Table 3 shows the chemical composition of baby food (Cerelac), these results are close to that reported by Egyptian Organization for Standardization and Quality Control (1990) which noted that the chemical composition's standard for weaning food mixture are as follows: moisture % not increases than 7% and ash not increases than 3%; but protein (7.10) was less than mentioned reference, being not less than 15%. Data of Table 3 for Cerelac, however similar to that reported by Sidky (1995). These results are in agreement with that reported by Bowes and Church (1983) and Thomokinson and Mathur (1985).

Chemical composition of baby food (Cerelac) as influenced by addition different levels of dried common carp fish surimi

The chemical composition of baby food (Cerelac) as influenced by addition different levels of dried common carp fish surimi is shown in Table 4. The obtained results indicated that the increasing fortification levels of dried common carp fish surimi in baby food (Cerelac) resulting a markedly increase of moisture content (%). In case of protein and ash contents (%), it could be noted that increasing the fortification levels of baby food (Cerelac) with dried common carp fish surimi result a significant increase in protein and ash contents. On the other hand, fat, fiber and carbohydrates content (%) showed a markedly reduction with increasing fortification levels by dried common fish surimi. Also, energy value recorded a

Table 4. Chemical composition of baby food (Cerelac) as influenced by addition different levels of dried common carp fish surimi.

Sample		Chemical composition (%)						
		Moisture	Protein	Fat	Ash	Fiber	Carbohydrates	Energy value (kcal/100 g)
Control (0%)	WW	2.95	7.10	1.70	1.66	2.10	84.49	381.66
	DW	-	7.32	1.75	1.71	2.16	87.06	-
With 10% dried	WW	4.85	11.50	1.59	1.90	1.79	78.37	373.79
common carp fish surimi	DW	-	12.09	1.67	2.00	1.88	82.36	-
With 20% dried	WW	5.67	15.52	1.45	2.17	1.58	73.61	369.57
common carp fish surimi	DW	-	16.45	1.54	2.30	1.68	78.03	-
With 30% dried	WW	6.10	18.45	1.32	2.43	1.37	70.33	367.00
common carp fish surimi	DW	-	19.65	1.45	2.59	1.46	74.90	-

WW, Wet weight basis; **DW.** dry weight basis.

Table 5. Nutritional evaluation of baby food (Cerelac) as influenced by addition different levels of dried common carp fish surimi.

Sample	Protein		Total calories	
	GDR (g)	PS /150%	GDR (g)	PS/150%
Control (0%)	197	25	223	22.45
With 10% dried common carp fish surimi	122	41	227	21.99
With 20% dried common carp fish surimi	90	55	230	21.24
With 30% dried common carp fish surimi	76	66	232	21.59

markedly reduction with increasing dried common carp fish surimi. These results are in agreement with the findings of Saad (2006).

Nutritional evaluation of baby food (Cerelac) for infant as influenced by addition different levels of dried common carp fish surimi.

Data given in Table 5 show the nutritional evaluation of baby food (Cerelac) for infant (6-12 months of age) as influenced by addition different levels of dried common carp fish surimi. It is clear to notice that grams daily requirement (GDR) is for protein decreased, while GDR for energy value increased with increasing fortification levels of dried common carp fish surimi in baby food (Cerelac).

On the other hand, PS/150 for protein increased, while PS/150 for energy value decreased with increasing fortification levels by dried common carp fish surimi in baby food (Cerelac).

Sensory evaluation of baby food (Cerelac) as influenced by addition different levels of dried common carp fish surimi

Data presented in Table 6 show the sensory evaluation of

baby food (Cerelac) as influenced by addition different levels of dried common carp fish surimi. It is clear to mentioned that the scores of all tested sensory evaluation (color, aroma, taste, texture and overall acceptability) of control baby food (0% dried common carp fish surimi). There is fortification of baby food (Cerelac) with 20% dried common carp fish surimi due to a markedly improvement of all tested sensory evaluation by different rates. On the other hand, the maximum sensory evaluation of baby food was recorded with 30% dried common carp fish surimi. .

Conclusions

In the current study, the TAPC was the only detected microorganisms in dried surimi from common carp fish. On the other hand, *Staphylococcus* spp., Coliform group, *Salmonella* and *Shigella,* anaerobic bacteria and mold and yeasts were not detected. The results also showed the increasing fortification levels of dried common carp fish surimi in (Cerelac) resulting a markedly increase of moisture, protein, ash contents (%), while fat, fiber and carbohydrates content (%) showed a markedly reduction with increasing fortification levels by dried common fish surimi. GDR for protein decreased; GDR for energy value increased with increasing fortification levels in (Cerelac).

Table 6. Sensory evaluation of baby food (Cerelac) as influenced by addition different levels of dried common carp fish surimi.

Panel test	Type of formulas			
	Control (0%)	With 10% dried common carp fish surimi	With 20% dried common carp fish surimi	With 30% dried common carp fish surimi
Colour	7[a]	8[ab]	9[b]	9[b]
Aroma	6[a]	7[ab]	8[bc]	9[c]
Taste	7[a]	8[ab]	9[b]	9[b]
Texture	9[a]	8[a]	8[a]	8[a]
Overall acceptability	7[a]	8[ab]	9[b]	9[b]

Mean under the same line bearing different superscript letters are different significantly ($p < 0.05$).

On the other hand, PS/150 for protein increased, while PS/150 for energy value decreased with increasing fortification levels in Cerelac. This work strongly recommend the fortification of Cerelac with 30% dried common carp fish surimi due to a maximum improvement of all tested sensory evaluation and nutritional value by different rates.

Conflict of Interests

The authors have not declared any conflict of interests.

ACKNOWLEDGMENT

This project was supported by King Saud University, Deanship of Scientific Research, College of Science, Research Center.

REFERENCES

Abd El–Aal HA, Ibrahim MM (2001). Koufta analog production from silver carp fish (Hypophthalmichthys molitrix) surimi. Assiut J. Agric. Sci. 32(1):17-35.

Abd El–Aal HA, Latif S (2002). Characteristics of surimi from Karmout fish (Claries lazera) and using it in sausage. The 3 rd Scientific Conference of Agricultural Science, Assiute, pp. 189-206.

AOAC (1995). Official Methods of Analysis, Association of Official Analytical Chemists, 16 th Ed., Verginia, U.S.A.

AOAC (2003). Official Methods of the Association of Official Analytical Chemists. Arlington, Virginia, U.S.A.

Bowes AD, Church CF (1983). Food Value of Portions Commonly Used 14th Ed, JB. Lippincott Company, Philadephia, London, Mexico City, New York, San Poulo, St. Louis, Sydney.

Bryan FL (1991). Teaching HACCP techniques to food processors and regulatory officials. Dairy Food Environ. Sant. 11(10):562-568.

Buffa A (1971). Food conservation on service; food technology and development. Part 1 – processing low cost nutritious native food for world's hungry children: factors, formulas, processes. UNICEF, Paris, France.

Campo-Deaño L, Tovar C, Pombo MJ, Solas MT, Borderias J (2009). Rheological study of giant squid surimi (Dosidicus gigas) made by two methods with different cryoprotectants added. J. Food Eng. 94:26–33.

Cort´es-Ruiz JA, Pacheco-Aguilar R, Lugo-S´anchez ME, Carvallo-Ruiz MG, Garc´ıa-S´anchez G (2008). Production and functional evaluation of a protein concentrate from giant squid (Dosidicus gigas) by acid dissolution and isoelectric precipitation. Food Chem. 110(2):486–92. http://dx.doi.org/10.1016/j.foodchem.2008.02.030

Difco manual (1970). Difco manual of dehydrated culture media and reagents for microbiological, clinical and laboratory produces. Detriot, Mich. USA.

Duncan's DB (1955). Multiple range and multiple F–test, Biometries, 11:1-42. http://dx.doi.org/10.2307/3001478

Egyptian Organization for Standardization and Quality Control (1990). Egyptian Standard. Cereal Based Foods, Egypt.

Garcia Zepeda CM, Kastner CL, Kropf DH, Hunt MC, Bkenney PB, Schwenk JR Schleusener DS (1993). Utilization of surimi – like products from pork with sex – odor in restructured, precooked pork roast. J. Food Sci. 58(1):53-83. http://dx.doi.org/10.1111/j.1365-2621.1993.tb03210.x

Ganesh A, Dileep AO, Shamasundar BA, Singh U (2006). Gel-forming ability of common carp fish (Cyprinus carpio) meat, effect of freezing and frozen storage. J. Food Biochem. 30(3):342–361.

Guenneugues P, Morrissey MT (2005). Surimi resources. In. Park JW, editor. Surimi and surimi seafood. 2nd ed. Boca Raton, Fla. Taylor & Francis Group. pp. 3–32.

Ibrahim MMM, Wally Fardus AA, El –Gendy Alia A (2005). Physical and panelists assessment of beef burger containing fish surimi. The Third International Conference for Food Science & Technology." Modernizing Food Industries, Egyptian Society of Food Science & Technology, February 22–24 th, Cairo, Egypt.

ICMSF (1996). Microorganisms in Food. 5. Microbiological Specification of Pathogens, International Commission of Microbiological Specification for Foods Blockie. Academic and Professional, an Imprint of Chapman & Hall, New York.

Lanier TC (1984). Surimi. A unique "new" food protein. Proc. Meat Industry Res. Corr. P. 80, American Meat Institute, Washington, DC.

Luo Y, Kuwahara R, Kaneniwa M, Murata Y, Yokyama M (2004). Effect of soy protein isolate on gel properties of Alaska Pollock and common carp surimi at different setting conditions. J. Sci. Food Agric. 84(7):663-671. http://dx.doi.org/10.1002/jsfa.1727

Niki H, Deya E, Kato T, Igarashi S (1982). The process of producing active fish powder. Nippon Swisan Gakkaishi, pp. 49-99.

Oxoid Manual (1979). The Oxoid Manual of Culture Media. Ingredients and other Laboratory Services, Fourth Ed., Oxoid Limited, Hamphire RG 24 0PW.

Park JW, Korhonen RW, Lanier, TC (1990). Effect of rigor mortis on gel–forming properties of surimi and unwashed mince prepared from Tilapia. J. Food Sci. 55:353-360. http://dx.doi.org/10.1111/j.1365-2621.1990.tb06761.x

Saad FM (2006). Utilization of Camel and Catfish Meats in Baby Food Formula. Ph.D. Thesis, Fac. of Home Economics, Minufiya University.

Sidky HMA (1995). Nutritional evaluation on soy bean extruded as weaning food. Egyptian J. Nutr. (2):125-137.

SPSS (1998). Statistical Package for Social Science, Computer Software, Ver. 10, SPSS Company, London, UK.

Thomokinson DK, Mathur BN (1985). Formulated infant food, a – prospective. National Dairy Res. 13(20):247-250.

Toyoda K, Kimura L, Fujita T, Noguchi SF, Lee CM (1992). The surimi manufacturing process: In " Surimi Technology ", Lanier TC and Lee CM (Eds) Marcel Dekker, Inc., N.Y. pp. 79-112.

Watts BM, Yamaki GL, Jeffery LE, Elias LG (1989). Sensory Methods for Food Evaluation,1stEd.,The International Development Research Center Pub., Ottawa, Canada.

WHO World Health Organization (1988). Health Education in Food Safety. WHO/88 (7):32.

15

Infection of *Hysterothylacium aduncum* (Namatoda: Anisakidae) in farmed rainbow trout (*Oncorhynchus mykiss* Walbaum, 1792)

Naim Saglam

Department of Aquaculture and Fish Diseases, Faculty of Fisheries, University of Fırat, 23119, Elazıg-Turkey.

Farmed rainbow trout (*Oncorhynchus mykiss*) were examined for anisakid nematodes at fish farms in Elazig city, Turkey. A total of 439 fish (246 from fish farm ponds and 193 from net-cages) were monthly investigated in the period from February 2000 to May 2003. Only the endoparasite, *hysterothylacium aduncum* (Nematoda: Anisakidae) was recorded in the digestive tract of the 91 cultured rainbow trout, *O. mykiss* fed with minced marine fish. Prevalence, mean intensity, and the abundance of *H. aduncum* on fish obtained from fish farm ponds were 36.99%, 16.00 ± 1.15, and 5.92 ± 0.15, respectively. However, this values in fish fed with freshly minced marine fish were 100%, 16.00 ± 1.15 and 16.00 ± 1.15, respectively. *H. aduncum* was found in the oesophagus, stomach, intestine, and pyloric caeca of fish. All *H. aduncum* were adult and was not found on fish fed with commercial pellets.

Key words: *Hysterothylacium aduncum*, rainbow trout, nematoda, anisakidae.

INTRODUCTION

Nematodes are usually considered the most economically important helminth parasites of fishes in the world (Dick and Choudhury, 1995). Most adult nematodes are found in the intestine of fish, but larval stages are sometimes found in the flesh and viscera which cause disease and economical loses. It is also the larval stages which are infective to humans and which have the greatest impact on consumer acceptance of fish as a source of protein (Dick and Choudhury, 1995; Moravec, 1994).

Members of the family Anisakidae parasitise are fish, mammals, birds and reptiles (Moravec, 1994; Zhu et al., 1998). Anisakids are among most common nematodes of fish. They cause patholagical symptoms and mortalities, and reduce the commercial value of fish (Dick and Choudhury, 1995). Larval and adult anisakids infect freshwater fishes (Cyprinidae, Ictaluridae, contrarchidae, percidae and salmonidae) (Hoffman, 1998; Ekingen, 1983).

The presence of larval and adult nematodes belonging to the genus *hysterothylacium* was reported in the freshwater and marine fish farms (Moravec, 1994; Hoffman, 1998; Gonzalez, 1998). Furthermore, it is known that species of marine fish can act as intermediate, paratenic or definitive host (Zhu et al., 1998). *Hysterothylacium aduncum* is mainly found in marine piscivorous fish (*Gadus morhua* Linnaeus, 1758) and its larvae occur in a variety of prey fishes including smaller cod (*G. morhua* Linnaeus, 1758) and mackerel (*Scomber scombrus* Linnaeus, 1758) (Dick and Choudhury, 1995).

Inoue et al. (2000) investigated the possibility of larval

anisakid infection in farmed salmon, *Oncorhynchus mykiss* (Walbaum, 1792) in Tokyo, Japan. The life cycle of *H. aduncum* was shown experimentally by Gonzalez (1998) and Yoshinaga et al. (1987). The third-stage larva of *H. aduncum* was defined in flounder (*Platichthys flesus* Linnaeus, 1758; Koie, 1999). Ismen and Bingel (1999) studied *H. aduncum* infection in the whiting, (*Merlangius merlangius* euxinus Nordmann, 1840), off Turkish coast of the Black Sea. Shih and Jeng (2002) observed *H. aduncum* infecting a herbivorous fish, (*Siganus fuscescens* Houttuyn, 1782), off the Taiwanese coast of the Northwest Pacific.

In this study, the existence of anisakid nematodes was investigated in organs and tissues of rainbow trout cultured with respect to a prevalence mean intensity of infection and mean abundance in freshwater ponds and net-cages in Elazıg city, in Turkey.

MATERIALS AND METHODS

Fish

A total of 439 cultured rainbow trout (246 from freshwater ponds and 193 from net-cages) were examined for endoparasites throughout the study period. Farmed rainbow trout (*O. mykiss*) (age 0+, weight, 200 to 250 g and total length, 20 to 26 cm) for this study were monthly obtained from two different commercial net-cages farms in the Keban Dame Lake and from a commercial fish farm ponds in Elazıg, Turkey in the period from February 2000 to May 2003. These were farmed in freshwater ponds and net-cages for about 10 to 12 months after hatching. Cultured fish in net-cages were fed with commercially prepared pellets. While some of fish obtained from fish farm ponds were fed with only commercially prepared pellets, some others were fed with only freshly minced marine fishes such as anchovy (*Engraulis engrasicholus* Linnaeus, 1758), whiting (*Merlangius merlangus* Linnaeus, 1758) and scad (*Trachurus trachurus* Linnaeus, 1758). All fishes were fed three times in a day.

Examination of rainbow trout for anisakid nematodes

The organs and tissues of rainbow trout were examined for anisakid nematodes using the methodology of Chubb and Powell (1966). The muscle of fish was sliced and carefully examined for the presence of nematodes (Inoue et al., 2000). The digestive tract, liver, spleen, kidney and heart were taken into petri dish separately, with a lancet and carefully investigated. The inner surface of the abdominal cavity was also checked according to the methods described by Inoue et al. (2000) and Chubb and Powell (1966).

Identification of nematode species

Parasites found were identified by the morphological characteristics given by Moravec (1994) and Bykhovskaya-povlovskaya et al. (1964). The number of *H. aduncum* in each fish was counted. The specimens were fixed in 70% alcohol and then transferred to lactophenol for becoming transparent (Kennedy, 1990; Pritchard and Kruse, 1982). The locations of *H. aduncum* in individual fish were also recorded. The data were analysed with respect to the infection prevalence (number of fish infected with *H. aduncum*/number of fish examined), mean intensity (the mean number of *H. aduncum* per infected fish) and abundance (the mean number of *H. aduncum* per studied fish) of worms (Bush et al., 1997).

RESULTS

The anisakids and other nematodes were not found in the muscle of 469 farmed rainbow trout between 2000 and 2003. Adult *H. aduncum* was only determined in the oesophagus, stomach, intestine and piloric-caeca of 91 trout fed with minced marine fish. Neither larvae nor adult *H. aduncum* was found in 348 farmed rainbow trout and fed with commercially prepared pellets (Table 1). No nematode was determined on the surface of the liver, spleen, kidney, muscle, heart or the inner surface of the abdominal cavity of examined fish.

The infection prevalence (%), mean intensity and mean abundance of *H. aduncum* are given in Tables 1 and 2. *H. aduncum* were found only in the digestive tract of infected 91 fishes, corresponding to a prevalence of 20.73%, a mean intensity of 16.00 ± 1.15 *H. aduncum* per infected fish, and an abundance of 3.32 ± 0.12 *H. aduncum* per examined fish. On the other hand, 91 rainbow trout fed with minced marine fish were infected by *H. aduncum* with the prevalence of 100%, the mean intensity of 16.00 ± 1.15 nematodes per infected fish, and the abundance of 16.00 ± 1.15 nematodes per fish (Table 1). Infection prevalence, mean intensity, and mean abundance of *H. aduncum* in farmed rainbow trout in the freshwater ponds were 36.99%, 16.00 ± 1.15 and 5.92 ± 0.15, respectively (Table 2). A total of 1456 *H. aduncum* (551 male and 905 female) were counted through the host of investigation period.

Morphology of *H. aduncum* (Rudolphi, 1802)

The body of *H. aduncum* is cylindrical. Females (Figure 1) are larger than males (Figure 2). Lips are approximately equal in size. Oesophagus is narrow, ending with a small ventriculus and the intestine is dark and straight. The nerve ring encircled the oesophagus approximately at the border of first, second and fifths of its length. Excretory pore is situated just below the nerve ring and rectum is short. Measurements of *H. aduncum* are given in Table 3.

DISCUSSION

Marine fishes are usually intermediate of anisakid nematodes. A few cases of transmission of these nematodes to humans have been reported (Dick and Choudhury, 1995; Moravec, 1994; Post, 1987). The *in vitro* culture and an experimental infection of *H. aduncum* demonstrated that the third larval stage hatching from the eggs easily infects their first intermediate host a calanoid

Table 1. Infection prevalence (%) of nematode, mean intensity, mean abundance and locality of *H. aduncum* in the rainbow trout (*O. mykiss*) fed with freshly minced marine fish in the freshwater ponds.

Source of food	No. of fish examined	No. of infected fish	No. of parasites (male/female)	Infection prevalence (%)	Mean intensity (±S.E.)	Mean abundance (±S.E.)	Locality of *H. aduncum*
Freshly minced marine fish	91	91	1456 (551/905)	100	16.00 ± 1.15	16.00 ± 1.15	Digestive tract*
Commercially prepared pellets	348	0	0	0	0	0	-
Total	439	91	1456 (551/905)	20.73	16.00 ± 1.15	3.32 ± 0.12	

(*) Oesophagus, stomach, intestine and piloric-caeca. No., number.

Table 2. Infection prevalence (%), mean intensity and mean abundance of H. aduncum in farmed rainbow trout (O. mykiss) in the freshwater ponds and net-cages.

Sample	No. of fish examined	No. of fish infected	No of parasites (male/female)	Infection prevalence (%)	Mean intensity	Mean abundance
Freshwater ponds	246	91	1456 (551/905)	36.99	16.00 ± 1.15	5.92 ± 0.15
Net-cages	193	0	0	0	0	0
Total	439	91	1456 (551/905)	20.73	16.00 ± 1.15	3.32 ± 0.12

or harpacticoid copepod, *Tisbe longisetosa* Gurney, 1927 (Gonzalez, 1998). In this study, *H. aduncum* infections were recorded in farmed rainbow trout in ponds fed with freshly minced marine fish such as; anchovy, whiting and scad. This result is the same with those reported by Yoshinaga et al. (1987). They experimentally demonstrated that rainbow trout could be the final host for *H. aduncum*, and are fed by marine fish infected with the 3rd stage larva of the nematode. They defined *H. aduncum* in the net-cages of Chilean marine farms in the inner ocean (Gonzalez, 1998) and *A. simplex* in the wild

salmons (Inoue et al., 2000). But in this study, infection of *H. aduncum* was not determined in rainbow trout fed with commercial pellets in the net-cages in the Keban Dame Lake in Elazig. *H. aduncum* were not observed in any wild fish species in the previous studies conducted in the Keban Dame Lake.

The larvae and adults of *H. aduncum* were not determined in the muscle of farmed rainbow trout. This result is in accordance with the finding of Inoue et al. (2000) and Koie (1999). In this study, *H. aduncum* was recorded in the digestive tract such as; oesophagus, stomach, intestine

and pilloric-caeca of rainbow trout, in contrast to be defined only in the intestine of sea trout (Byrne et al., 1999).

The infection prevalence of *H. aduncum* was recorded as 8 to 44% in the intestine of flounder *P. flesus* (Koie, 1999) and 21.8 to 54.8% in the visceral organs of whiting, *M. merlangus euxinus* (Ismen and Bingel, 1999). According to this study, the infection prevalence of *H. aduncum* in farmed rainbow trout in the freshwater ponds was 36.99%. Ninety-one rainbow trout fed with minced marine fish were infected by *H. aduncum* with the prevalence of 100%. These results are

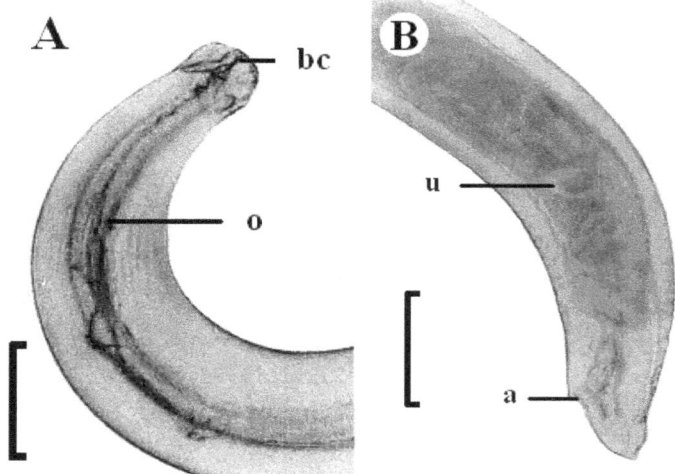

Figure 1. The anterior (A) and posterior (B) view of female *Hysterothylacium aduncum.* Scale bar = 0.5 mm. a, anus; bc, buccal capsule; o, oesophagus; u, uterus.

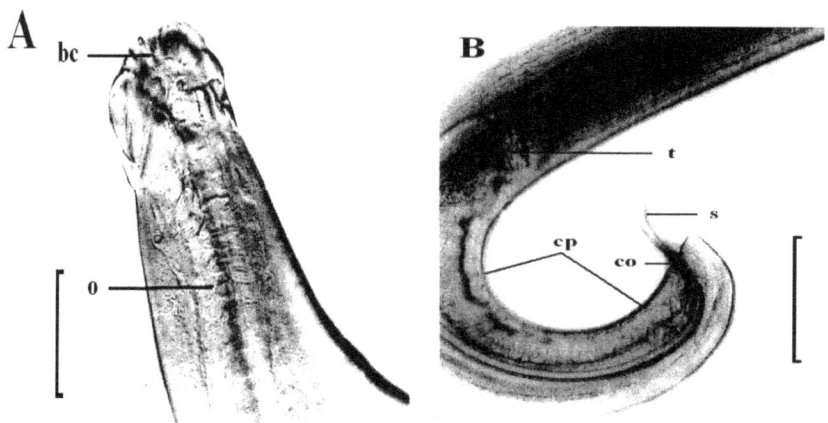

Figure 2. The anterior (A) and posterior (B) view of male *Hysterothylacium aduncum.* Scale bar = 0.5 mm. bc, buccal capsule; co, cloacal opening; cp, caudal pa -1921. pillae, o, oesophagus; s, spicules; t, testis.

Table 3. Measurements of *H. Aduncum.*

Measurement	Male (Mean ± SE) (range)	Female (Mean ± SE) (range)
Length of body	14.0 ± 2.7 mm (8.0 to 18.0)	25.7 ± 2.6 mm (15.2 to 42.0)
Width of body	0.40 ± 0.01 mm (0.20 to 0.50)	0.76 ± 0.06 mm (0.45 to 1.25)
Length of dorsal lip	0.14 ± 0.01 mm (0.13 to 0.15)	0.14 ± 0.05 mm (0.14 to 0.15)
Width of dorsal lip	0.13 ± 0.01 mm (0.12 to 0.14)	0.15 ± 0.02 mm (0.14 to 0.17)
Length of oesophagus	0.83 ± 033 mm (0.81 to 1.79)	1.83 ± 0.43 mm (1.41 to 3.54)
Ventriculus	0.12 to 0.14 x 0.09 to 0.11 mm	0.16 to 0.17 x 0.15 to 0.16 mm
Length of intestinal caecum	0.45 to 0.75 mm	0.91 to 0.99 mm
Nerve ring	0.43 to 0.44 mm	0.56 to 0.80 mm
Long of spicules 1 and 2	0.97 to .91 mm	to
Eggs	to	0.059 to 0.065 x 0.040 to 0.042 mm

SE, standard error.

similar to the infection prevalence of *H. aduncum* to that of researchers in *P. flesus* and *M. merlangus* euxinus.

The abundance of anisakids observed in the chum salmon by Inoue et al. (2000) was nearly equal to that of the present study (16.0 ± 1.15). The prevalence mean intensity and mean abundance of *H. aduncum* on the sea trout were 61.2, 9.3 and 6.3%, respectively (Byrne et al., 1999). Our results were found rather higher on the rainbow trout fed with freshly minced marine fish in the freshwater ponds.

Conclusion

In conclusion, in order to prevent farmed fish from this parasites infection. It is not advisable to use the freshly minced fish for farmed fish feeding. If it is necessary, this kind of food has to pass some processes to eliminate the parasites.

REFERENCES

Bush AO, Lafferty KD, Lotz JM, Shostak AW (1997). Parasitology Meets Ecology on its Own Terms: Margolis et al. Revisited. J. Parasitol. 83:575-583.

Bykhovskaya-Povlovskaya IE, Gusev AV, Dubinina MN, Izyumova NA, Smirnova TS, Sokolovskaya IL, Shtein GA, Shul'man SS, Epshtein VM (1964). Key to parasites of freshwater fishes of the USSR I (Translated from Rusian by. Birrow A, Cole ZS). Isr. Prog. For. Sci. Jeruselam, Israel. pp. 615-887.

Byrne CJ, Holland C, Tully O (1999). Metazoan parasite community structure of sea trout on the west coast of Ireland. J. Fish Biol. 55:127-134.

Chubb JC, Powell AM (1966). The examination of fish parasites. pp. 87-90. Department of Zoology University of Liverpool. Liverpool, UK.

Dick TA, Choudhury A (1995). Phylum Nematoda. In: Fish Diseases and Disorders Volume I Protozoon and Metazoon Infection (ed. Woo PTK Cambridge University Press. Cambridge, UK). pp. 415-446.

Ekingen G (1983). Freshwater Fish Parasites. University of Firat Press., Elazig, Turkey.

Gonzalez L (1998). The life cycle of *Hysterothylacium aduncum* (Nematoda: Anisakidae) in Chilean marine farms. Aquaculture, 162:173-186.

Hoffman GL (1998). Parasites of North American Freshwater Fishes. University of California Press. New York.

Inoue K, Oshima SI, Hirata T, Kimura I (2000). Possibility of anisakid larvae infection in farmed salmon. Fisheries Sci. 66:1049-1052.

Ismen A, Bingel F (1999). Nematodes infection in the whiting *Merlangius merlangius euxinus* of Turkish Coast of the Black Sea. Fish. Res. 42:183-189.

Kennedy MJ (1990). Basic Methods of Specimen Preparation in Parasitology. Canada.

Koie M (1999). Metazoan parasites of flounder *Platichthys flesus* (L.) along a transect from the southwestern to the northeastern Baltic Sea. ICES J. Mar. Sci. 56:157-163.

Moravec F (1994). Parasitic Nematodes of Freshwater Fishes of Europe. Kluwer Academic Publishers. London, UK.

Post G (1987). Animal Parasites of Fishes. In: Textbook of Fish Health. T.F.H. Publications Inc. USA. pp. 159-214.

Pritchard MH, Kruse GOW (1982). The collection and preservation of animals parasites. Illustrations by M. Marcuson, Technical Bulletin No. 1 University of Nebraska Press, USA.

Shih HH, Jeng MS (2002). *Hysterothylacium aduncum* (Nematoda: Anisakidae) Infecting a Herbivorous Fish, *Siganus fuscescens*, off the Taiwanese Coast of the Northwest Pacific. Zoological Studied-Taipei, 41:208-215.

Yoshinaga T, Ogawa K, Walcabayashi K (1987). Experimental life cycle of *Hysterothylacium aduncum* (Nematoda: Anisakidae) in freshwater. Fish. Pathol. 22:243-251.

Zhu X, Gasser RB, Podolska M, Chilton NB (1998). Characterisation of anisakid nematodes with zoonotic potential by nuclear ribosomal DNA sequences. Int. J. Parasitol. 28:1911.

16

Comparative study of the concentration of mercury and lead and the chemical characteristics of Japanese and Korean chub mackerel (*Scomber japonicas* Houttuyn, 1782) in the East China Sea

Jin Han Bae and Sun Young Lim*

Division of Marine Environment and Bioscience, Korea Maritime University, Busan 606-791, Korea.

We compared heavy metals and proximate composition of Japanese and Korean chub mackerel from the East China Sea. J-mackerel showed higher mercury content than K-mackerel, but no difference in lead content was observed. J-mackerel had significantly lower ash content than K-mackerel. There were no significant differences in saturated, monounsaturated, and polyunsaturated fatty acids, but J-mackerel showed significant differences in 22:0, 24:0 and total lipids. The primary amino acid in J- and K-mackerel was glutamic acid, followed by aspartic acid, lysine, and leucine in decreasing amounts; no differences in the quantities of essential amino acids were found among the fish.

Key words: Mercury, lead, chemical characteristic, chub mackerel, fatty acid.

INTRODUCTION

Chub mackerel (*Scomber japonicas* Houttuyn, 1782) is a cosmopolitan pelagic species inhabiting tropical and subtropical regions of the Atlantic, Indian, and Pacific Oceans, and adjacent seas. Chub mackerel in the East China Sea is an important fishery resource for Japan and Korea. The Asian countries with the largest catches of chub mackerel (2009) were China and Japan, with 0.52 and 0.47 million tons respectively (MAFF, 2009; FAO, 2012). In 2009, productions of chub mackerel in Korea reached 0.18 million ton, comprising 14.3% of the coastal and offshore capture (MIFAFF, 2010). The spawning ground of chub mackerel is considered to be located in the central and southern part of the East China Sea, and the spawning period spans February to June (Yukami et al., 2009). Chub mackerel provides omega-3 (n-3) fatty acids that reduce cholesterol levels and the incidence of heart disease, stroke, and preterm delivery (Daviglus et

al., 2002; Patterson, 2002).

Environmental pollution occurs from both anthropogenic sources and natural weathering. The major anthropogenic sources of surface-water pollution include discharge of industrial effluent, atmospheric deposition of pollutants, and accidental spills of toxic chemicals. Harmful substances including heavy metals accumulate in marine organisms through the food chain. Among the heavy metals, mercury and lead merit special attention due to their ability to cause serious health problems in various life forms (Loux, 1998). Since chub mackerel is a favorite food in Korea and Japan, the accumulation of metals in this species in excess of permissible limits is a serious health concern. Concern about heavy metals in edible fish in general has been increasing, as fish form a major part of the human diet in much of the world. Knowledge of the chemical composition of fish can be used to estimate the nutritional value of various types of fish and to report their nutrient composition from a public health perspective. For this reason, the objective of this study was to compare heavy metal contents and chemical characteristics between Japanese (J-mackerel)

---

*Corresponding author. E-mail: sylim@hhu.ac.kr.

Table 1. Comparison of mercury and lead content (mg/kg) between J- and K-mackerel.

Heavy metals[1]	J-mackerel	K-mackerel
Mercury	0.08 ± 0.01	0.05 ± 0.01*
Lead	0.02 ± 0.01	0.01 ± 0.00

[1]Data were presented as Mean ± Standard deviation (n=5); *p<0.05, significantly different as compared to J-mackerel.

and Korean mackerel (K-mackerel) in order to determine the safety and nutritional information of chub mackerel from the East China Sea.

MATERIALS AND METHODS

Samples collection

J-mackerel and K-mackerel (n = 5) were purchased from the Jagalchi fish market of Korea. J- and K-mackerel were captured near Nagasaki in Japan (129°0'–129°5'E, 32°5'–33°0'N, J-mackerel), and near Jeju in Korea (126°0'–126°5'E, 32°5'–33°0'N, K-mackerel). The average length of J- and K-mackerel were 35.1 ± 0.4 and 31.5 ± 2.2 cm and average weights were 431.7 ± 42.2 and 371.7 ± 75.2 g, respectively. Mackerel were gutted, eviscerated, and filleted on two sides. Mackerel fillets were cut into small pieces, blended in a homogenizer (HMF-985; Hanil, Seoul, Korea), packed separately in polyethylene bags, and stored at -20°C until analyses of heavy metals and chemical composition were conducted.

Measurement of mercury and lead

Mercury concentration was measured using a direct mercury analyzer (DMA-80; Milestone Srl, Bergamo, Italy) without chemical pre-treatment. The analysis of lead was weighed and ashed in a muffle furnace at 500°C for 24 h. Ash was dissolved in a small amount of nitric acid, and the resulting solution was diluted to 50 ml with double-distilled water. Lead was determined using inductively coupled plasma (Optima 3000DV; Perkin-Elmer, Norealk, CT, USA).

Analysis of proximate composition

Moisture content was determined by drying samples in an oven at 105°C until constant weight was obtained. Crude fat was determined using the Soxhlet extraction method. Crude protein content was determined by the Kjeldahl nitrogen method using a 6.25 conversion factor. Ash content was determined by incineration in a muffle furnace at 550°C for 24 h (KFN, 2000).

Measurement of fatty acid composition

Sample lipids were extracted with chloroform-methanol (2:1 v/v) according to the method of Bligh and Dyer (1959). Fatty acid methyl esters were prepared with 14% BF3/methanol and analyzed with a gas chromatograph (CP-3380; Varian Inc., Palo Alto, CA, USA) using a flame-ionization detector, as described previously (Salem et al., 1996).

Analysis of amino acids

The sample was hydrolyzed with 6 N HCl at 110°C for 24 h. The hydrolyzed sample was dried in a rotary vacuum evaporator. The residue was then dissolved in distilled water and filtered through a 0.2 μm glass filter. The amino acid profiles of an aliquot were determined using an amino acid analyzer (L-8800; Hitachi, Tokyo, Japan).

Statistical analysis

Data were presented as Means ± Standard Deviation. Significant differences between J-mackerel and K-mackerel were tested using the independent samples t-test. Analyses were conducted using SPSS software package (version 10.0; SPSS Inc., Chicago, IL, USA).

RESULTS AND DISCUSSION

Content of mercury and lead

The average contents of mercury and lead in J-mackerel were 0.08 ± 0.01 and 0.02 ± 0.01 mg/kg respectively, while those in K-mackerel were 0.05 ± 0.01 and 0.01 ± 0.00 mg/kg respectively (Table 1). J-mackerel showed a higher content of mercury than K-mackerel (p<0.05). The Food Code of Korea specifies that the content of mercury and lead in seafood should be <0.5 mg/kg wet weight (KFDA, 2010). The Japanese government has recommended that fish with mercury levels ≥ 0.3 mg/kg should not be sold because of the high consumption of fish in that country (Nakagawa et al., 1997). The content of mercury and lead in canned fish from Georgia and Alabama (USA) were 0.02 to 0.74 and 0.0 to 0.03 mg/kg respectively and the mercury contents exceeded a recommended action level in some tuna samples (Ikem and Egiebor, 2005). This present study showed that heavy metal contents of J- and K-mackerel were below action levels specified in the Food Code of Korea. Concentrations of heavy metals in fish tissue reflect past or present exposure and are influenced by ecological factors such as place of development, season, nutrient availability, temperature, and salinity of the water (Barghigiani and De Ranieri, 1992; Kagi and Schaffer, 1998; Deshpande et al., 2009).

Proximate composition

Body composition is a good indicator of the physiological condition of a fish, but it is relatively time consuming to measure. The average contents of moisture, crude fat, crude protein, and ash in J-mackerel were 69.37 ± 1.82, 8.01 ± 3.20, 22.16 ± 1.37, and 0.99 ± 0.10%, respectively; those in K-mackerel were 71.95 ± 1.52, 4.32 ± 2.07, 20.61 ± 1.10, and 1.49 ± 0.06%, respectively. K-mackerel showed lower levels of crude fat and protein but these differences were not significant. There was a significant difference in ash content between J- and K-mackerel (p<0.05). Previously, we reported that Atlantic mackerel had 64.84% moisture, 13.82% crude fat, 20.50% crude

Table 2. Comparison of fatty acid composition (area %) between J- and K-mackerel.

Fatty acids[1]	J-mackerel	K-mackerel
14:0	4.75 ± 0.19[2]	4.57 ± 0.49
16:0	21.47 ± 2.26	25.80 ± 1.47
18:0	7.59 ± 0.23	8.15 ± 0.33
20:0	3.34 ± 1.66	1.33 ± 0.32
22:0	4.32 ± 0.93	1.76 ± 0.61*
24:0	2.10 ± 0.77	3.69 ± 0.50*
Total SFA	43.57 ± 1.17	45.30 ± 2.21
16:1n-7	5.19 ± 0.40	3.97 ± 0.89
18:1n-9	17.60 ± 1.33	19.14 ± 2.05
18:1n-7	3.73 ± 0.44	3.69 ± 1.34
20:1n-9	2.59 ± 0.17	0.26 ± 0.23
24:1n-9	1.20 ± 0.38	1.55 ± 0.37
Total MUFA	30.30 ± 1.22	28.60 ± 2.39
18:2n-6	1.53 ± 0.22	2.32 ± 0.77
18:3n-3	0.56 ± 0.20	0.53 ± 0.01
20:2n-6	0.30 ± 0.02	ND
20:4n-6	1.71 ± 0.82	2.11 ± 0.75
20:5n-3	6.40 ± 0.67	5.25 ± 0.90
22:4n-6	0.44 ± 0.25	ND
22:5n-3	1.60 ± 0.39	1.59 ± 0.45
22:6n-3	13.61 ± 2.31	14.17 ± 1.29
Total PUFA	26.13 ± 2.35	26.10 ± 2.50
Total n-3 PUFA	22.17 ± 2.17	21.55 ± 2.65
Total n-6 PUFA	3.96 ± 0.66	4.43 ± 0.20
Total lipid (g/100 g wet wt)	4.54 ± 0.53	1.55 ± 0.62**

[1]SFA, saturated fatty acid; MUFA, monounsaturated fatty acid; PUFA, polyunsaturated fatty acid. [2]Data were presented as Mean ± Standard deviation (n=5); ND, not detected; wt, weight. *$p<0.05$, **$p<0.01$ significantly different as compared to J-mackerel.

protein, and 1.23% ash; chub mackerel had 58.29% moisture, 18.62% crude fat, 18.12% crude protein, and 1.47% ash (Bae et al., 2011). Our results indicated that moisture content was proportional to lipid content, and a negative correlation was observed. The crude lipid content of horse mackerel caught offshore from Tsushima of Japan was higher than that of the Nagasaki and East China Sea, and there were no significant differences among the other catches. The reason for this could be that there are many eddies and upwelling streams offshore from Tsushima that create ocean currents which are more complicated than those in the East China Sea or offshore from Nagasaki (Osako et al., 2002). It is known that variation in the proximate composition of marine fish is closely related to nutrition, living area, fish size, catch season, and seasonal and sexual variations. (Luzia et al., 2003; Zlatanos and Laskaridis, 2007).

Fatty acid composition

The fatty acid profiles of fish muscle tissue tends to

reflect their profile in the diet, but the extent to which this occurs depends on a variety of factors including concentration and profile of fatty acids in feed, fish species, specific muscle, lipid class isolated, and physiological state. There were no significant differences in saturated (SFA), monounsaturated (MUFA), and polyunsaturated fatty acids (PUFA) between J- and K-mackerel (Table 2). It indicated that there was little variation in both mackerel since fatty acid contents are closely related to feed intake, migratory swimming, and sexual changes in connection with spawning. J-and K-mackerel showed significant differences in the percentages of 22:0, 24:0, and total lipids ($p<0.05$). The major fatty acids were palmitic acid (16:0, 21 to 26%), oleic acid (18:1n-9, 18 to 19%), docosahexaenoic acid (22:6n-3, 14%) and eicosapentaenoic acid (20:5n-3, 5 to 6%). The compositional dominance of palimitic acid in marine animals has been typically observed, which was independent of species, season, temperature, diet or geographical region (Suriah et al., 1995). It could be attributed to the fact that the palmitic acid is a key metabolite and utilized as an energy source, for which *de novo* synthesis of the fatty acid occurs in fish (Kluytmans et al., 1985). Osako et al. (2003) reported that 22:6n-3 ratios of total fatty acids in summer-caught horse mackerel were lower than those in winter-caught horse mackerel; this tendency was significantly more pronounced in smaller-sized fish. In contrast, 22:6n-3 levels in fish tissues varied little throughout the year. The main fatty acid classes were MUFA (37.9 to 39.3%), SFA (33.3 to 35.2%), and PUFA (26.7 to 27.2) in cultured mackerel; in wild mackerel the distribution was SFA (36.8 to 39.5%), MUFA (31.8 to 33.4%), and PUFA (28.7 to 29.6%) (Moon et al., 2009). J- and K-mackerel are excellent sources of MUFA, PUFA, and n-3 fatty acids. It is known that n-3 PUFA reduces cholesterol levels and the incidence of heart disease, prevents cardiovascular diseases, and improves learning ability (Larsen et al., 2011).

Amino acid composition

Fish muscle protein is characterized by a very desirable composition of amino acids. The total amount of amino acid in J-mackerel was 217.74 ± 8.11 mg/g, and that in K-mackerel was 204.21 ± 6.69 mg/g (Table 3). The major amino acid in J- and K-mackerel was glutamic acid, followed by aspartic acid, lysine, and leucine in decreasing amounts, with the exception of glycine. Similarly, the major amino acids found in Atlantic, blue, and chub mackerel were glutamic and aspartic acid (Bae et al., 2011). This present study demonstrated that the essential amino acids (EAA) comprised 50% of total amino acids (TAA), and the ratio of EAA to non-essential amino acids (NEAA) was 0.9. The FAO-WHO (1973) recommended that the percentage of EEA in TAA, and the ratio of EAA to NEAA approximate reference values of 40% and 0.6 respectively. Our results demonstrated

Table 3. Comparison of amino acid composition (mg/g) between J- and K-mackerel.

Amino acid	J-mackerel	K-mackerel
Essential		
Arginine	13.03 ± 0.39[1]	12.35 ± 0.28
Histidine	13.62 ± 1.82	9.93 ± 1.04
Isoleucine	10.32 ± 0.36	10.34 ± 0.88
Leucine	18.12 ± 0.65	17.48 ± 0.55
Lysine	21.15 ± 0.42	20.15 ± 0.86
Methionine	6.89 ± 0.11	6.19 ± 0.78
Phenylalanine	8.59 ± 0.54	8.18 ± 0.38
Threonine	10.03 ± 0.36	9.79 ± 0.42
Valine	12.24 ± 0.56	12.07 ± 1.22
Total EAA	114.00 ± 5.05	106.49 ± 4.60
Non-essential		
Alanine	13.32 ± 0.69	12.49 ± 0.42
Aspartic acid	21.47 ± 0.68	20.87 ± 0.81
Glutamic acid	34.52 ± 0.19	33.64 ± 2.32
Glycine	11.38 ± 0.72	9.58 ± 0.25*
Proline	7.37 ± 0.41	6.62 ± 0.25
Serine	8.67 ± 0.28	8.32 ± 0.32
Tyrosine	6.98 ± 0.21	6.19 ± 0.82
Total NEAA	103.73 ± 3.05	97.72 ± 3.71
TAA	217.74 ± 8.11	204.21 ± 6.69
EAA/TAA 100	52.35 ± 0.37	52.14 ± 1.21
Total NEAA/total EAA	0.91 ± 0.01	0.92 ± 0.04

[1]Data were presented as Mean ± Standard deviation (n=5). *$p<0.05$ significantly different as compared to J-mackerel. EAA, essential amino acids; NEAA, non-essential amino acids; TAA, total amino acids.

that there were no significant differences in the relative amount of amino acids between J- and K-mackerel. It suggested that the environments surrounding the fish might not be sufficiently different to influence on amino acid composition.

Chub mackerel is considered a favorite traditional food in Japan and Korea that offers nutritional, cultural, and economic benefits. The effects of PUFAs on human health have been reported, and the intake of fish lipids has increased steadily. Although PUFAs in fish have been associated with health benefits, there is increasing evidence that the accumulation of heavy metals exceeding permissible limits in some fish can lead to adverse health effects. According to this present study, the heavy metal contents of J- and K-mackerel were below limits specified by the Food Code of Korea. The major finding of this study was that the amounts of fats were significantly different between J- and K-mackerel, but their fatty acid compositions remained constant. We assumed that food availability in the respective habitants might be a factor that could explain the regional discrepancies in the absolute contents of fats in chub mackerel. We believe that it could give world-widely

applicable information, in that this study established that the contents of heavy metals and chemical characteristics in chub mackerel might vary with collection regions even within the same species.

ACKNOWLEDGMENT

This work is the outcome of a Manpower Development Program for Marine Energy by the Ministry of Land, Transport and Maritime Affairs (MLTM).

REFERENCES

Bae JH, Yoon SH, Lim SY (2011). Heavy metal contents and chemical compositions of Atlantic (Scomber scombrus), blue (Scomber australasicus), and chub (Scomber japonicus) Mackerel Muscles. Food Sci. Biotechnol. 20:709-714.

Barghigiani C, De Ranieri S (1992). Mercury content in different size classes of important edible species of the Northern Tyrrhenian Sea. Mar. Poll. Bull. 24:114-116.

Bligh EG, Dyer WJ (1959). A rapid method of total lipid extraction and purification. Can. J. Biochem. Phys. 37:911-917.

Daviglus M, Sheeshka J, Murkin E (2002). Health benefits from eating fish. Comments Toxicol. 8:345-374.

Deshpande A, Bhendigeri S, Shirsekar T, Dhaware D, Khandekar RN (2009). Analysis of heavy metals in marine fish from Munbai Docks. Environ. Monit. Assess. 159:493-500.

Food and Agriculture Organization (FAO) (2012). FAO fishfinder. Available at: http://www.fao.org /fishery/species/3277/en.

Food and Agriculture Organization-World Health Organization (FAO-WHO) (1973). Energy and protein requirements. FAO-WHO Press, Rome, Italia.

Ikem A, Egiebor NO (2005). Assessment of trace elements in canned fishes (mackerel, tuna, salmon, sardines and herrings) marketed in Georgia and Alabama (United States of America). J. Food Compos. Anal. 18:771-787.

Kagi JH, Schaffer A (1998) Biochemistry of metallothionein. Biochemistry 27:8509-8515.

Kluytmans JH, Boot JH, Oudejans RCHM, Zandee DI (1985). Fatty acid synthesis in relation to gametogenesis in the mussel Mytilius edulis L. Comp. Biochem. Physiol. 81B:959-963.

Korea Food and Drug Administration (KFDA) (2010). Food Code. Korea Foods Industry Association (KFIA) Press, Seoul, Korea. pp. 20-21.

Korean Society of Food Science and Nutrition (KFN) (2000). Handbook of experiments in food science and nutrition. Hyoil Press, Seoul, Korea. pp. 96-128.

Larsen R, Eilertsen KE, Elvevoll EO (2011). Health benefits of marine foods and ingredients. Biotechnol. Adv. 29:508-518.

Loux NT (1998). An assessment of mercury-species-dependent binging with natural organic carbon. Chem. Speciation Bioavailability 10:127-136.

Luzia LA, Sampaio GR, Castellucci CMN, Toreres EAFS (2003). The influence of season on the lipid profiles of five commercially important species of Brazilian fish. Food Chem. 83:93-97.

Ministry for Food, Agriculture, Forestry and Fisheries (MIFAFF) (2010). Agricultural and forestry statistical yearbook 2010. MIFAFF Press, Seoul, Korea. p. 284.

Ministry of agriculture, forestry and fisheries (MAFF) (2009). Statistic database for Fisheries production. Available at: http://www.maff.go.jp.

Nakagawa R, Yumita Y, Hiromoto M (1997). Total mercury intake from fish and shellfish by Japanese people. Chemosphere 35: 2909-2913.

Osako K, Yamaguchi A, Kurokawa T, Kuwahara K, Saito H, Nozaki Y (2002). Chemical components and body color of horse mackerel caught in different areas. Fish. Sci. 68:587-594.

Osako K, Yamaguchi A, Kurokawa T, Kuwahara K, Saito H, Nozaki Y (2003). Seasonal variation in docosahexaenoic acid content in horse mackerel caught in the East China Sea. Fish. Sci. 69:589-596.

Patterson J (2002). Introduction-comparative dietary risk: balance the risks and benefits of fish consumption. Comments Toxicol. 8:337-344.

Salem N, Reyzer M, Karanian J (1996). Losses of arachidonic acid in rat liver after alcohol inhalation. Lipids 31:153-156.

Suriah AR, The SH, Osman H, Nik-Mat D (1995). Fatty acid composition of some Malaysian freshwater fish. Food Chem. 54:45-49.

Zlatanos S, Laskaridis K (2007). Seasonal variation in the fatty acid composition of three Mediterranean fish–sardine (*Sardina pilchardus*), anchovy (*Engraulis encrasicholus*) and picarel (*Spicara smaris*). Food Chem. 103:725-728.

Yukami R, Ohshimo S, Yoda M, Hiyama Y (2009). Estimation of the spawning grounds of chub mackerel *Scomber japonicus* and spotted mackerel *Scomber australasicus* in the East China Sea based on catch statistics and biometric data. Fish. Sci. 75:167-174.

17

Effects of dietary soybean oil inclusion to replace fish oil on growth, muscle fatty acid composition, and immune responses of juvenile darkbarbel catfish, *Pelteobagrus vachelli*

Xueqin Jiang[1], Liqiao Chen[1], Jianguang Qin[2], Chuanjie Qin[1], Haibo Jiang[1] and Erchao Li[1]

[1]School of Life Science, East China Normal University, Shanghai, 200062 China.
[2]School of Biological Sciences, Flinders University, Adelaide, SA 5001, Australia.

An 80 day feeding trial was conducted to evaluate the effects of dietary inclusion of soybean oil to replace fish oil in the diet on growth, muscle fatty acids and immune responses of *Pelteobagrus vachelli*. Isonitrogenous and isocaloric diets were formulated with four fish oil to soybean oil ratios at 8:0 (FO, control), 6:2 (MO1), 2:6 (MO2) and 0:8 (SO) in triplicate. Each diet was fed to juvenile darkbarbel catfish (1.0 ± 0.02 g) twice daily. No significant differences were found in weight gain, special growth rate, hepatosomatic index and intraperitoneal fat ratio among all the treatments. Incorporation of soybean oil significantly modified the fatty acid composition and n-3 and n-6 PUFAs ratio in muscle of fish. Fish fed MO1 showed significantly higher serum lysozyme activity, complement C3 and C4 contents and total IgM content than those fed other diets. These results indicate that partial replacement of fish oil with soybean oil does not compromise fish growth, but improve fish immunity. This study suggests that 25% fish oil replacement with soybean oil can be used in the practical diet of this fish.

Key words: Darkbarbel catfish, *Pelteobagrus vachelli*, soybean oil, growth, fatty acid composition, immune responses.

INTRODUCTION

Fish oils are considered the main source of lipid in aquaculture feeds to promote growth and development of farmed species by providing essential polyunsaturated fatty acids (PUFAs), especially high unsaturated fatty acids (HUFAs) (Sargent et al., 2002). However, in recent years, the increasing demand with limited supply of fish oil (Barlow, 2000; Petropoulos et al., 2009) necessitates the search for alternative lipids to replace fish oil in aquaculture feeds (Mourente and Bell, 2006). Thus,

vegetable oils are potential and sustainable candidates to partially replace fish oils in aquaculture feeds (Montero et al., 2003; Lin and Shiau, 2007). Among available vegetable oils, soybean oil has been sought after due to its availability, affordable price and rich content of essential fatty acids (Bell et al., 2001; Caballero et al., 2003).

Like in mammals, nutritional status can affect immune system and disease resistance in fish (Blazer, 1992;

Calder, 2001). Studies on dietary lipid and fish immune system have been emphasized on because incorporation of vegetable oils in diets may result in the potential imbalance of n-3 to n-6 fatty acids and thereafter directly or indirectly affect immune system and disease resistance in fish (Pablo et al., 2002; Montero et al., 2003). Recent research has shown that fish immunity is compromised when fish oil is replaced with vegetable oils in *Epinephelus malabaricus* (Lin and Shiau, 2007), *Ictalurus punctatus* (Fracalossi and Lovell, 1994), *Psetta maxima* (Regost et al., 2003), *Salmo salar* L. (Brandsen et al., 2003) and *Sparus aurata* (Montero et al., 2008).

Darkbarbel catfish, *Pelteobagrus vachelli* is an important freshwater species in aquaculture in Asia because of its high nutritional and market values (Xu et al., 2012). Currently, nutrition research on this species has been limited to define the environmental and nutritional requirements for optimal growth (Lu et al., 2008; Huang et al., 2009; Tan et al., 2009; Ye et al., 2009; Zheng et al., 2010). However, information on nutritional modulation of fish immunity by dietary lipid is scarce, especially on the immune response of fish when fish oil is replaced with vegetable oils. The purpose of this study was to investigate the effect of inclusion of soybean oil at various levels in feed on growth, muscle fatty acid composition and immune responses of this species.

MATERIALS AND METHODS

Experimental diets

Four isonitrogenous and isocaloric diets (Table 1) were formulated with four levels of fish oil and soybean oil at 8:0 (FO, control), 6:2 (MO1), 2:6 (MO2) and 0:8 (SO). Fish meal and soybean meal were used as a dietary protein source, and wheat starch was used as a dietary carbohydrate source. Diet samples were analyzed for crude protein, lipid, dry matter and ash content (Table 1) according to the standard methods of AOAC (1995), followed by the analysis of fatty acid composition (Table 2). The diet containing 51% crude protein and 11% lipid is sufficient to support the optimal growth of darkbarbel catfish (Ye et al., 2009). Diet ingredients were ground into fine powder and water was added to produce stiff dough. The dough was then pelleted with an experimental feed mill and dried in a ventilated oven at 40°C to moisture less than 10%. After being dried, the diets were processed into pellets (1 mm dia×2 mm length), and stored at -20°C until use.

Experimental fish, feeding and sampling

Juvenile darkbarbel catfish were obtained from a fish hatchery in Shanghai, China. Fish were fed with the control diet for two week prior to the growth trial. Fish (1.0 ± 0.02 g) were randomly distributed into 12 tanks (0.8 × 0.6 × 0.6 m) with 30 fish in each tank. Fish were fed each experimental diet in triplicate at 5% fish weight twice a day at 0800 and 1800 h, and were bulk weighed every two weeks to adjust the daily ration. All tanks were configured in a recirculation system with flow-through dechlorinated water. Dissolved oxygen in water was maintained at about 8.0 mg/L by continuous aeration; ammonia-N was less than 0.1 mg/L; pH varied from 7.0 to 7.3; water temperature ranged from 25 to 28°C. A light/dark period was set as 12 h:12 h during the feeding trial. Fish

were fasted for 24 h before sampling. The trial lasted 80 days. At the end of the feeding trial, all fish were weighed, and five fish per tank were taken for blood sampling from the caudal vein with a 1.0 ml heparinized syringe, subsequently allowed to clot for 1 h in microtubes at room temperature and centrifuged at 1500×g for 5 min at 4°C to recover serum which was frozen at -20°C until use (Montero et al., 2003). While livers and intraperitoneal fats were obtained from these fish and weighed, dorsal muscle was taken for fatty acid analysis (Piedecausa et al., 2007). Growth variables were calculated as follows:

Weight gain (WG) = $(Wt-W_0)/W_0 \times 100$
Specific growth rate (SGR) = $(LnWt-LnW_0) \times 100/t$
Hepatosomatic index (HSI) = (liver wet weight/body wet weight)×100
Intraperitoneal fat ratio (IPF) = (intraperitoneal wet weight/body wet weight)×100

where W_0 and Wt were initial and final fish weight (g), respectively; t is duration of experiment (day).

Biochemical composition analysis

All diets were analyzed for proximate composition following the standard methods (AOAC, 1995). Moisture was determined by drying in an oven at 105°C to a constant weight. Then dry matter was digested with nitric acid and incinerated in a muffle furnace at 600°C overnight for ash content. Crude protein was measured with the Kjeldahl method and crude lipid was determined by the ether extraction method using the Soxhlet System (2055 Soxhlet Avanti; Foss Tecator, Hoganas, Sweden).

Diets and fish muscles were dried by lyophilization for fatty acid analysis. The analysis was performed using gas chromatography (GS, HP6890, USA) with minor modifications as described by Mourente et al. (1999) and Satoh et al. (1989). Briefly, total lipid was extracted with chloroform: methanol (2:1, v/v) according to the method of Folch et al. (1957). The capillary gas chromatography (GC) method was employed to determine the fatty acid profile. The HP6890 (FID detector) and SPTM-2380 column (30 m×0.25 mm×0.20 ìm) were used on a GC machine. Separation was carried out with nitrogen as carrier gas. The column temperature was programmed from 140 to 240°C at 4°C/min, held for 5 min at 140°C and 10 min at 240°C, with a detector at 260°C. A split injector (50:1) at 260°C was used. Fatty acids were identified by comparing their retention time to the chromatographic standard (Sigma). Peak areas were determined using Varian software.

Immune parameters assay

Serum lysozyme activity

The serum lysozyme activity was determined as described by Ellis (1990) and Alcorn et al. (2002), using a detection kit (Nanjing Jiancheng Bioengineering Institute, China). One unit of enzyme activity was defined as the amount of enzyme causing a decrease in absorbance of 0.001 per min per ml serum.

Total immunoglobulin M (IgM) in serum

Total IgM was determined according to the method of Siwicki and Anderson (1993). As modified by Tang et al. (2008), the assay was based on the measurement of total protein contents in plasma using a micro protein determination method (C-690; Sigma) prior to and after precipitating down the IgM molecules employing a 12%

Table 1. Formulation and proximate composition of the experimental diets.

Ingredients (g·kg[-1])	Diets[*]			
	FO	MO1	MO2	SO
Fish meal[a]	500	500	500	500
Soybean meal	280	280	280	280
Wheat starch	64	64	64	64
Fish oil	80	20	60	
Soybean oil		60	20	80
Mineral mix[b]	50	50	50	50
Vitamin mix[c]	5	5	5	5
Choline chloride	1	1	1	1
CMC	20	20	20	20
Proximate composition of diets (g·kg[-1] dry matter)				
Dry matter	910.0	920.8	920.1	910.2
Crude protein	510.0	510.1	520.9	520.3
Crude fat	110.1	110.0	110.4	110.0
Ash	90.0	90.6	90.4	90.2

*Diet abbreviations are as follows: FO=fish oil; MO1=mixture oil 1 (fish oil:soybean oil=1:3); MO2= mixture oil 2 (fish oil:soybean oil=3:1); SO= soybean oil. [a]Made in New Zealand. [b]Mineral mixture (mg/Kg diet): MnSO4.7H$_2$O, 399; Ca (H$_2$PO$_4$)$_2$, 3; AlCl$_3$.6H$_2$O, 21; ZnSO$_4$ 7H$_2$O, 60; KI, 0.15; K$_2$HPO$_4$, 0.6; FeSO$_4$.7H$_2$O, 105; CuSO$_4$.5H$_2$O, 30; NaCl, 150 ; cellulose, 231.25. [c]Vitamin mixture (mg/kg diets, NRC 1977): retinol acetate, 5500 IU; cholecalciferol, 1000 IU; a-tocopherol acetate, 50 IU; menadione, 10 ; choline chloride, 550; nicotinic acid,100; riboflavin, 20; pyridoxine hydrochloride, 20; thiamin hydrochloride, 20; biotin,0.1; folic acid,, 5; vitamin B12, 20; L-ascorbyl-2-monophosphate Mg, 100; myo-inositol, 100. All ingredients were diluted with a-cellulose to 1 g.

Table 2. Fatty acid composition of experimental diets for darkbarbel catfish *Pelteobagrus vachelli* (% total fatty acids).

Fatty acids[a]	Diets			
	FO	MO1	MO2	SO
C8:0	0.08	0.61	1.67	2.20
C12:0	0.10	0.17	0.30	0.36
C14:0	7.19	6.04	3.54	3.13
C15:0	0.65	0.63	0.34	0.35
C16:0	18.77	19.11	19.67	20.12
C17:0	0.87	0.83	0.51	0.54
C18:0	3.44	4.53	6.70	7.78
$\sum$SFA[1]	31.95	32.62	34.93	36.12
C16:1	10.98	6.33	3.92	3.98
C18:1n-9	15.64	19.45	18.78	20.72
C20:1n-9	5.09	1.76	4.21	0.84
$\sum$MUFA[2]	33.31	31.43	28.39	26.08
C18:2n-6	4.01	10.44	19.21	24.27
20:4n-6	3.59	3.01	1.38	0.50
$\sum$n-6[3]	8.11	13.65	21.58	25.26
C18:3n-3	1.43	1.75	2.19	2.70
C20:5n-3	12.8	10.72	6.45	5.01
C22:6n-3	12.4	9.83	5.46	3.80
$\sum$n-3[4]	26.64	23.52	15.1	12.1
$\sum$PUFA	34.74	35.95	36.68	37.8
n-3/ n-6	3.28	1.72	0.69	0.47

[a]SFA, saturated fatty acid; MUFA, monounsaturated fatty acid; PUFA, polyunsaturated fatty acid. [1]Total SFA includes *C4:0, C6:0, C8:0, C10~17:0, C18:0, C21:0, C22:0,C23:0* and *C24:0*. [2]Total MUFA includes *C14:1, C15:1, C16:1, C17:1, 18:1n-9* and *C24:1n-9*. [3]Total *n-6* includes *C18:3n-6, C18:2n-6* and *C22:4n-6*. [4]Total *n-3* includes *C18:3n-3, C20:3n-3, C20:5n-3, C22:5n-3* and *C22:4n-6*.

Table 3. Growth performance* of darkbarbel catfish *P. vachelli* fed diets with different lipid sources.

Growth performance[2]	Diets[1]			
	FO	MO1	MO2	SO
IBW(g)	1.00±0.12	1.00±0.12	1.00±0.13	1.00±0.17
FBW(g)	7.68±1.46	8.24±1.70	8.31±0.64	7.47±0.83
WG	6.51±1.44	7.06±1.66	7.14±0.65	6.32±0.82
SGR	2.50±0.22	2.59±0.24	2.62±0.11	2.48±0.14
HSI	2.66±0.42	3.04±0.76	2.93±0.62	2.54±0.60
IPF	3.52±0.39	3.70±0.42	3.92±0.17	4.49±0.22

*Values are means ± S.D. of three replicates. [1]Diet abbreviations refer to Table 1. IBW (g), initial mean body weight; FBW (g), final mean body weight; WG, weight gain; SGR, specific growth rate ; IPF, intraperitoneal fat ratio; HSI, hepatosomatic index.

(w/v) solution of polyethyleneglycol (Sigma). The difference in the protein contents was considered the IgM content.

Serum complements C3 and C4 contents

The kit of immunoturbidimetry (Nanjing Jiancheng Bioengineering Institute, China) was adopted for the detection of serum complement C3 and C4 contents as described by Tang et al. (2008). The serum was mixed with the antibody from the kit, and then an antigen-antibody complex was produced. The optical density (OD) was measured at 340 nm with UV-visible spectrophotometer. Compared with the values of the standards from the kit, the C3 and C4 contents were calculated as mg/L.

Statistical analysis

All data were analyzed by one-way analysis of variance (ANOVA) and Duncan's multiple range tests using software SPSS 16.0. The results are presented as mean ± SD, and probabilities of $P < 0.05$ were considered significant.

RESULTS

In the 80-day growth trial, no significant differences were found in weight gain (WG), special growth rate (SGR), hepatosomatic index (HSI) and intraperitoneal fat ratio (IPF) of fish fed different diets ($P > 0.05$, Table 3). But WG, SGR and HSI were slightly higher in fish fed MO1 or MO2 than the control, while IPF increased progressively with the increase of soybean oil in feed.

Total saturated fatty acid (SFA) in the muscle of fish fed FO and MO1 were significantly higher than those in fish fed SO and MO2 ($P < 0.05$, Table 4). Total monounsaturated fatty acid (MUFA) in fish fed FO was significantly higher than that in fish fed other diets ($P < 0.05$, Table 4). Similarly, MUFA in fish fed MO1 was significantly higher than that in fish fed MO2 and SO ($P < 0.05$). The highest total n-6 fatty acids was found in fish fed SO and lowest in fish fed FO. In contrast, the total n-3 fatty acids in fish fed FO were highest, but lowest in fish fed SO ($P < 0.05$). Inclusion of soybean oils in feed significantly increased the content of linoleic acid (18: 2n-

6, $P < 0.05$) and thus significantly modified the content of arachidonic acid (ARA), eicosapentaenoic acid (EPA) and docosahexaenoic acid (DHA) and the ratio of n-3/n-6 fatty acids in the muscle ($P < 0.05$).

Table 5 presents the significant effects of the inclusion of soybean oil in feed on the activities or contents of some serum immune parameters. The lysozyme activity, contents of complement C3, C4 and total IgM in fish fed MO1 were higher than those in fish fed other diets ($P < 0.05$), but were lower in fish fed SO than those in fish fed other diets ($P < 0.05$). However, there was no significant difference between the treatments of FO and MO2 ($P > 0.05$).

DISCUSSION

In this study, juvenile darkbarbel catfish fed diets with different inclusions of soybean oil showed no significant differences in growth after 80 days. This is in agreement with the results previously reported in *Acanthopagrus schlegeli* (Peng et al., 2008), *Diplodus puntazzo* (Piedecausa et al., 2007), *E. malabaricus* (Lin and Shiau, 2007) and *Oncorhynchus mykiss* (Caballero et al., 2002), representing successful replacement of dietary fish oil with vegetable oils in fish species without compromising their growth. Despite statistical insignificance obtained in the study, fish fed the diets with partial inclusion of soybean oil (MO1 and MO2) showed numerically higher growth rate (WG, SGR, HSI and IPF) than fish fed FO or SO. This pattern of growth may probably resulted in from 11% lipid content in formulation of the experimental diets satisfying fish growth and development based on the previous studies of dietary lipid requirement for darkbarbel catfish (Han et al., 2005; Huang et al., 2009). In addition, the higher growth performance was observed on juvenile darkbarbel catfish fed diets MOs (MO1 and MO2), probably due to the n-3 to n-6 ratio of 1.72-0.69 from diets MOs was suitable for improving growth of darkbarbel catfish. The fatty acid (FA) ratio can affect fish growth by the interaction between dietary n-3 and n-6 fatty acids on endogenous FA elongation and

Table 4. Muscle fatty acid composition* of darkbarbel catfish *P. vachelli* fed diets with different lipid sources (% total detectable FA).

Fatty acid	Diets[1]			
	FO	MO1	MO2	SO
C14:0	3.49±0.04[a]	2.94±0.03[b]	2.03±0.16[c]	1.79±0.02[c]
C15:0	0.42±0.05[a]	0.39±0.08[ab]	0.25±0.04[bc]	0.21±0.03[c]
C16:0	17.45±0.71[a]	17.05±0.49[a]	15.29±0.26[b]	15.02±0.68[b]
C17:0	0.63±0.08	0.60±0.12	0.40±0.07	0.37±0.10
C18:0	4.43±0.20a	4.97±0.36b	4.80±0.11b	4.85±0.23b
∑SFA	26.89±0.39[a]	26.52±0.41[a]	23.40±0.27[b]	22.94±0.33[b]
C16:1	6.44±0.19[a]	4.88±0.12[b]	3.22±0.30[b]	3.12±0.28[b]
C18:1n-9	27.71±0.29[b]	28.64±0.37[a]	28.03±0.46[b]	28.02±0.15[b]
C20:1n-9	4.17±0.02[a]	3.30±0.02[ab]	2.15±0.04[bc]	2.41±0.03[c]
∑MUFA	38.32±0.82[a]	36.82±0.65[b]	33.41±0.91[c]	33.55±0.47[c]
C18:2n-6	9.9±0.12[a]	12.79±0.06[b]	25.84±0.47[c]	27.39±0.61[d]
20:4n-6	0.51±0.01[a]	0.57±0.08[a]	0.74±0.10[b]	0.87±0.23[b]
∑n-6	10.47±0.15[a]	13.80±0.14[b]	27.22±0.62[c]	28.91±0.83[d]
C18:3n-3	1.70±0.10[a]	1.96±0.20[a]	2.92±0.22[b]	3.21±0.19[b]
C20:5n-3	4.85±0.08[a]	4.48±0.11[b]	2.52±0.43[c]	2.14±0.23[d]
C22:6n-3	13.45±1.06[a]	12.68±1.21[b]	8.18±0.97[c]	7.21±0.75[d]
∑n-3	23.63±1.24[a]	22.16±1.52[b]	15.78±1.33[c]	13.81±1.09[d]
∑PUFA	34.80±0.79[a]	36.67±0.95[b]	43.2±1.58[c]	43.51±0.88[c]
n-3/ n-6	2.26±0.05[a]	1.61±0.02[b]	0.58±0.04[c]	0.48±0.06[c]

*Values are means of three replicates ± SD. Means in a row with the same letter superscript are not significantly different ($P < 0.05$).

Table 5. Serum immune parameters* of darkbarbel catfish *P. vachelli* fed diets with different lipid sources.

Immune parameter	Diets			
	FO	MO1	MO2	SO
Lysozyme activity (U/ mL)	97.65±14.02[b]	132.72±16.16[a]	106.40±22.32[b]	69.43±16.16[c]
Complement C3 content (mg/L)	0.22±0.03[b]	0.33±0.08[a]	0.23±0.09[b]	0.13±0.05[c]
Complement C4 content (mg/L)	0.08±0.01[b]	0.16±0.01[a]	0.10±0.03[b]	0.06±0.01[c]
Total IgM level (mg/L)	0.28±0.02[b]	0.47±0.07[a]	0.25±0.08[b]	0.22±0.01[b]

*Values are means ± S.D. of three replicates and values with different letter superscripts within the same row are significantly different at $P <0.05$.

desaturation enzyme systems, especially by the Δ6 and Δ5 desaturase enzymes (Tan et al., 2009). Therefore, the higher HSI and IPF in fish fed MOs with n-3 to n-6 ratio of 1.72 and 0.69, probably resulted in from the increasing activities of elongase and desaturase in hepatocytes of fish due to their higher affinity towards the n-3/n-6 fatty acids (Tocher et al., 2002; Stubhaug et al., 2005; Blanchard et al., 2008). Our result indicated that inclusion of soybean oil in feed does not compromise the species growth, suggesting the possible replacement of fish oil with soybean oil in diet formulation for darkbarbel catfish. The positive relationship between tissue fatty acid composition and dietary fatty acid contents has been well demonstrated in previous studies (Bransden et al., 2005; Zia-Ul-haq et al., 2007a, b, 2008, 2010, 2011a, b; Luo et

al., 2008; Rezek et al., 2010). In this study, the muscle fatty acid composition in fish generally reflected the lipid composition in diets similar to the report of Ng et al. (2003) that the high fatty acids in feed can lead to high fatty acids in muscle. Moreover, there were differences in fatty acid composition between fish muscle and the diets in our study.

For instance, lower content of SFA in muscle relative to the diets indicates that darkbarbel catfish has limited ability to deposit SFA into the tissue and metabolize it to meet up energy demands. This claim is supported by the research in *Mystus nemurus* (Ng et al., 2000) and *Pelteobagrus fulvidraco* (Tan et al., 2009). Freshwater fish has an innate ability to convert these C18 fatty acids to the long-chain n-3 and n-6 fatty acids (e.g.

EPA, DHA and ARA) *in vivo* by an alternating sequence of desaturation and elongation (Garg et al., 1988; Blanchard et al., 2008), which have been observed in some freshwater fish species such as *M. nemurus* (Ng et al., 2000), *Oncorhyncus mykiss* (Buzzi et al., 1996), *Perca fluviatilis* (Blanchard et al., 2008) and *Pelteobagrus fulvidraco* (Tan et al., 2009). Therefore linolenic acid (LnA, C18:3n-3) and linoleic acid (LA, C18:2n-6) are considered as the essential fatty acids for freshwater fish (Turchini et al., 2006). Darkbarbel catfish probably possesses the similar ability to convert LnA and LA to 20:5n-3 (EPA), 22:6n-3 (DHA) and 20:4n-6 (ARA), respectively (Nakamura and Nara, 2004; Sprecher et al., 1995). Therefore in this study, the dietary inclusion of soybean oil (rich in C18:2n-6) finally led to accumulation of n-6 fatty acids and reduction of n-3 fatty acids in fish muscle. Piedecausa et al. (2007) also reported the similar result that the higher level of 18:2n-6 fatty acid in soybean oil can promote lipid accumulation in fish body. Therefore, an optimal replacement (25 to 75%) of fish oil by soybean oil in feed with the n-3 to n-6 PUFA ratios of 0.69-1.72 may provide better growth for juvenile darkbarbel catfish.

The change of dietary fatty acid composition and the ratio of n-6/n-3 fatty acids can also affect fish immunity and disease resistance (Blazer, 1992; Kiron et al., 1995; Yildirim-Aksoy et al., 2007; Montero et al., 2008). In this study, the MO1 diet with the n-3/n-6 fatty acid ratio of 1.72 significantly increased the activity of serum lysozyme, the contents of serum complement C3, C4 and total IgM, suggesting that such an n-3/n-6 fatty acid ratio in diet was appropriate to darkbarbel catfish and stimulated its immune function. This result was probably attributed to, on the one hand, the ratio of n-3 and n-6 PUFA can affect immune cell functions, cell signaling and humoral immunological processes by regulating the production of eicosanoids, derived from DHA, EPA and ARA as their precursors (Calder, 2006; Yaqoob and Calder, 2007; Montero et al., 2010). On the other hand, the balanced n-3/n-6 fatty acid ratio is a requirement for fish health and good immunity (Simopoulos, 2008). An excess of n-6 PUFA adversely affects antibody production and serum immune response parameters, whereas excessive level of n-3 PUFA reduce macrophage killing ability and disease resistance in fish (Bell et al., 1996; Yildirim-Aksoy et al., 2007). In this study, in comparison to the control FO (total fish oil) and SO (total soybean oil), a well-balanced n-3/n-6 fatty acids ratio of 1.72 (MO1) significantly improved the immunity of the fish.

However, despite the crucial correlation between dietary lipid composition and fish immunity, the functional effects of dietary lipid on fish immune response are controversial. Positive effects of n-3 fatty acids on the immune response were found in *O. mykiss* (Kiron et al., 1995) and *S. aurata* (Montero et al., 2008), whereas negative effects of high levels of dietary n-3 PUFAs on immunity are found in *I. punctatus* (Fracalossi and Lovell,

1994) and *O. mykiss* (Ashton et al., 1994). In this study, although we found that partial inclusion (25%) of soybean oil in diet can enhance immunity, it still warrants further study to investigate the mechanism why inclusion of vegetable oil can improve fish immunity.

ACKNOWLEDGEMENTS

This work was supported by grants from the National Basic Research Program (973 Program, No. 2009CB118702), National Key Technology Support Program (2012BAD25B03), National Natural Science Foundation of China (No. 31172422, 31001098), Special Fund for Agro-scientific Research in the Public Interest (No. 201003020, 201203065), Shanghai Committee of Science and Technology, China (10JC1404100), Shanghai technology system for Chinese mitten-handed crab industry, and partially by the E-Institute of Shanghai Municipal Education Commission (No. E03009).

REFERENCES

Alcorn SW, Murray AL, Pascho RJ (2002). Effects of rearing temperature on immune functions in sockeye salmon (*Oncorhynchus nerka*). Fish Shellfish Immun.12:303-334.

AOAC (Association of Official Analytical Chemists). (1995) Official Methods of Analysis, 16th edn. AOAC, Arlington, VA.

Ashton I, Clements K, Barrow SE, Secombes CJ, Rowley AF (1994). Effects of dietary fatty acids on eicosanoid-generating capacity, fatty acid composition and chemotactic activity of rainbow trout (*Oncorhynchus mykiss*). Leucocytes. Biochim. Biophys. Acta. 1214:253-262.

Barlow S (2000). Fish meal and oil: sustainable feed ingredients for aquaculture. Global Aquacult. Advocate 4:85-88.

Bell JG, Ashton I, Secombes CJ, Weitzel BR, Dick JR, Sargent JR (1996). Dietary lipid affects phospholipid fatty acid compositions, eicosanoids production and immune function in Atlantic salmon (*Salmo salar*). Prostag. Leukotr. Ess. 54:173-182.

Bell JG, McEvoy J, Tocher DR, McGhee F, Campbell PJ, Sargent JR (2001). Replacement of fish oil with rapeseed oil in diets of Atlantic salmon (*Salmo salar*) affects tissue lipid compositions and hepatocyte fatty acid metabolism. Nutrition 131:1535-1543.

Blanchard G, Makombu JG, Kesternont P (2008). Influence of different dietary 18:3n-3/18:2n-6 ratio on growth performance, fatty acid composition and hepatic ultrastructure in Eurasian perch, Perca fluviatilis. Aquaculture 284:144-150.

Blazer VS (1992). Nutrition and disease resistance in fish. Annu. Rev. Fish. Dis. 2:309-323.

Bransden MP, Butterfield GM, Walden J, McEvoy LA, Bell JG (2005). Tank colour and dietary arachidonic acid affects pigmentation, eicosanoid production and tissue fatty acid profile of larval Atlantic cod (*Gadus morhua*). Aquaculture 250:328-340.

Brandsen MP, Carter CG, Nichols PD (2003). Replacement of fish oil with sunflower oil in feeds for Atlantic salmon (*Salmo salar* L.): effect on growth performance, tissue fatty acid composition and disease resistance. Comp. Biochem. Physiol. 135:611-625.

Buzzi M, Henderson RJ, Sargent JR (1996). The desaturation and elongation of linolenic acid and eicosapentaenoic acid by hepatocytes and liver microsomes from rainbow trout (*Oncorhyncus mykiss*) fed diets containing fish oil or olive oil. Biochim. Biophys. Acta. 1299:235-244.

Caballero MJ, Obach A, Rosenlund G, Montero D, Gisvold M, Izquierdo MS (2002). Impact of different dietary lipid sources on growth, lipid digestibility, tissue fatty acid composition and histology of rainbow trout,

Oncorhynchus mykiss. Aquaculture 214:253-271.

Caballero MJ, Izquierdo MS, Kjorscik E, Montero D, Socorro J, Fernandez AJ, Rosenlund G (2003). Morphological aspects of intestinal cells from gilthead seabream (*Sparus aurata*) fed diets containing different lipid sources. Aquaculture 225:325-340.

Calder PC (2001). Polyunsaturated fatty acids, inflammation and immunity. Lipids 36:1007-1024.

Calder PC (2006). Polyunsaturated fatty acids and inflammation. Prostag. Leukotr. Ess. 75:197-202.

Ellis AE (1990). Lysozyme assays. In: Stolen JS, Fletcher TC, Anderson DP, Kaattari SL, Rowley AF (eds) Techniques in Fish Immunology, SOS Publications, Fair Haven, NJ. pp. 101-103.

Fracalossi DM, Lovell RT (1994). Dietary lipid sources influence responses of channel catfish (*Ictalurus punctatus*) to challenge test with the pathogen *Edwardsiella ictaluri*. Aquaculture 119:287-298.

Folch J, Lees M, Sloane-Stanley GH (1957). A simple method for the isolation and purification of total lipids from animal tissues. J. Biol. Chem. 226:497-509.

Fountoulaki E, Alexis MN, Nengas I, Venou B (2003). Effects of dietary arachidonic acid (20:4n-6), on growth, body composition and tissue fatty acid profile of gilthead bream fingerlings (*Sparus aurata* L.). Aquaculture 225:309-323.

Garg ML, Sebokava E, Thomson ABR, Clandinin MT (1988). Δ6-desaturase activity in liver microsomes of rats fed diets enriched with cholesterol and/or ω3 fatty acids. Biochemistry 249:351-356.

Han Q, Tian ZC, Xia WF (2005). Dietary lipid requirement of darkbarbel catfish *Pelteobagrus vachelli*. Fisheries Sci. 24:8-11 (in Chinese).

Huang J, Feng J, Sun T, Huang XY, He L, Du WP (2009). Macronutrient composition of formulated diets for juvenile yellow catfish (*pelteobagrus fulvidraco* richardson). Oceanol. Limnol. Sin. 40:437-445. (in Chinese)

Kiron V, Fukuda H, Takeuchi T, Watanabe T (1995). Essential fatty acid nutrition and defence mechanisms in rainbow trout *Oncorhynchus mykiss*. Comp. Biochem. Physiol. 111:361-367.

Lin YH, Shiau SY (2007). Effects of dietary blend of fish oil with corn oil on growth and non-specific immune responses of grouper, *Epinephelus malabaricus*. Aquac. Nutr. 13:137-144.

Lu SF, Zhao N, Zhao AY, He RG (2008). Effect of soybean phospholipid supplementation in formulated microdiets and live food on foregut and liver histological changes of *Pelteobagrus fulvidraco* larvae. Aquaculture 278:119-127.

Luo Z, Li X, Bai H, Gong S (2008). Effects of dietary fatty acid composition on muscle composition and hepatic fatty acid profile in juvenile *Synechogobius hasta*. J. Appl. Ichthyol. 24:116-119.

Montero D, Grasso V, Izquierdo MS, Ganga R, Real F, Tort L, Caballero MJ, Acosta F (2008). Total substitution of fish oil by vegetable oils in gilthead sea bream (*Sparus aurata*) diets: Effects on hepatic Mx expression and some immune parameters. Fish Shellfish Immun. 24:147-155.

Montero D, Kalinowski T, Obach A, Robaina L, Tort L, Caballero MJ, Izquierdo MS (2003). Vegetable lipid source for gilthead seabream (*Sparus aurata*): effects on fish health. Aquaculture 225:353-370.

Montero D, Mathlouthi F, Tort L, Afonso JM, Torrecillas S, Fernández-Vaquero A, Negrin D, Izquierdo MS (2010). Replacement of dietary fish oil by vegetable oils affects humoral immunity and expression of pro-inflammatory cytokines genes in gilthead sea bream *Sparus aurata*. Fish Shellfish Immun. 29:1073-1081.

Mourente G, Bell JG (2006). Partial replacement of dietary fish oil with blends of vegetable oils (rapeseed, linseed and palm oils) in diet for European sea bass (*Dicentrarchus labrax* L.) over a long term growth study: effects on muscle and liver fatty acid composition and effectiveness of a fish oil finishing diet. Comp. Biochem. Physiol. 145:389-399.

Mourente G, Tocher DR, Diaz-Salvago E, Grau A, Pastor E (1999). Study of the n-3 highly unsaturated fatty acids requirement and antioxidant status of *Dentex dentex* larvae at the Artemia feeding stage. Aquaculture 179:291-307.

Nakamura MT, Nara TY (2004). Sctructure, function, and dietary regulation of delta-6, delta-5, and delta-9 desaturases. Annu. Rev. Nutr. 24:345-376.

Ng WK, Campbell PJ, Dick JR, Bell JG (2003). Interactive effects of dietary palm oil concentration and water temperature on lipid

digestibility in rainbow trout, *Oncorrynchus mykiss*. Lipids 38:1031-1038.

Ng WK, Tee MC, Boey PL (2000). Evaluation of crude palm oil and refined palm olein as dietary lipids in pelleted feeds for a tropical bagrid catfish *Mystus nemurus* (Cuvier and Valenciennes). Aquaculture 31:337-347.

Pablo MA, Puertollano MA, Cienfuegos GV (2002). Biological and clinical significance of lipids as modulators of immune functions. Clin. Diagn. Lab. Immun. 9:94-950.

Peng SM, Chen LQ, Qin JG, Hou JL, Yu N, Long ZQ, Ye JY, Sun XJ (2008). Effects of replacement of dietary fish oil by soybean oil on growth performance and liver biochemical composition in juvenile black seabream, *Acanthopagrus schlegeli*. Aquaculture 276:154-161.

Petropoulos IK, Thompson KD, Morgan A, Dick JR, Tocher DR, Bell JG (2009). Effects of substitution of dietary fish oil with a blend of vegetable oils on liver and peripheral blood leucocyte fatty acid composition, plasma prostaglandin *E*2 and immune parameters in three strains of Atlantic salmon (*Salmo salar*). Aquac. Nutr. 15:596-607.

Piedecausa MA, Mazo'n MJ, Garcy'a B, Herna'ndez MD (2007). Effects of total replacement of fish oil by vegetable oils in the diets of sharpsnout seabream (*Diplodus puntazzo*). Aquaculture 26:211-219.

Regost C, Arzel J, Robin J, Rosenlund G, Kaushik SJ (2003). Total replacement of fish oil by soybean oil with return to fish oil in turbot (*Psetta maxima*) I. Growth performance, flesh fatty acid profile, and lipid metabolism. Aquaculture 217:465-82.

Rezek TC, Watanabe WO, Harel M, Seaton PJ (2010). Effects of dietary docosahexaenoic acid (22:6n-3) and arachidonic acid (20:4n-6) on the growth, survival, stress resistance and fatty acid composition in black sea bass *Centropristis striata* (Linnaeus 1758) larvae. Aquac. Res. 41:1302-1314.

Sargent J, Tocher DR. Bell JG (2002). The lipids. In: Halver JE(ed) Fish Nutrition, 2nd, Academic Press, London. pp. 181-257.

Satoh S, Poe WE, Wilson RP (1989). Effect of dietary n-3 fatty acids on weight gain and liver polar lipid fatty acid composition of fingerling channel catfish. Nutrition 119:23-28.

Simopoulos AP (2008). The omega-6/omega-3 fatty acid ratio, genetic variation, and cardiovascular disease. Asia. Pac. J. Clin. Nutr. 17:131-4.

Siwicki AK, Anderson DP (1993). Nonspecific Defense Mechanisms Assay in Fish. II. Potential Killing Activity of Neutrophils and Macrophages, Lysozyme Activity in Serum and Organs and Total Immunoglobulin Level in Serum. In: Siwicki AK, Anderson DP, Waluga J (eds) Fish Disease Diagnosis and Prevention Methods. Olsztyn, Poland: pp. 105-112.

Sprecher H, Luthria DL, Mohammed BS, Baykousheva SP (1995). Reevaluation of the pathways for the biosynthesis of polyunsaturated fatty acids. J. Lipid Res. 36:2471-2477.

Stubhaug I, Frøyland L, Torstensen BE (2005). β-oxidation capacity of red and white muscle and liver in Atlantic salmon (*Salmo salar* L.) — effects of increasing dietary levels of rapeseed oil (0-100%) and olive oil (50%) to replace capelin oil. Lipids, 40: 39-47.

Tan XY, Luo Z, Xie P, Liu XJ (2009). Effect of dietary linolenic acid/linoleic acid ratio on growth performance, hepatic fatty acid profiles and intermediary metabolism of juvenile yellow catfish *Pelteobagrus fulvidraco*. Aquaculture 296:96-101.

Tang HG, Wu TX, Zhao ZY, Pan XD (2008). Effects of fish protein hydrolysate on growth performance and humoral immune response in large yellow croaker (*Pseudosciaena crocea* R.). J. Zhejiang Univ. Sci. 9:684-690.

Tocher DR, Agaba M, Hastings N, Bell JG, Dick JR, Teale AJ (2002). Nutritional regulation of hepatocyte fatty acid desaturation and polyunsaturated fatty acid composition in zebrafish (*Danio rerio*) and tilapia (*Oreochromis niloticus*). Fish. Physiol. Biochem. 24:309-320.

Turchini GM, Francis DS, De Silva SS (2006). Fatty acid metabolism in the freshwater fish Murray cod (*Maccullochella peelii peelii*) deduced by the whole body fatty acid balance method. Comp. Biochem.Physiol. 144:110-118.

Xu MH, Long LN, Chen LQ, Qin JG, Zhang L, Yu N, Li EC (2012). Cloning and differential expression pattern of pituitary adenylyl cyclase-activating polypeptide and the PACAP-specific receptor in darkbarbel catfish Pelteobagrus vachelli. Comp. Biochem. Physiol.

161:41-53.

Yaqoob P, Calder PC (2007). Fatty acid and immune function: new insights into mechanisms. Br. J. Nutr. 98:41-45.

Ye WJ, Tan XY, Chen DY, Luo Z (2009). Effects of dietary protein to carbohydrate ratios on growth and body composition of juvenile yellow catfish, Pelteobagrus fulvidraco (Siluriformes, Bagridae, Pelteobagrus). Aquaculture 40:1410-1418.

Yildirim-Aksoy M, Lim C, Davis DA, Shelby R, Klesius PH (2007). Influence of Dietary Lipid Sources on the Growth Performance, Immune Response and Resistance of Nile Tilapia, Oreochromis niloticus, to Streptococcus iniae Challenge. J. Appl. Aquac. 19:27-49.

Zheng KK, Zhu XM, Han D, Yang YX, Lei W, Xie SQ (2010). Effects of dietary lipid levels on growth, survival and lipid metabolism during early ontogeny of Pelteobagrus vachelli larvae. Aquaculture 299:121-127.

Zia-Ul-Haq M, Ahmad S, Chiavaro E, Mehjabeen, Ahmed S (2010). Studies of oil from cowpea (Vigna unguiculata (l) walp.) cultivars commonly grown in Pakistan. Pak. J. Bot. 42(2):214-220.

Zia-Ul-Haq M, Ahmad M, Iqbal S, Ahmad S, Ali H (2007a). Characterization and compositional studies of oil from seeds of desi chickpea (Cicer arietinum L.) cultivars grown in Pakistan. J. Am. Oil Chem. Soc. 84:1143-1148.

Zia-Ul-Haq M, Ahmad S, Shad MA, Iqbal S, Qayum M, Ahmad A, Luthria DL, Amarowicz R (2011a). Compositional studies of some of lentil cultivars commonly consumed in Pakistan. Pak. J. Bot. 43(3):1563-1567.

Zia-Ul-Haq M, Ćavar S, Qayum M, Imran I, Feo V (2011b). Compositional studies: antioxidant and antidiabetic activities of Capparis decidua (Forsk.) Edgew. Int. J. Mol. Sci. 12(12):8846-8861.

Zia-Ul-Haq M, Iqbal S, Ahmad M (2008). Characteristics of oil from seeds of 4 mungbean (Vigna radiate L. wilczek) cultivars grown in Pakistan. J. Am. Oil Chem. Soc. 85:851-856.

Zia-Ul-Haq M, Iqbal S, Ahmad S, Imran M, Niaz A, Bhanger MI (2007b). Nutritional and compositional study of desi chickpea (Cicer arietinum L.) cultivars grown in Punjab, Pakistan. Food Chem. 105:1357-1363.

18

Length weight relationship and ponderal index of rainbow trout (*Oncorhynchus mykiss* W., 1792) from Dachigam stream in Kashmir

Tasaduq Hussain Shah, Masood Ul Hassan Balkhi, Oyas Ahmad Asimi and Imran Khan

Faculty of Fisheries, Sher-e-.Kashmir University of Agricultural Sciences and Technology of Kashmir, Rangil, Ganderbal, C/o Shuhama Campus, Alusteng, Srinagar, Kashmir, Jammu and Kashmir, 190006, India.

The rainbow trout (*Oncorhynchus mykiss*), introduced in Kashmir in 1912, has thrived well since then and is now established in almost all the cold water streams, lakes and rivers of the valley. This study was aimed to describe the length-weight relationship and ponderal index (condition factor) of the fish from Dachigam stream in Kashmir. The investigation was carried out between April and December, 2005 and the data were recorded from anaesthetized fish, which ranged from 110 to 488 mm in length and 20 to 1425 g in weight. The length-weight relationship was estimated as LogW = - 4.9216 + 2.9618LogL and the *b* value (2.9618) did not differ significantly from 3. The coefficient of correlation (r) for the length-weight relationship was estimated at 0.9968 which showed a high degree of positive correlation between the length and weight of the fish. The condition factor (K) value was estimated as 1.15 ± 0.013.

Key words: *Oncorhynchus mykiss*, rainbow trout, length weight relationship, ponderal index, Dachigam stream, Kashmir.

INTRODUCTION

Rainbow trout are members of the genus *Oncorhynchus*, which also includes Pacific salmon and members of the family Salmonidae, such as, Atlantic salmon, trout, char, graylings, white fish and several other groups. Rainbow trout are native to coldwater environments in the north temperate zones and are distributed from Southern California through Alaska, the Aleutians, and the Western Pacific areas of the Kamchatka Peninsula and Okhotska sea drainages. Rainbow trout are, thought of as freshwater fish, but in the eastern Pacific, seawater forms called steelhead trout exhibit an anadromous life history, meaning that they spend a part of their life in the ocean, but return to the lakes and rivers to spawn. Rainbow trout

have been widely transplanted around the world and are established in South America, Japan, China, Europe, parts of Africa, India, Pakistan, Australia and New Zealand. Trout and salmon are the only exotic game fishes introduced in India. According to Mitchell (1918), the first attempt at shipment of trout ova to the country was in the year 1900 when the Duke of Bedford sent trout ova as a present to the Maharaja of Kashmir. This whole consignment, however, perished on the way on account of heat. Subsequently, eyed eggs of brown trout were imported in 1900 as well as in 1901. This attempt was successful in yielding fish which were fortuitously distributed far and wide in the valley due to a heavy flood in 1903. Rainbow trout was introduced in Kashmir in 1913 when nearly 1000 alevins hatched out from a consignment of eyed ova presented by the Bristol Waterworks from their head works at Blagdon, England (Mitchell, 1918). Length-weight relationship and condition factor are extremely useful tools for understanding the biological changes in fish stocks (Le Cren, 1951; Bagenal and Tesch, 1978). For more efficient fishery management, the knowledge of growth in fish is of paramount importance.

The length-weight relationship is a very useful tool in fisheries assessment. It is usually easier to measure the length of a specimen than the weight, and weight can be predicted later on using the length-weight relationship. Furthermore, standing crop biomass can be estimated (Morey et al., 2003) and seasonal variations in fish growth can be tracked in this way (Richter et al., 2000). The length-weight relationship also helps in predicting the condition, reproductive history and life history of fish species (Nikolsky, 1963; Wooton, 1992) and in morphological comparison of species and populations (King, 1996). The ponderal index or condition factor is often associated with fitness, that is, a poor condition can manifest as a number of negative fitness consequences for the individual fish and fish populations. Somatic growth potential of fish can be reduced (Danzmann et al., 1988). Reproductive success can be reduced through a number of factors like lower fecundity, poor quality eggs and sperms (Kjesbu et al., 1991, 1992; Rakitin et al., 1999). Additionally, poor condition may also lower the chances of survival (Wilkins, 1967). Of the hill stream fishes reported from Kashmir, the rainbow trout (*Oncorhynchus mykiss*), being a transplanted exotic fish, has established very well; almost in every nook and corner of the valley. Rainbow trout supports an important coldwater fishery resource of this northernmost state of the country. Biological aspects of the indigenous schizothoracids have been studied in detail by a number of authors. However, not much information is available as far as rainbow trout is considered. This work determined the length-weight relationship and condition factor of rainbow trout (*O. mykiss*) from Dachigam stream of Kashmir.

MATERIALS AND METHODS

A total of 359 specimens of rainbow trout (*O. mykiss* W.) in the length range of 110 to 488 mm and weight range of 20 to 1425 g were caught from the Dachigam stream, Harwan area of Srinagar, Kashmir, India from April, 2005 to December, 2005. The total length of the fish was taken from the tip of the snout to the end of the caudal fin, measured to the nearest millimetre and the weight to the nearest gram. All the fish were anaesthetized using clove oil (50 mgl^{-1}) and then were measured in length and weight. After the recovery from the anaesthetic effect, the fish were released in the vicinity of the site from where they were caught. Care was taken to remove the moisture from the fish by blotting prior to the measurement of weight. The general formula adopted for the evaluation of the length-weight relationship was given by Le Cren

(1951) as:

$$W = aL^b$$

where, W is the weight of the fish specimen (g); L is the length of the fish specimen (mm) and a and b are constants. The formula is expressed logarithmically as:

$$Log\ W = Log\ a + b\ Log\ L$$

The coefficient of correlation (r) for length-weight relationship was calculated as per the method described by Snedecor and Cochran (1967) to show the degree and nature of the correlation between length and weight of the fishes.

Condition factor or ponderal index (K) meant for studying the degree of the well being of the fish was determined by using the formula of Hile (1936), which is:

$$K = [W/L^3] \times 10^5$$

The number 10^5 is a factor to bring ponderal index (condition factor) to near unity (Carlander, 1950). The study of condition, a standard practice in fisheries ecology, is based on the analysis of length-weight data and assumes that heavier fish of a given length are in better condition (Bolger and Connolly, 1989). A fish is said to be in better condition when the value of K is more than 1 and in worse condition than an average individual with the same length, when K value is less than 1. At the population level, the average K indicates whether a population is in better (K > 1) or worse (K < 1) condition than an average population.

RESULTS AND DISCUSSION

The equation obtained for the length-weight relationship in this study is given by LogW = - 4.9216 + 2.9618 LogL. The coefficient of correlation (r) for the length-weight relationship was estimated at 0.9968, indicating a high degree of positive correlation between the two parameters. The value of exponent b (2.9618) did not differ significantly from 3.0. The length-weight exponents (b) for most animals fall roughly around 3.0 (Siegfried, 1980; Uye, 1982; Hopcroft et al., 1998). An exponent above 3.0 indicated that the fish become wider or deeper as they grow, while an exponent below 3.0 indicates that the fish become more slender. The value of the condition factor (K) was estimated as 1.15 ± 0.013.

The nearness of the K value to 1.0 clearly indicated the suitability of the environment in the Dachigam stream for good growth of the fish. The results are in consonance with those of Kumar et al. (1979) where the length-weight relationship and ponderal index of brown trout catches from five Kashmir streams were studied and analyzed. According to the authors, the values of the exponents in the length-weight equations estimated for males and females indicated that females departed more from the cube law. The general relationship between LogW and LogL was LogW = -5.2844 + 3.14862 LogL. The value of the ponderal index (K) ranged between 1.19 and 1.31.

Reimers et al. (1955) studied the fisheries of some lakes in Mono County, California and reported the value of K factor for 61 rainbow trout ranging from 125 to 325 mm to be between 0.881 and 1.023. Rabe (1967) reported the value of condition factor (K) between 0.859 and 1.104 for rainbow trout in Alpine lakes. Zimmerman (1999) reported the length-weight relationship of rainbow trout from Portal lake as $W = 0.00004 \times L^{2.72}$ (r = 0.99). The author estimated the value of the condition factor as 1.09 ± 0.14. Maia and Valente (1999) worked on the brown trout populations in the river Lima and obtained *b* value between 2.92 and 3.00. The authors reported a high degree of correlation (r = 0.99) for the length-weight relationship.

The value of condition factor was estimated to range between 1.13 and 1.25. Arslan et al. (2004) studied the length-weight relationship of brown trout from Coruh Basin, Turkey and reported the *b* value for the fish to be 2.97. Kimmerer et al. (2005) opined that the length-weight relationships for a single species of fish may differ substantially from one study to the next. Reasons for these differences can generally be categorised as biological, procedural and statistical. The biological causes of different length-weight relationships consist of real differences in weight and length among data sets. The weight of fish varied as a result of feeding history and the allocation of energy to growth and reproduction, so weight at a given length may vary spatially (especially between regions) and temporally (particularly between seasons). Ahmet et al. (2005) studied the brown trout populations in Firniz stream of the river Ceyhan, Turkey and reported that the value of exponent *b* in the length-weight relationship of *Salmo trutta macrostigma* was 2.971 for females and 3.009 for males. The authors reported the mean condition factor of 1.521 ± 0.010 for the population.

REFERENCES

Ahmet A, Kara C, Murat H (2005). Age, growth and diet composition of the resident brown trout, *Salmo trutta macrostigma* Dumeril 1858, in Firniz stream of the river Ceyhan, Turkey. Turk. J. Vet. Anim. Sci. 29:285-295.

Arslan M, Yildirim A, Bektas S (2004). Length-weight relationship of brown trout, *Salmo trutta* L., inhabiting Kan stream, Coruh Basin, North-Eastern Turkey. Turk. J. Fish. Aquat. Sci. 4:45-48.

Bagenal TB, Tesch FW (1978). Age and growth. *In*. T. B. Bagenal (ed.) Methods for the assessment of fish production in freshwaters. Oxford, Blackwell Scientific Publication. pp. 101-136.

Bolger T, Connolly PL (1989). The selection of suitable indices for the measurement and analysis of fish condition. J. Fish. Biol. 34:171-182.

Carlander KD (1950). Handbook of freshwater fishery biology. William C. Brown. Dubuque, Iowa.

Danzmann, RG, Ferguson, MM, Allendorf FW (1988). Heterozygosity and components of fitness in rainbow trout. Biol. J. Linn. Soc. 33:285-304.

Hile R (1936). Age and growth of the cisco, *Leucichthys artedi* Le Sueur, in the lakes of the Northeastern Highlands, Wisconsin. Bull. Bur. Fish. U.S. 48(19):211-317.

Hopcroft RR, Roff JC Bouman HA (1998). Zooplankton growth rates: the larvaceans *Appendicularia, Fritillaria,* and *Oikopleura* in tropical waters. J. Plankton Res. 20:539-55.

Kimmerer W, Avent SR. Bollens SM (2005). Variability in Length-weight relationships used to estimate biomass of estuarine fish from survey data. Trans. Am. Fish. Soc. 134:481-495.

King RP (1996). Length-weight relationships of Nigerian freshwater fishes. *Fish Byte.* 19(4):53-58.

Kjesbu OS, Klungsoyr J, Kryvi H, Whitthames P, Greer R Walker M (1991). Fecundity, atresia and egg size of captive Atlantic cod (*Gadus morhua*) in relation to proximate body composition. Can. J. Fish Aquat. Sci. 48:2333-2343.

Kjesbu OS, Kryvi H, Sundby S, Solemdal P (1992). Buoyancy variations in eggs of Atlantic cod (*Gadus morhua*) in relation to chorion thickness and egg size: theory and observations. J. Fish Biol. 41:581-599.

Kumar K, Sehgal KL, Sunder S (1979). Length weight relationship and ponderal index of brown trout *Salmo trutta fario* (Linnaeus) catches in the streams of Kashmir. J. Inland Fish. Soc. India 11(1):156-61.

Le Cren ED (1951). The length-weight relationship and seasonal cycle in gonad weight and condition in the perch (*Perca fluviatilis*). J. Anim. Ecol. 20:201-219.

Maia CFQ, Valente ACN (1999). The brown trout *Salmo trutta* L. populations in the river Lima catchment. *Limnetica* 17:119-126.

Mitchell FJ (1918). How trout were introduced into Kashmir. J. Bombay Nat. Hist. Soc. 26(1):295-99.

Morey G, Moranta J, Massuti E, Grau A, Linde M, Riere F, Morales-Nin B (2003). Weight-length relationship of littoral to lower slope fishes from the Western Mediterranean. Fish. Res. 62:89-96.

Nikolsky GW (1963). The ecology of Fishes. Academic Press, London and New York. p. 352.

Rabe FW (1967). Rainbow trout in Alpine lakes. *Northwest. Sci.* 41(1):12-22.

Rakitin A, Ferguson, MM, Trippel EA (1999). Sperm competition and fertilization success in Atlantic cod (*Gadus morhua*): effect of sire size and condition factor on gamete quality. Can. J. Fish. Aquat. Sci. 56:2315-2323.

Reimers N, Maciolek JA. Pister EP (1955). Limnological study of the lakes in Convict Creek Basin, Mono County, California. U.S. Fish and Wildlife Service. Fisheries Bull. 103:437-503.

Richter HC, Luckstadt C, Focken U, Becker K (2000). An improved procedure to assess fish condition on the basis of length-weight relationships. Arch. Fish. Mar. Res. 48:255-264.

Siegfried CA (1980). Seasonal abundance and distribution of *Crangon franciscorum* and *Palaemon macrodactylus* (Decapoda, Carida) in the *San Francisco Bay-Delta*. Biol. Bull. 159:177-192.

Snedecor GW, Cochran W G (1967). Statistical Methods. Oxford and IBH Publishing Company. p. 435.

Uye S (1982). Length-weight relationship of important zooplankton from the Inland sea of Japan. J. Oceanogr. Soc. Japan 38:149-158.

Wilkins NP (1967). Starvation of the herring, *Clupea harengus* L.: survival and some gross biochemical changes. Comp. Biochem. Physiol. 23:503-518.

Wooton RS (1992). Fish Ecology. Printed in Great Britain by Thomson Litho Ltd. Scotland.

Zimmerman T (1999).Recreational Fishery Stock Assesment 1999 Final report of Portal Lake. Ministry of Environment, British Columbia.

19

An implementation of the hazard analysis and critical control points (HACCP) system in the cage culture of *Siniperca scherzei* in Zhelin Lake, China

Wei Song[1,4], Zhiqiang Wu[2], Nan Wu[3,4] and Lingbo Ma[1]

[1]Key Laboratory of East China Sea and Oceanic Fishery Resources Exploitation, Ministry of Agriculture, East China Sea Fisheries Research Institute, Chinese Academy of Fishery Science, 200090, Shanghai, China.
[2]College of Environmental Science and Engineering, Guilin University of Technology, 12 Jiangan Road, 541004, Guilin, China.
[3]College of Life Science, Huaiyin Normal University, 111 West Changjiang Road, 223300, Jiangsu, China.
[4]Jiangsu Engineering Laboratory for Breeding of Special Aquatic Organisms, Huaiyin Normal University, 111 West Changjiang Road, 223300, Jiangsu, China.

The quality and safety of aquatic products has always been the first priority in food quality control in China. The Hazard Analysis and Critical Control Points (HACCP) is recognized as an effective system designed to improve food safety. To promote the establishment and implementation of the HACCP system in the aquaculture industry in China, the paper analyzed potential hazards in the cage aquaculture of *Siniperca scherzei* and defined the critical control point. Meanwhile, the table of HACCP plan was established.

Key words: Hazard analysis and critical control points (HACCP), cage culture, *Siniperca scherzei*, quality.

INTRODUCTION

The Hazard Analysis and Critical Control Points (HACCP) system was introduced in the 1960s by NASA for the design and manufacture of food for space flights (Mayes, 1998; Panisello and Quantick, 2001; Song and Wu, 2002; Mul and Koenraadt, 2009). Subsequently, HACCP has been recognised internationally as a logical tool for the adaptation of traditional inspection methods to a modern, science-based, food safety system (Mayes, 1993; Maldonado et al., 2005; Sun and Ockerman, 2005). The application of HACCP in aquaculture aims to ensure the comprehensive monitoring and management of factors relevant to the operation of an aquaculture facility. These factors can include the environment surrounding the culture site, the water quality, the water body used for the

culture, and the feedstuff and pesticide used throughout the whole production cycle.

Zhelin Lake (308 km^2) is located on the upper reaches of the Xiu River in Jiangxi Province of China. It occupies the three counties of Yongxiu, Wuning, and Xiushui. The cage culture site investigated in this research is located along the upstream reaches of Zhelin Lake. This site is an environmentally friendly aquaculture system in Jiangxi province. *Siniperca scherzei* is the most popular of the various species (*S. scherzei, Lateolabrax japonicus, Silurus meridionalis*) that are raised there in more than 260 cages. *S. scherzeri* is considered to be valuable freshwater species that are widely cultured in China.

This paper further elaborates on the identification of the

risk factors and critical control points (CCPs) for the cage culture of *S. scherzei* in Zhelin Lake and presents suggestions for corrective actions. We have developed a checklist based on this information and on evaluative inputs by the manager of the cage culture site that can be used to control hazards that have a potential impact on the cage culture of *S. scherzei*, so as to promote the establishment and implementation of the HACCP system in the aquaculture industry in China.

HACCP STEPS

Based on the seven principles of the HACCP system, the following steps or principles provide a HACCP model for the cage culture of *S. scherzei*.

Hazard analysis

Design and construction of the base

Faulty design and construction of cages may present chemical and biological hazards at the *S. scherzei* culture site. These factors can degrade water quality by obstructing water exchange, and they can readily induce disease. Once disease has appeared, it can easily progress to an outbreak and cause great economic losses. According to the governing ordinance, the cages are of the net structure type. They are constructed of polyethylene. The mesh size of the net is 2 cm. The cages are 3 × 4 × 4 m in size.

Open frame floating cages are adopted because this design is easy to manage. However, this national design standard does not regulate the layout of the cages. The high density of the cages at the site may produce decreased water exchange and may therefore induce cross-infection with disease as well as the accumulation of residual amounts of drugs. The process is identified as a critical control point.

Water quality

The water quality associated with the culture system directly affects the safety of the aquaculture product (Hamblin and Gale, 2002). The potential hazards resulting from poor water quality include chemical and biological pollution. The primary sources of chemical pollution are near the cage culture site. Runoff from the nearby farms may contain agrochemicals, insecticides and heavy metals. Moreover, these harmful chemical contaminants can migrate through the food chain, accumulate in the body of *S. scherzei*, and subsequently threaten the health of the human consumers. The potential biological hazards to the culture system are mainly associated with pathogens and parasites. Pathogens infect *S. scherzei* and cause disease. Some

pathogens can also infect humans. For example, during the initial aquaculture period, the pathogens of Chilodenella and Trichodina can infect *S. scherzei*, which can also infect humans. Owing to the low exchange rate of the water, any remaining excess feed in the system will deteriorate rapidly and may trigger infections by pathogens. Moreover, waterfowl may carry pathogens and may therefore represent a potential source of biological hazards. The process is identified as a critical control point.

Supply of fry and domestication

The fish fry are obtained from wild-caught fish. Before the wild fish fry placed in the culture cages, they are put in a small tank for acclimatization for approximately a month. They are then housed in standard cages for feeding. In this process, the culturist does not make strict demands on the quality of fish fry. The fish fry may carry pathogens and parasites. Biological pollution is the main hazard. The process is identified as a critical control point.

Supply of feedstuffs

Biological and chemical pollution are the main causes of the hazards that generate the need for proper attention and control. The origin of these hazards is that, the feed used for *S. scherzei* in the cages consist of bait (small dead fish and shrimp). The bait may be contaminated with a variety of pathogens and parasites that if not treated, can induce outbreaks of diseases that are particularly high during May to September. Therefore, the process is a critical control point.

Disinfection of cages

Prior to stocking and during the breeding process, disinfection of cages must be performed in a way that does not introduce hazards to the culture. The culture cages are treated by using bleaching powder twice a month. In addition, allitridi (at 100 g allitridi per 200 small fish) is added to the cages at a rate of 10 times per year. Chemical compounds or disinfectants that are used in compliance with good aquaculture practises will not produce chemical contamination in the water and will therefore not harm the consumer (Song et al., 2011). Hence, the proper use of the abovementioned disinfectant during the breeding process will not harm the culture. The process is not a critical control point.

Use of drugs to prevent and cure disease

In this aspect of *S. scherzei* aquaculture, the drugs used for disease treatment represent the main hazard that requires control. The primary sources of this hazard are that, the culturist lacks knowledge of the use of drug and that antibiotics may be abused in the culture process. The antibiotics then accumulate inside the body of *S. scherzei*. This situation may lead to the appearance of drug-resistant pathogens. The human body is susceptible to the pathopoietic functions that these pathogens induce. For example, chloromycetin is used to treat human and fish diseases. But if this antibiotic reaches humans through the food chain, it may induce the appearance of drug-resistant pathogens. The use of this drug in fish culture may therefore be potentially harmful to human health. Furthermore, malachite green and methylene blue can prevent and cure wheel verminosis and water mildew if these compounds are used properly. However, malachite green is highly toxic and is carcinogenic to humans. The process is identified as a critical control point.

Daily management regime

The potential hazards in the culture system are pathogenic fungi and helminthes. These potentially harmful organisms need to be controlled through good management practises. The source of this hazard is that, small fishes used as feed are not disinfected before they are introduced to the cage culture in Zhelin Lake. If they are contaminated, these small fishes may cause pathogen infection in *S. scherzei*. Rapid growth of *S. scherzei* is observed from June to September when the temperature is high. Many diseases and pathogens are particularly likely to thrive when the temperature is high and when the fish are growing rapidly. Poor management of aquaculture and of poultry farms may cause disease and death to *S. scherzei* by introducing potentially harmful organisms to the cage culture. The process is identified as a critical control point.

Transportation

The harvest is transported to Qingdao and sold to middlemen. The product is subsequently exported to South Korea. Hazards that may arise during transport are physical injury and the growth of pathogenic fungi. However, these hazards are not CCPs because good practise can result in effective control. Professional carrier vehicles are used to transport aquatic products, and the workers possess good technical skill.

Establishment of critical control points (CCPs)

This study has identified the CCPs for the *S. scherzei* cage culture (Table 1). The study conducted a hazard analysis and investigated cage dimensions, stocking densities, the environment surrounding the culture site, supply of fish fry, use of drugs to prevent and cure

disease and the supply of feedstuffs. These investigations indicated that, design and construction of the base, water quality, supply of fry and domestication, supply of feedstuffs, use of drugs to prevent and cure disease, and daily management regime were the CCPs associated with the process of culturing *S. scherzei*.

Establishment of critical limits for each CCP

In accordance with relevant regulations, applicable criteria and technical literature and practical experience, critical limits were defined in a way that would result in effective control of the hazards. The critical limits can be defined by a limitation index for product security or by the implementation of control factors in the production process to ensure product safety. The critical limits for the cage aquaculture of *S. scherzei* were established based on the regulations governing aquatic products established by China and by South Korea (Table 2).

Implementation of a monitoring system

Monitoring measures, which focus on the CCPs are one of the most important components of the HACCP plan for *S. scherzei* cage aquaculture. The monitoring system of the cage aquaculture of *S. scherzei*, including monitoring method and frequency, are established, and the personnel responsible are identified on the HACCP worksheet (Table 2). Effective monitoring measures can be adopted to determine whether the situation at a critical control point is out of control or deviates from the critical limits.

Establishment of corrective actions

When the monitoring indicates that, a particular CCP is not under control, an established corrective action must be taken. An overview of possible corrective actions is provided in Table 2.

Establishment of effective record keeping procedures

Record keeping is one important aspect of the implementation of the HACCP system. The preparation and the implementation of the HACCP plan should be recorded completely, and the records should include all the amendments and rectifications that have been adopted. All records should be kept for two years in document form, and they should be made available for evaluation and checking by the relevant monitoring department of the government (Table 2).

Establishment of procedures for verification

The technician feeding *S. scherzei* in the breeding

Table 1. Hazard analysis of cage culture of *S. scherzeri*.

The processing step	Potential hazards	Significant hazard, or not	Primary causes of the hazard	Preventive measures	CCP status of this processing step
Design and construction of the site	Biological: potential pathogenic microbes and parasites Chemical: heavy metals and drug residues	Significant hazard	The national design standards do not regulate the layout of the cage, However, density of the cages in the site is high	The choice and construction of cages should meet the state standard	CCP
Water quality	Biological: potential pathogenic microbes and parasites Chemical: heavy metals and drug residues	Significant hazard	The water quality will be degraded. Pathogenic organisms will be bred. Pesticides, fishery drugs, and heavy metals are introduced from the adjacent pollution sources and by the cultivation processes. These chemical contaminants may cause the appearance of resistant pathogens	Choose a satisfactory water source, purify and disinfect the water, clean up the feed-borne trematodes, intermediate hosts and pathogens	CCP
The supply of fish fry and domestication	Biological: potential pathogenic microbes and parasites	Significant hazard	The fish fry and bait are not effectively disinfected. The cultivation density is overly high	Adopt strict disinfection measures for the fish fry and bait, use moderate cultivation densities	CCP
Feed supply	Biological: potential pathogenic microbes and parasites Chemical: heavy metals	Significant hazard	The bait might be polluted by heavy metals. It may introduce massive infestations of pathogenic bacteria or parasites. The bait might be not fresh during the hot season	Apply strict disinfection procedures for the fish bait and supply fresh fish bait	CCP
Disinfection of cages	Chemical: chemical residues	Not significant	The proper use of the abovementioned disinfectant during the breeding process will not harm the culture		Not
Use of drugs for prevention and cure of disease	Chemical: drug residues	Significant hazard	The abuse of the fish drug	Purchase and use the medication permitted by national or international standards. Do not use illicit drugs, such as malachite green and the like. Read the instructions carefully before applying the treatment	CCP
Daily management regime	Biological: potential pathogenic microbes and parasites Chemical: drug residues	Significant hazard	Unscientific management of cultivation	Use appropriate timing and quantitative feeding. Remove all surplus bait in a timely manner. Isolate fish showing signs of illness in time to prevent an outbreak	CCP
Transportation	Physical: abrasion	Not significant	Professional carrier vehicles are used to transport aquatic products, and the workers possess good technical skill		Not

Table 2. Identification and monitoring of important process phases (ranked as CCPs) in the cage culture of *S. scherzei*.

Critical control points (CCPs)	Significant hazard	Specifications/critical limits	Monitoring method and frequency	Corrective action	Recordkeeping	Proof process
Design and construction of the site	Biological: potential pathogenic microbes and parasites Chemical: heavy metals and drug residues	NY/T5167-2002 (environmentally friendly food, the technological norm of the breeding process for mandarin fish)	After the site has been built, checking whether the design and construction are unreasonable	Rebuilding the site	Drawing(s) of the site made for planning purposes	Checking the plan drawing of the site and investigating the construction conditions present at the site
Water quality	Biological: potential pathogenic microbes and parasites Chemical: heavy metals and drug residues	NY5051-2001 (environmentally friendly food, water quality for fresh water aquaculture)	Laboratory tests of water quality every month	Improving the protocol used for water quality analysis protocol, elimination of helminths, cleaning of cages to prevent blinding sieve	Results of the tests used to assess the water quality of the samples	Check the water quality test results, sample to verify the test results
The supply of fish fry and domestication	Biological: potential pathogenic microbes and parasites	NY5071-2001 (environmentally friendly criteria for the use of drugs in food fish) NY5073-2006; (environmentally friendly food for ensuring restricted level of poisonous and harmful materials in aquatic products); Items and criteria for the inspection and quarantine for the entry and exit of aquatic animals for South Korea	Microbiological laboratory analyses every month	Discarding poor-quality fry, selection of high-quality fry	Test results from fry sampling	Check the test results, sample to verify the test results
Feed supply	Biological: potential pathogenic microbes and parasites Chemical: heavy metals	NY5073-2006 (environmentally friendly food for ensuring restricted levels of poisonous and harmful materials in aquatic products)	Microbiological and chemical laboratory analyses every month	Selection of high-quality bait if sampled fry is found to be of poor quality and is rejected	Test results from bait sampling	Check the results of the tests made on the bait, sample to verify the test results
Use of drugs for prevention and cure of disease	Chemical: drug residues	NY5071-2001 (environmentally friendly criteria for use of fish drugs); NY5073-2006 (environmentally friendly food for ensuring restricted levels of poisonous and harmful materials in aquatic products); NY5070-2002 (Restricted levels of fish medicine remain in aquatic product); Items and criteria for inspection and quarantine for the entry and exit of aquatic animals for South Korea	Chemical laboratory analyses every month	Purchasing drugs that comply with national or international standards and using dosage strictly as recommended, and confirming decontamination after use of the drug	Test results from fish sampling	Check the results of the tests made on fish, sample to verify the tests results
Daily management regime	Biological: potential pathogenic microbes and parasites Chemical: heavy metals	NY/T5167-2002 (environmentally friendly food, the technological norm of the breeding process for mandarin fish)	Checking cages regularly every day	Adopting a standard procedure for feed management	The records obtained from the procedures used for checking cages	Checking the cage records and reviewing the management operations

facility generates an appropriate summary of the breeding technique. The purpose of this report is to confirm that, each CCP has been controlled, that the situation at the critical point is sufficient to ensure food safety in the process of breeding *S. scherzei* and that, the HACCP system is functioning effectively. Validation of the result should reveal whether or not the measure adopted for the purpose is correct (Table 2).

Conclusion

There is interest worldwide in the implementation of the HACCP system by the food industry, especially for high-risk foods such as meat, poultry or fishery products (Lupin et al., 2010; Wang et al., 2010). The HACCP system is based on a scientific, systematic, rational, multi-disciplinary, and cost-effective approach to the control of safety problems (Song et al., 2011). The countrywide establishment of HACCP and of a traceability system in China has gradually enhanced the security of aquatic products across the whole production chain from aquaculture, processing, and distribution to consumption. The safety and quality of aquatic products should be ensured beginning at the headstream.

This study applied the HACCP system to the cage aquaculture of *S. scherzei*. Chemical and biological contaminants were two of the main hazards identified by the analysis. A variety of factors could lead to pollution by these agents. The chemicals of concern consisted of antibiotics, hormones and minerals. Antibiotic use is of concern because the antibiotic may affect non targeted species and thus, produce antibiotic resistance and other toxic effects (Cole et al., 2009). Microbiological contaminants are of concern because diseases and parasites can be deleterious. Parasites may serve as a vector for other lethal diseases, such as infectious salmon aenemia (ISA).

ISA has been detected in farmed and wild fish (Cole et al., 2009). The CCP decision tree in the HACCP system was activated and executed. These computations identified six CCPs. The CCPs identified by the decision tree were design and construction of the site, water quality, supply of fish fry and domestication, feed supply, use of drugs for prevention and cure of disease and culture system. Corrective actions were immediately taken at each CCP. Implementation of the HACCP system helped to improve the chemical and biological quality of the cage aquaculture of *S. scherzei*. However, it should be pointed out that, the success of the system does not depend exclusively on the fact that, the HACCP process produced promising results regarding chemical and microbiological contaminants. The significance of the results is that, they help to indicate whether or not the HACCP system is working effectively.

ACKNOWLEDGEMENTS

Special thanks are due to the manager Chuan Yong Li for the two-year secondment that permitted the completion of this work. This research was supported by the opening foundation of the Jiangsu Engineering Laboratory for Characteristic Aquatic Species Breeding, grant NO. CASB1301.

REFERENCES

Cole DW, Cole R, Gaydos SJ (2009) Aquaculture: Environmental, toxicological, and health issues. Int. J. Hyg. Environ. Heal 212(4):369-377.

Hamblin PF, Gale P (2002) Water quality modeling of caged aquaculture impacts in lake wolsey, north channel of lake huron. J. Great Lakes Res. 28(1):32-43.

Lupin HM, Parin MA, Zugarramurdi A (2010). HACCP economics in fish processing plants. Food Control. 21(8):1143-1149.

Maldonado ES, Henson SJ, Caswell JA (2005) Cost-benefit analysis of HACCP implementation in the Mexican meat industry. Food Control. 16(4):375-381.

Mayes T (1993). The application of management systems to food safety and quality. Trends Food Sci. Tech. 4(7):216-219.

Mayes T (1998). Risk analysis in HACCP: Burden or benefit? Food Control 9(2-3):171-176.

Mul M, Koenraadt C (2009). Preventing introduction and spread of *Dermanyssus gallinae* in poultry facilities using the HACCP method. Exp. Appl. Acarol. 48(1):167-181.

Panisello PJ, Quantick PC (2001) Technical barriers to Hazard Analysis Critical Control Point (HACCP). Food Control 12(3):165-173.

Song W, Wu ZQ (2002) Application example of HACCP in aquaculture in China. J. Nanchang Univer. (Chinese) 26:117-118.

Song W, Ma CY, Ma LB (2011) Application and development of HACCP in aquaculture. Chinese Fishery Quality and Standards (Chinese) 1(3):73-79.

Sun YM, Ockerman HW (2005) A review of the needs and current applications of hazard analysis and critical control point (HACCP) system in foodservice areas. Food Control 16(4):325-332.

Wang D, Wu H, Hu X, Yang M, Yao P, Ying C, Hao L, Liu L (2010). Application of hazard analysis critical control points (HACCP) system to vacuum-packed sauced pork in Chinese food corporations. Food Control. 21(4):584-59.

20

Training needs of the freshwater fish growers in Assam, India

Uttam Kumar Baruah, Jyotish Barman, Hitu Choudhury and Popiha Bordoloi

Krishi Vigyan Kendra Goalpara, National Research Centre on Pig, Indian Council of Agricultural Research, Dudhnoi – 783124, Assam, India.

A study was conducted in five rural development blocks of India to investigate the training needs of freshwater fish growers. A total of fifty fish growers having training exposures were interviewed using a structured questionnaire. Independent variables included thirteen socio-economic parameters and dependant variables included seven critical technical areas. Frequency and percentage, mean score, standard deviation (SD), co-efficient of variance (CV) and simple correlation were analyzed. The study revealed that majority of the respondents need trainings on water quality management (80%). Only 16 and 10% respondents expressed training need on fish seed handling and transportation, and fish nutrition and feeding, respectively. Negatively significant correlations were observed between interest and attitude towards fish farming age and education. Individual independent characteristic and training needs of the farmers had negative correlations with education (X_2) $(p<0.01)$ and attitude (X_{13}) $(p<0.05)$. Positively significant correlation between possession of pond (X_5) and age (X_1) $(p<0.01)$, income (X_4) and age (X_1) $(p<0.01)$, economic motivation (X_7) and age (X_1) $(p<0.01)$, decision-making ability (X_8) and age (X_1) $(p<0.05)$, attitude (X_{13}) and education (X_2) $(p<0.05)$, income (X_4) and main occupation $(X_3)(p<0.05)$. The study concluded that before assessing the training needs, the fish growers of the district should be made aware of the latest technologies in aquaculture.

Key words: Assessment, training needs, freshwater fish growers.

INTRODUCTION

The district has a population of 8,22,306 with a density of 451 km^{-2} (Anon, 2009). Almost all the people consume fish in their daily diet. The district has 0.039 million ha water spread area, comprising 0.038 million ha (97.44%) of lentic and 0.001 million ha (2.56%) of lotic waters. This amounts to 10.43% of state's fisheries resources and 0.53% of the inland fisheries resources of the country (ARDB, 2011). Most farm families have water bodies in their homesteads in the form of seasonal ponds, sumps, roadside ditches, etc (NRCP, 2007). From all these resources, the district produces 5,347 tonnes of fish

against the biophysical potential of 40,969.25 tonnes per annum (Anon, 2009). The present demand for fish in the district is 9,045.37 tonnes per annum, considering the per capita requirement of fish at 11 kg/person/year as recommended by the World Health Organisation (WHO) for the country. There are number of proven package of practices for fish farming (AAU, 1997), which are also suitable for the agro-ecological situations of the district (Mandal et al., 1981).

Fishes reared in the district under composite culture of carps are *Catla catla* (Catla), *Cirrhinus mrigala* (Mrigal),

Labeo rohita (Rohu), *Hypopthalmichthys molitrix* (Silver carp), *Ctenopharyngodon idella* (Grass carp) and *Cyprinus carpio* (Common carp). All these species need s lightly alkaline water and pH ranging between 6.5 and 8.0. The soil and water of the district are acidic in reaction, which is the major limiting factor for development of carp aquaculture. As per the recommended package of practices, 2,100 kg/ha of agricultural lime ($CaCO_3$) is needed per year to check acidity. The lime is applied in split doses. Most of them operate aquaculture in old ponds, where production is limited by anaerobic condition due to non removal of silt from the bottom (88%). Ponds (80%) are generally well impounded and do not allow entry of wild water. However, 80% farmers do not take erosion control measures. No farmer applies fertilizers in their ponds regularly. They stock their ponds arbitrarily and do not follow any norms in terms of species composition, fingerlings' size and even stocking period and time. Since, carps require water temperature above 28°C and optimum temperature prevails during the period from mid April to mid September, it is recommended that ponds should be stocked during April to May with stunted yearlings at a density of 5000 numbers/ha preferably in the morning hours. Recommended species composition includes silver carp (20%), catla (15%), rohu (15%), grass carp (10%), mrigal (20%) and common carp (20%). No farmer follows this practice. Carps require concentrated feed at the rate of 1% of the body weight/day for maintenance (Paulraj, 1997). For maximum growth at a declining the requirement of concentrated feed increases to 3 to 4% of the body weight/day and for the highest growth quantity of feed required is equivalent to 6 to 7% of the body weight/day. The recommended practice for feeding is 3 to 5% of the body weight/day with 1:1 mixture of rice bran and mustard oil cake or with formulated feed as per manufacturer's prescription. No farmer supplies sufficient feed in their fishponds. There is a wide gap between recommended practices and farmers practice, which has resulted in lower yield (SREP, 2006). This situation underlines the needs of appropriate measures for building farmers' capacity for scaling up aquaculture to increase fish production and to improve their livelihood.

METHODOLOGY

This study was conducted in Assam, India. Five rural development blocks having population of different tribes were purposively selected for the study. A multi-stage sampling design was used to select the respondents using random sampling technique. A total of fifty (50) farmers having training exposure were selected purposefully. The survey was conducted over a period of nine months during 2011. Data were collected using a semi structured questionnaire. Thirteen (13) independent variables were selected to investigate the socio-economic characteristics of the farmers. The independent variables were age (X_1), education (X_2), main occupation (X_3), annual income (X_4), possession of pond area (X_5), localiteness-cosmopoliteness (X_6), economic motivation (X_7),

decision making ability (X_8), scientific orientation (X_9), interest (X_{10}), information seeking behaviour (X_{11}), knowledge (X_{12}) and attitude (X_{13}). A total of seven critical technical areas of training, namely; (i) integrated livestock and fish farming, (ii) water quality management, (iii) fish nutrition and feeding, (iv) integrated rice-fish farming, (v) seed handling and transportation, (vi) pen culture in open water, and (vii) fish disease diagnosis and prevention, were selected based on the observations recorded during the number of village transects made across the district before this study. The test schedules were developed to determine the training needs of the farmers. Weight of the technical areas of the training was decided by judge's rating and extent of need was measured as most essential (ME), essential (E), slightly essential (SE) and not essential (NE) with assigned scores of 3, 2, 1 and 0, respectively. Final scores were attained by multiplying the weights of an area with the corresponding extent of need score. Various descriptive and inferential statistical methods were employed to analyze the data following Panse and Sukhatme (1985).

The main statistical techniques and tools used were (1) frequency and percentage analysis, (2) mean score, (3) standard deviation (SD), (4) co-efficient of variance (CV), and (5) simple correlation.

RESULTS

Findings on socio-economic characteristics are summarized in Table 1. Majority (76%) of the respondents belonged to middle age category (29 to 58 years) followed by old (>58 years) which is 22% and young (<29 years) which is 2%. Most of the farmers (92%) belonged to high category of educational status, that is, above high school standard. Only 8% of the respondents belonged to medium education level, that is, between primary and high school standard. Only 18% of the respondents were fully engaged in fish farming. Others took fish farming as subsidiary occupation, while 48% of the respondents had agriculture as major occupation, and 30% had other business. Negligible section of respondents (4%) had government job as major occupation with fish farming as subsidiary occupation. Data on annual income revealed that 56% of the respondents had middle level of annual income (INR 150000 to 250000) followed by high category (24%) and low category (20%) with annual income of more than INR 250000 and less than INR 150000, respectively. Amongst the respondents, 54% had medium level possession of ponds (0.5 to 1.00 ha) followed by 38% low level of possession of ponds (<0.5 ha) and 4% high (> 1.00 ha). All the respondents belonged to medium level (52%) and low level (48%) of localiteness-cosmopoliteness. Economic motivation of the respondents showed that 90% were in medium level and 10% were in low level category. None of the respondents had high level decision-making ability. While majority (84%) of the respondents was in medium level, 16% were in low level in terms of decision making ability. Analysis of data on scientific orientation revealed mean score of 22.420, standard deviation (SD) of 1.907 and co-efficient of variation (CV) of 8.51% indicating quite homogeneity amongst the respondents, while 68% of the respondents

Table 1. Socio-economic characteristics of the fish growers.

Variable	Frequency	Percentage	Mean	SD	CV
Age (X1)					
Young (up to 28)	38	76			
Middle aged (29-58)	11	22	2.2	0.452	20.53
Old (above 58)	1	2			
Education (X2)					
Low	-	-			
Medium	4	8	2.92	0.274	9.39
High	46	92			
Main occupation (X3)					
Aquaculture	9	18			
Agriculture	24	48			
Business	15	30	2.2	0.782	35.57
Govt. service	2	4			
Annual income (X4)					
Low (< 1.50 Lakh)	10	20			
Medium (1.50 - 2.50 Lakh)	28	56	2.04	0.669	32.79
High > 2.50 Lakh)	12	24			
Pond area (X5)					
Low	19	38			
Medium	27	54	1.7	0.614	36.14
High	4	8			
Localiteness-cosmopoliteness (X6)					
Low	24	48			
Medium	26	52	12.52	2.493	19.91
High	-	-			
Economic motivation (X7)					
Low	-	-			
Medium	45	90	21.8	2.491	11.43
High	5	10			
Decision making ability (X8)					
Low	8	16			
Medium	42	84	22.22	3.61	16.25
High	-	-			
Scientific orientation (X9)					
Low	-	-			
Medium	3	6	22.42	1.907	8.51
High	47	94			
Interest (X10)					
Low	-	-			
Medium	16	32	14.74	1.736	11.78
High	34	68			
Information seeking behaviour (X11)					
Low	38	76	11.4	1.948	17.09

Table 1. Contd.

Medium	12	24			
High	-	-			
Knowledge (X12)					
Low	-	-			
Medium	17	34	17.98	1.744	9.7
High	33	66			
Attitude (X13)					
Less favourable	12	24			
Favourable	38	76	26	3.264	12.27
More favourable	-	-			

Table 2. Frequency and percentage distribution of respondents in different response categories against water quality management (N = 50).

Area of training	Distribution of respondents				Mean	SD
	ME (3)	E (2)	SE (1)	NE (0)		
Integrated livestock and fish farming	0 (0.00)	2 (4.00)	4 (8.00)	44 (88.00)	0.160	0.46773
Water quality management	40 (80.00)	4 (8.00)	3 (6.00)	3 (6.00)	2.620	0.85452
Fish nutrition and feeding	5 (10.00)	3 (6.00)	2 (4.00)	40 (80.00)	0.460	0.99406
Integrated rice-fish farming	2 (4.00)	6 (12.00)	5 (10.00)	37 (74.00)	0.460	0.86213
Seed handling and transportation	8 (16.00)	12 (24.00)	10 (20.00)	20 (40.00)	1.160	1.13137
Pen culture	0 (0.00)	0 (0.00)	0 (0.00)	50 (100.00)	0.000	0.00000
Fish disease diagnosis and prevention	0 (0.00)	0 (0.00)	0 (0.00)	50 (100.00)	0.000	0.00000

Data in parentheses are percentage of frequencies.

exhibited a medium level and 32% respondents exhibited high level in respect of interest. None of the respondents exhibited high level of information seeking behaviour. The study revealed that 66% of the respondents exhibited high level of knowledge on the existing practice followed by 34% in the medium level. As high as 94% respondents had favourable attitude, while 6% of the respondents were found to be more favourable.

The study revealed that the farmers in the district operate aquaculture in an easy going manner and they lack the entrepreneurial approach. Training needs as expressed by the farmers are summarized in Table 2. The farmers of Assam normally keep livestock at their homestead. Since feed is the major input in fish aquaculture and it can be replaced by recycling livestock wastes, they were asked whether training on integrated livestock fish farming is essential for them. Most of the farmers (88%) opined that such training is not essential for them. The farmers (80%) were aware of the effect of water quality on fish production and opted for training on water quality management. However, 80% farmers were not cautious about fish nutrition and feed. Only 10% farmers felt it most essential. Agriculture, especially rice cultivation is the major farm operation in the state. The rice ecosystem offers tremendous scope for integration of fish culture in rice field. The farmers (74%), however, did not express concurrence to the need; only 2% farmers felt the requirement of training in this area. Farmers were not aware of the impact of mishandling and wrong transportation of fish seed on growth of fish in grow out ponds. Only 16% farmers felt the training on fish seed handling and transportation most essential, while 24% felt essential, 20% felt slightly essential and 40% felt not essential. There is a vast area of non-impounded water bodies suitable for fish farming using pen or enclosure. The farmers were asked about training requirement on pen culture. No requirement of training on fish disease diagnosis and prevention was also opined by the farmers.

Analysis of simple correlations amongst the independent variables was done and results are presented in Tables 3 and 4. Table 4 revealed positively significant correlation between possession of pond (X_5) and age (X_1) (r=0.3676, p<0.01), income (X_4) and age (X_1) (r=0.3782, p<0.01), economic motivation (X_7) and age (X_1) (r=0.3990, p<0.01), decision-making ability (X_8) and age (X_1) (r=0.3479, p<0.05), attitude (X_{13}) and education (X_2) (r=0.2829, p<0.05), income (X_4) and main occupation (X_3) (r=0.3354, p<0.05). Table 3 revealed positively significant correlations between economic motivation and income (X_4) (r=0.6909, p<0.01), decision-making ability (X_8) and income (X_4) (r=0.6018, p<0.01), information seeking behaviour (X_{11}) and income

Table 3. Correlations amongst the independent variables.

Variable	Age (X₁)	Education (X₂)	Main occupation (X₃)	Annual income (X₄)	Pond size/area (X₅)	Localiteness-cosmopoliteness (X₆)	Economic motivation (X₇)	Decision-making ability (X₈)	Scientific orientation (X₉)	Interest (X₁₀)	Information seeking behaviour (X₁₁)	Knowledge (X₁₂)	Attitude (X₁₃)
Age (X₁)	1	-0.5275**	-0.0577	0.3782**	0.3676**	0.1051	0.3990**	0.3479*	-0.0284	0.1718	0.0232	0.057	-0.1523
Education (X₂)	-0.5275**	1	0.0761	-0.0935	-0.1454	-0.0275	-0.0837	-0.1881	-0.0125	-0.1304	0.0994	-0.0888	0.2829*
Main occupation (X₃)	-0.0577	0.0761	1	0.3354*	-0.1273	-0.1695	0.0838	0.0058	-0.0711	-0.2314	0.174	0.1675	0.1598
Annual Income (X₄)	0.3782**	-0.0935	0.3354*	1	0.2284	-0.0984	0.6909**	0.6048**	0.0506	0.1498	0.4573*	0.2107	-0.0299
Pond size/area (X₅)	0.3676**	-0.1454	-0.1273	0.2284	1	0.2105	0.2	0.1316	0.0749	0.0976	0.1193	0.1467	0.0204
Localiteness-cosmopoliteness (X₆)	0.1051	-0.0275	-0.1695	-0.0984	0.2105	1	-0.0947	-0.1853	0.2279	0.013	-0.2706	-0.129	-0.1043
Economic motivation (X₇)	0.3990**	-0.0837	0.0838	0.6909**	0.2	-0.0947	1	0.5769**	-0.0894	0.1907	0.1977	0.1212	-0.2159
Decision-making ability (X₈)	0.3479*	-0.1881	0.0058	0.6048**	0.1316	-0.1853	0.5769**	1	-0.1175	0.2699	0.4341**	0.1174	-0.2401
Scientific orientation (X₉)	-0.0284	-0.0125	-0.0711	0.0506	0.0749	0.2279	-0.0894	-0.1175	1	-0.0835	-0.1945	-0.0956	0.1653
Interest (X₁₀)	0.1718	-0.1304	-0.2314	0.1498	0.0976	0.013	0.1907	0.2699	-0.0835	1	0.0072	0.0589	-0.3393*
Information seeking behaviour (X₁₁)	0.0232	0.0994	0.174	0.4573*	0.1193	-0.2706	0.1977	0.4341**	-0.1945	0.0072	1	0.3268*	0.0321
Knowledge (X₁₂)	0.057	-0.0888	0.1675	0.2107	0.1467	-0.129	0.1212	0.1174	-0.0956	0.0589	0.3268*	1	0.1743
Attitude (X₁₃)	-0.1523	0.2829*	0.1598	-0.0299	0.0204	-0.1043	-0.2159	-0.2401	0.1653	-0.3393*	0.0321	0.1743	1

**Correlation is significant at the 0.01 level (2-tailed). *Correlation is significant at the 0.05 level (2-tailed).

(X₄) (r=0.4573, p<0.01), economic motivation (X₇) and decision-making ability (X₈) (r=0.5769, p<0.01), knowledge (X₁₂) and information seeking behaviour (X₁₁) (r=0.3268, p<0.05).

Negatively significant correlations were observed between interest and attitude towards fish farming (r=(-) 0.3393, p<0.05), age and education (r=(-) 0.5275, p<0.01). Individual independent characteristic and training needs of the farmers had negative correlations with education (X₂) at p<0.01 level (r=(-) 0.4247) and attitude (X₁₃) at p<0.05 level (r=(-) 0.2863). Table 4 revealed the details of correlation amongst the independent variables and pond management practices.

DISCUSSION

Training is a planned process to modify attitude, knowledge or skill behavior through a learning experience to achieve effective performance in an activity or range of activities and education is an activity which aim at developing the knowledge, skills, and moral values (Smith, 1992). Both training and education are essential for building capacity in man which is the key for human resource development (HRD). UNDP (1995) defined capacity as the ability of actors (individuals, groups, organisations, institutions, and countries) to perform specified functions (or pursue specified objectives) effectively, efficiently and sustainably. Capacity building is the efforts made by the actors themselves to achieve or strengthen their ability to perform the functions in question. Capacity can be built through a wide range of activities undertaken by the various actor involved, and is not a restrictive or prescriptive term. It may involve activities the actors organise themselves, or activities organised by others in which they participate. HRD is defined as 'the process of increasing the capacity of the human resources through development. It is thus a process of adding value to individuals, teams or an organization as a human system' (McGlagan, 1989). Training need analysis refers to the learning needs of individuals to enable them to reach the required standard of performance in their current or future jobs. Wilson (1999) suggested that the conventional and simpler methods such as interviews, questionnaires, observations, and focus groups to gather information for HRD needs analysis. Skill and training needs are a special category of needs that can be identified through a formal training needs analysis or a skills audit, or through more informal means. Training needs analyses and skills audits are usually done within formal organisations, corporations or highly structured workplaces that have clearly defined tasks and outputs. Wilson (1999) described that they are not easy to apply to community- based organisations and voluntary groups involved in integrated

Table 4. Correlations amongst independent variables and training needs.

Variable	Training need
Age (X_1)	0.2781
Education (X_2)	-0.4247**
Main occupation (X_3)	-0.0930
Income (X_4)	0.0540
Operational holding (X_5)	0.0879
Localiteness-cosmopoliteness (X_6)	0.1885
Economic motivation (X_7)	0.0381
Decision-making ability (X_8)	0.2215
Scientific orientation (X_9)	0.1095
Interest (X_{10})	-0.0306
Information seeking behaviour (X_{11})	-0.1369
Knowledge (X_{12})	-0.0325
Attitude (X_{13})	-0.2863*

**Correlation is significant at the 0.01 level (2-tailed). *Correlation is significant at the 0.05 level (2-tailed).

natural resource management (INRM) planning where there is considerable flexibility in people's roles and contributions, and where the details of groups and tasks may vary considerably from region to region and community to community and therefore the method suggested by Wilson (1999) was followed.

Scientific and technological revolution (STR) is characterized by deep interconnection and interaction of processes and fundamental changes in all the areas of science, technology and production, with science playing the leading role as the productive force (Marinko, 1989). Science and technology constitute the means of enhancing men's strength and the potentiality of his hands and brain. Drawing on the actor oriented perspectives in rural sociology (Long and Long, 1992), it was advocated that success of adoption of a technology at higher level are not merely a function of the technology, nor of the research and extension methodology, but result from a complex conjunction of people and events with outcomes. The present study revealed an apathetic approach of the farmers towards their works. Their decision making ability is negatively correlated with their educational qualification. Moreover, interest and attitude are negatively correlated. According to Wilkening (1953), adoption of a specific practice is not the result of a single decision to act but series of action and meaningful decisions. Rogers (2003) explains that the adoption decision and its timing depend on decision maker's perception and inherent characteristics, with innovators at one extreme and laggards at the other.

The farmers of the study area in general live under uncertain, harsh social and environmental conditions with heterogeneity in terms of social, economic and psychological characters. People are basically small holders (80.12%) and 75.25% people live below the poverty line (Anon, 2009). Gini co-efficient of the district is 0.488 (AHDR, 2003). They operate their farms with little access to land, water, extension service and credit. Farming in the district itself is fraught with the uncertainties of floods, drought and anthropo-political conflicts. The fish farmers are normally repelled to high input farming technologies owing to (1) adoption does not sustain due to high cost involved; (2) low access of the household to technology extension and credit; and (3) vulnerability of the households to risk involved such as floods, drought and societal problems (Lightfoot et al., 1992).

Farmers in the same environment with same livelihood resource base have different objectives and livelihood strategies and therefore respond differently to a given technological areas. Only 18% of the respondents have adopted fish farming as major occupation, the rest 82% were engaged in agriculture, other business and government jobs. Biot et al. (1995) suggested that 'different behaviour is as much a function of different opportunities and constraints as of different perception'. Even within the farm households, the ability to make decision on resource use and technology adoption varies according to age, gender and other category and actual decision can depend on a complex bargaining process amongst the members (Ellis, 1993; Jackson, 1995; Biot et al., 1995). Beyond the household group processes and ability to harness them can play a crucial role in adoption decision (Chamala and Mortiss, 1990; Frank and Chamala, 1992; Pretty and Shah, 1994).

While Wozniak (1984) opined that education increases ones' ability to receive, decode and understand information relevant to making innovative decisions; Clay et al. (1998) found that education is an insignificant determinant of adoption decision. In this study, it was hypothesized that high level of institutional education increases the probability of interest on and adopting a new technology. But the results revealed that individual independent characteristic and training needs of the farmers had negative correlations with education level and attitude (X_{13}).

Au and Enderwick (2000) explained that six beliefs, namely, compatibility, enhanced value, perceived benefits, adaptive experiences, perceived difficulties and suppliers' commitments, affect the cognitive process that determines the farmers' attitude towards technology adoption. This study revealed positive correlation with main occupation, education, scientific orientation, information seeking behaviour, knowledge and possession of ponds. This study suggests a change in farmers' attitude for development of aquaculture in the district.

Fisheries have been a caste based activity in Assam. Flood plain lakes, which were once unmanaged natural water bodies, were the main source of fish. Historically, there have been three distinct groups of people involved in fisheries activity: (1) those who catch fish for their own daily consumption; (2) those belonging to the fisher community and (3) the rural fisher entrepreneurs (leaseholders). Ordinary people usually used to catch fish daily for food, while fishers were full-time operators. Aquaculture is comparatively a new sector of food production and it is undergoing continuous change in Assam. During the last 20 years, it has been mastering a driving force that has propelled aquaculture to the forefront. However, pond productivity is limited to 2800 kg/ha in Goalpara, Assam; although, much higher yields (5000 kg/ha) have been recorded by Luu et al. (2003) in China and Vietnam. Biophysical potential of aquaculture in Assam reveals that same production could be achieved if new technology/practices are adopted. For which farmers should be substantially trained on the latest technological innovations in aquaculture for a paradigm shift from the current perception on aquaculture as a poverty alleviation programme to a prestigious income generating enterprise. Before analysing the training needs, farmers must be made aware of the technologies through frequent awareness camps on recent advances on aquaculture technologies.

ACKNOWLEDGEMENTS

The authors are grateful to Dr. Anubrata Das, Director, National Research Centre on Pig (ICAR), Guwahati-781 131, Assam, for his inspiration, encouragements and supervision. They are also thankful to the Department of Fisheries and Department of Agriculture, Government of Assam for providing the secondary data incorporated in this paper. The authors also extend thanks to Dr. A. K. Gogoi, Zonal Project Director, Zone-III, Indian Council of Agricultural Research, Barapani, Meghalaya for his unstinted initiatives for training need analysis. The authors acknowledge the helps received from their fellow colleagues working in the KVK Goalpara, Assam.

REFERENCES

AAU (1997). Package of Practices for Fish Farming in Assam. Assam Agricultural University, Jorhat, Assam, India.
AHDR (2003). Assam Human Development Report, Planning and Development Department, Govt. of Assam, India. (http://planningcommission.nic.in/plans/stateplan/sdr_pdf/shdr_assam03.pdf).
Anon (2009). Statistical Handbook of Assam. Directorate of Economic and Statistics. Government of Assam, India.
ARDB (2011). Online Agricultural Research Data Book. (http://www.iasri.res.in/agridata/11data/HOME_11.HTML.)
Au AK, Enderwick P (2000). A cognitive model on attitude towards technology adoption. J. Managerial Psychol. 15(4):266-282.
Biot Y, Blaikie PM, Jackson C, Palmer-Jones R (1995). Rethinking research on land degradation in developing countries. World Bank discussion World Bank: Washington DC. P. 289.
Chamala S, Mortiss P (1990). Working Together for Landcare: Group Management Skills and Strategies. Academic Press, Brisbane. P. 369.
Clay D, Reardon T, Kangasniemi J (1998). Sustainable intensification in the highland tropics: Rwandan farmers' investments in land

conservation and soil fertility. Econ. Dev. Cult. Change. 46(2):351-378.
Ellis F (1993). Peasant Economics: Farm Households and Agrarian Development. 2nd Edition, Cambridge, Cambridge University Press, England. P. 309.
Frank BR, Chamala S (1992). Effectiveness of extension strategies. In: Lawrence, G., Vanclay F, Furze B. (Eds.). Agriculture, Environment and Society: Contemporary Issues for Australia, Macmillan, Melbourne. pp. 122-140.
Jackson C (1995). Environmental reproduction and gender in the Third World. In: Morse, S. and Stocking, M. (Eds.). People and Environment, London UCL Press. pp. 109-130.
Lightfoot C, Gupta MV, Ahmed M (1992). Low external input sustainable aquaculture for Bangladesh – An operational framework. Naga, ICLARM Quarterly July. pp. 9-12.
Long N, Long A (1992). Battlefields of Knowledge: The Interlocking of Theory and Practice in Social Research and Development. London, Routledge. P. 306.
Luu LT, Trang PV, Cuong NX, Demaine H, Edwards P (2003). Promotion of small-scale aquaculture in the Red River Delta, Vietnam. In: Edwards, P., Little, D.C. and Demaine H. (Eds.). Rural Aquaculture, CAB Publishing, Oxford. pp. 55-75.
Mandal SC, Singh P, Borthakur BC, Mahanta K, Pande HK (1981). Report on ICAR Research Review Committee on National Agricultural Research Project. NARP Document No. 39. ICAR, New Delhi. P.167.
Marinko G (1989). What is the scientific and technological revolution? Progress Publishers, Moscow, USSR. P. 318.
McGlagan P (1989). Models for HRD Practice," 1989, American Society for Training and Development. St. Paul, Minnesota.
National Research Centre on Pig NRCP (Annual Report) (2006-2007). Annual report of Krishi Vigyan Kendra Goalpara, National Research Centre on Pig, ICAR, Dudhnoi, Assam, India. Report submitted to the Zonal Coordinating Unit, Zone-III, ICAR, Umiam, Meghalaya, India (http://icarzcu3.gov.in).
Panse VG, Sukhatme PV (1985). Statistical Methods for Agricultural Workers. Indian Council of Agricultural Research, New Delhi. India. p. 359.
Paulraj R (1997). Aquaculture feed- Handbook of Aquafarming. The Marine Product Export Development Authority (Ministry of Commerce, Govt. Of India), Kochi, India.
Pretty JN, Shah P (1994). Soil and Water Conservation in the Twentieth Century: A History of Coercion and Control. Research Series No. 1, Rural History Centre, University of Reading.
Rogers EM (2003). The Diffusion of Innovations. 4th Edition, New York, Free Press. P. 512.
Smith A (1992). Training and Development in Australia. Sydney, Butterworth. P. 279.
SREP (2006). Strategic Research and Extension Plan, Goalpara District, Assam, Agricultural Technology Management Agency, Assam. P. 169.
UNDP (1995). Human Development Report 2005: International cooperation at a crossroads – Aid, trade and security in an unequal world. United Nations Development Programme, New York, USA. P. 372.
Wilkening EA (1953). Adoption of improved farm practices as related to family factors. Wisconsin Experiment Station Research Bulletin. 183 Wisconsin, USA.
Wilson JP (1999). Human resource development. Learning and training for individuals and organizations. London, Kogan Page.
Wozniak GD (1984). The adoption of interrelated innovations: A human capital approach. Rev. Econ. Stat. 66:70-79.

21

Study on the quality and safety aspect of three sun-dried fish

Mohammad Abul Mansur[1], Shafiqur Rahman[2], Mohammad Nurul Absar Khan [3] Md. Shaheed Reza[4], Kamrunnahar[5] and Shoji Uga[6]

[1]Department of Fisheries Technology, Faculty of Fisheries, Bangladesh Agricultural University, Mymensingh, Bangladesh.
[2]Faculty of Earth Science, University Malaysia Kelantan, Jeli Campus, 17600 Jeli, Malaysia.
[3]Faculty of Fisheries, Chittagong Veterinary and Animal Science University, Chittagong, Bangladesh.
[4]Department of Fisheries Technology, Faculty of Fisheries, Bangladesh Agricultural University, Mymensingh-2202, Bangladesh.
[5]Department of Fisheries Technology, Faculty of Fisheries, Bangladesh Agricultural University, Mymensingh-2202, Bangladesh.
[6]Department of Parasitology, Graduate School of Health Sciences, Kobe University, Japan.

Market samples of three sun-dried freshwater fish species namely Indian major carp (*Labeo rohita*), snake headed fish (*Channa striatus*), and a type of eurasian catfish (*Wallago attu*) were included in this study. The quality of the sun-dried fish samples were evaluated by examining physical properties, biochemical composition, and reconstitution behaviour. The safety aspect of the sun-dried fish samples were studied by the detection of heavy metal, total viable bacterial count (TBC), aerobic plate count (APC) and Total volatile base-Nitrogen (TVB-N). According to the physical characteristics such as colour, odour, texture, it appeared that the *L. rohita* and *C. striatus* were of better quality than that of the *W. attu* which had already developed rancid odour, and bitter taste. The examined samples were brownish to light brown compared to the attractive cream colour of a freshly prepared samples. The products had moisture content ranging from 19.17 to 23.12%. The protein content ranged from 49.23 to 62.85%. Lipid and ash content were in the range of 4.92 to 11.0% and from 11.11 to 18.89%, respectively. *C. striatus* contained the highest protein content (62.85%) and *W. attu* contained lowest protein (49.23%) on moisture free basis. The result of heavy metal analysis of the sun-dried fish samples showed that, in sun-dried *L. rohita*, Arsenic (As) was 0.001 µg/g, Cadmium (Cd) was 0.53 µg/g, and Chromium (Cr) was 0.025 µg/g. In the sample of sun-dried *C. striatus* As was 0.003 µg/g, Cd was 0.089 µg/g and Cr was 0.054 µg/g; while in the sample of sun-dried *W. attu* As was 0.004, Cd 0.097, and Cr 0.068, which were within the acceptable limit for human consumption. The TBC of the experimental samples ranged from 1.84×10^4 to 5.3×10^6 per g of the dried samples. The total volatile base Nitrogen (N) content of the dried fish samples ranged from 7.54 to 8.32%. Reconstitution rate was found to be faster in *C. striatus* and *L. rohita* but slower in *W. attu*.

Key words: Total volatile base-Nitrogen, sun-dried, trace elements, fish.

INTRODUCTION

Drying along with salting and smoking as a fish preservation technique had been practiced perhaps longer than any other food preservation technique. Drying and other two curing methods have all continued as preservation techniques virtually unaltered from prehistory to the present day.Modern developments have centered around understanding and controlling the process to achieve the standard product demanded by today's market (Horner, 1992). Drying is one of the important methods of preserving fish throughout the world. It is still a vital method in the developing regions of the world.

Fish is the major source of animal protein in the diet of the people of Bangladesh contributing 58% of the total animal protein supply (DoF, 2008). It contributes 3.74% of the Gross Domestic Product (GDP) and 4.04% of the foreign exchange earnings. A sizeable quantity of fish is preserved by sun drying in Bangladesh from inland water fish as well as from sea fish. Domestic consumers as well as the ethnic community in developed countries eat dried fish. Bangladesh earns a good quantity of foreign exchange by exporting dried fish every year. The food value of dried fish is well established by the scientists (Cutting, 1962; Qudrat-I-Khuda et al., 1962; De, 1967; Bhattacharya et al., 1985; Humayun, 1985). A major problem associated with sun drying of fish in Bangladesh is the infestation of the product by fly and insect larvae during drying and storage (Ahmed et al., 1978).

In general, bacteriological problem associated with quality loss is not significant in properly dried and well-packaged fish products. The problem is the contamination during different stages of handling and improper packaging. Connell (1957) reported that, fresh fish species dried in sun reabsorbs water to comparatively small extent and once reconstituted, are very tough, almost rubber-like, fibrous, compact, and dry in mouth. Freshly prepared dry fish will have an attractive cream colour. On long storage, they become brownish yellow or brown which indicates varying degree of spoilage (Connell, 1957). It is highly desirable that, the requirement of a safe dried product should be available to all those who are concerned with the expansion and development of fish processing and preservation particularly in those regions of the world where an improvement in the fishery can have a marked effect on the standard of living of the people. Poulter et al. (1988) described losses of fish, which have been cured by salting, drying, smoking or a combination of these processes. Undue delay in processing and poor processing methods may lead to low value poor quality products; this represents a financial loss to fishermen or processor. The causes and extent of different types of quality reduction is described. Excessive heat treatment is known to impair the availability of amino acids such as lysine in fish protein (Carpenter and Booth, 1973). Hoffman et al. (1977) found a significant reduction in lysine availability and net protein utilization in tropical fish dried at 75°C and smoked at 100°C. Post harvest loss in dried fish product is estimated as 25% in Bangladesh

(Doe et al., 1977).Under very humid conditions cured fish reabsorb moisture. Rao et al. (1962) found that, a relative humidity of over 70% was conductive to mould attack. Fish damaged by mould has a lower price thereby resulting in economic loss.

Good quality raw material supply is essential for both domestic consumption and value-added product development for export market. It needs proper research support to produce safe and quality product for export. Global climate change has lead to an increasing concern in recent years regarding the abundant entry of heavy metal into the water and their probable adverse effect that might be reflected on aquatic animal like fish, and finally, on human health through the food chain.

Attention to environmental pollution in Bangladesh was not adequate which is now affecting its fisheries resources. Fish living in the polluted water may accumulated toxic trace metals via their food chain. High level of As, Lead (Pb), Copper (Cu), and Iron (Fe) have been found to cause rapid physiological changes in fish (Tarrio et al. 1991). As is identified as a toxic environmental pollutant because, it causes chronic and epidemic effects on human health through widespread water and crop contamination due to the natural release of this toxic element from aquifer rock in Bangladesh (Fazal et al., 2001). Cd is a known teratogen and carcinogen, probable mutagen and has been implicated as the cause of serious deleterious effect on fish. Trace elements can be accumulated by fish, both through the food chain and water (Hadson, 1998). Mercury (Hg), Cd, Cr, Pb, Selenium (Se), are known to be potentially harmful pollutants contaminated in fish, but so far only Hg has been implicated in disease to consumers caused by eating fish. Hg above 0.5 to 1.0 mg/kg cause a disease in human which affects central nervous system (Connell, 1980). For a sufficient margin of safety and a fairly high level of weekly fish consumption is 4 to 8 meals of 150 g each.

In Bangladesh, very little work has been done on the presence of heavy metals in freshwater fish, despite such data are important to assess the quality and safety of fish and fishery product for domestic consumption as well as for export. Some research, on the As on ground water relevant to human health are done by foreign scientists. But research on As, Cd, and Cr detection in fish and fishery products is still imperative for human health concern. Examination of fish and fishery products on specific metal and elements (heavy metal) may be necessary because sometimes fish are caught from suspected area as for example near effluent discharges or waste dumps. These heavy metals are cumulative poison that cause injury to health through progressive and irreversible accumulation in the body as a result of eating repeated small amounts. In the present study, we have attempted to create data information, which will be helpful in producing a quality product as well as a safe product for domestic consumption as well as for export.The present study was undertaken with the aim to have a clear idea about the quality and safety aspects of sun-dried fish. The nutritive value of dried fish is already established. Quality of dried fish in term of physical properties, bio-chemical composition and reconstitution behaviour was studied.At the same time the safety aspect in terms of heavy metal, total viable bacterial count, Aerobic plate count and Total volatile base Nitrogen was studied.

A better knowledge on quality and safety of sun-dried fish is important because a reasonable quantity of sun-dried fish is exported to International market every year. To continue export of this fishery product the quality and safety of the product should be assured. At the same time, the product should have desired quality and it should be safe for health of the domestic consumers. At the first step, the result of the present investigation is expected to provide a clear idea on the quality and safety of the sun-dried fish under present study.

MATERIALS AND METHODS

Species selection, sample collection, and storage

Three traditionally sun-dried freshwater fish species namely *Labeo rohita*, *Channa striatus,* and *Wallago attu* were included in this study. These sun-dried fishes were purchased from the local market of Mymensingh town and brought to the Laboratory of the Department of Fisheries Technology, Bangladesh Agricultural University, Mymensingh, Bangladesh. Dried fish samples were packed tightly in polyethylene bags and stored at -20°C until further analysis for subsequent studies. A quantity of 2 kg sun-dried fish of each species were been purchased for the study. The samples were been subjected to laboratory analyses within 2 weeks of purchase.

Physical characteristics

Physical characteristics such as colour, odour, and texture of the traditionally sun-dried fishes were examined by organoleptic test/sensory test on the basis of the method described by Howgate et al. (1992). All determinations were done in triplicate and the mean value was reported.

Biochemical composition

Biochemical composition of the traditionally sun-dried fishes were determined according to the methods proposed by Analytical Methods Committee AMC (1979). Biochemical analysis included determination of crude protein, lipid, ash, and moisture determinations. All determinations were done in triplicate. The mean values have been reported.

Total volatile base-Nitrogen (TVB-N)

Total volatile base-Nitrogen (TVB-N) was determined by steam distillation method proposed by Analytical Methods Committee AMC (1979). All determinations were done in triplicate and the mean value was reported.

Quantitative bacteriological analysis

Aerobic plate count (APC) was done by consecutive decimal dilution technique. Sample for the APC was accurately weighed and added with required amount of water and liquefied in a sterilized blender jar and consecutive ten fold dilution were prepared in test tubes. From all of the dilutions spread plate, cultures were made in duplicate and incubated at 35°C for 24 to 48 h. Colonies developed on the plates having 30 to 300 colonies were selected for APC. APC was calculated by the following formula:

$$APC/g = C \times D \times 10/S \; CFU/g$$

Where C = number of colonies found, D = dilution factor, S = weight of sample in grams, CFU = colony forming unit.

Experimental media

The media used in this experiment were as follows

Plate count agar

Plate count agar was a commercial preparation (Hi media, India) that was used for enumeration of viable bacterial count in sample. Accurately weighed 23.5 g media was suspended in 1000 ml distilled water and boiled until the ingredients were completely dissolved. The media was then sterilized at 121°C for 15 min under 15 lb/inch^2 pressure in an autoclave.

Peptone diluents (0.2%)

Peptone diluents were used as diluents in determining APC.

Calculation of microbial load

The microbial load of dried fish product was calculated by using the following formula:

$$\text{Colony Forming Unit (CFU/g)} = \frac{\text{No. of Colony} \times 10^n \times 10 \times \text{Vol. of Soln.}}{\text{Weight of Sample}}$$

Detection of heavy metal

The collected sun-dried fish samples were subjected to analysis for the detection of heavy metals namely As, Cd, and Cr. A known quantity of dried fish sample was weighed by an electronic balance and 5 ml of diacid mixture (5 ml conc. HNO_3; 3 ml 60% $HClO_4$) was added to each sample. The content was mixed for overnight. Samples were then digested initially at 80°C temperature and later at 150°C for 2 h. The completion of digestion was indicated by almost colourless condition of the material. The brown fumes also cease to exist at completion of digestion. The samples were separately filtered by using an ash less filter paper and volume made up to 25 ml with 0.5% HNO_3 prepared for the determination of As, Cd and Cr (Eboh et al., 2006). The samples were subjected to analysis by Atomic Absorption Spectrophotometer (HG-AAS, PG-990, PG Instruments, UK) at Professor Muhammad Hossain Central Laboratory, Bangladesh Agricultural University, Mymensingh according to the method of Clesceri et al. (1989). The wave length of As, Cr, and Cd were 193.7, 127 and 217 nm, respectively. The concentration of As, Cr, and Cd were calculated by the following

Table 1. Physical characteristics and organoleptic characteristics of sun-dried fishes

Species	Colour	Odour	Texture		Taste	Overall quality
C. striatus	Brown	Good	Tough and springy	Fibrous	Slightly salty and bitter	Satisfactory
L. rohita	Brown	Good	Tough and springy	Fibrous	Good	Satisfactory
W. attu	Dark brown	Rancid	Soft	Slightly fibrous	Bitter	Poor

Table 2. Biochemical and bacteriological characteristics of sun-dried fishes.

Species	Moisture (%)	Protein (%)	Lipid (%)	Ash (%)	TVB-N mg/100 g	TBC cfu/g
C. striatus	19.17	62.85	4.92	11.11	7.54	1.84×10^4
L. rohita	20.27	59.32	10.83	12.89	7.73	2.32×10^4
W. attu	23.12	49.23	11.00	18.89	8.32	5.3×10^6

formula:

$$\text{Metal concentration} = \frac{\text{ppm conc. observed} \times \text{final vol. of sample in ml}}{\text{Weight of tissue taken in g}}$$

Water reconstitution behaviour

Accurately weighed 5 g of dried fish flesh/tissue was kept soaked in 1 L of water at room temperature for 150 min and in hot water at 80°C for 60 min with occasional stirring. Then water was dried off. All the flesh were then transferred to the strainer and the water on surface was wiped off by a piece of blotting paper and the flesh tissue was weighed again. During the soaking time, the flesh could reabsorb maximum amount of water. Result in this respect has been expressed in terms of weight of water absorbed by 5 g of moisture frees sample (Jason, 1965).

RESULTS AND DISCUSSION

The physical (organoleptic) characteristics of the sun-dried fish samples are presented in Table 1. The colour, odour, texture, and taste revealed that the *C. straitus* and *L. rohita* were comparatively of good quality, while the *W. attu* had already lost the shelf life. Rancid odour and bitter taste was developed in the dried *W. attu* samples. It appears that, if moisture content of the product is comparatively high, then the deteriorative changes may result in browning reactions and development of rancid and off odour. The samples examined were slightly brown, brown or dark brown compared with freshly prepared samples which were expected to have an attractive cream colour. Rancid odour and off odour was developed in some of the samples. Some of the samples contained high quantity of broken pieces, which might be the result of using poor quality raw material, excess drying or improper drying and handling or due to moisture reconstitution.

The result of biochemical analysis on the sun-dried fish samples are presented in Table 2. Moisture content of *L.*

rohita and *C. straitus* was 20.07 and 19.17%, whereas the moisture content of *W. attu* was 23.12%. The protein content which is most important from nutritional point of view ranged from 49.23 to 62.85%. Lipid and ash content were in the range of 4.92 to 11.01% and 11.11 to 18.89%, respectively. For better evaluation of data, the content of protein, total lipid and ash of all the samples have been recalculated on moisture free basis. There was an inverse relationship between protein and fat content of the dried fish products where the relationship is markedly evident by the data calculated on moisture free basis. As shown in Table 2, the sun-dried *C. straitus* contained the highest protein content (62.85%) while the sundried *W. attu* contained the lowest (49.23%) protein content. From the analytical data, it is evident that, the proximate composition of fish varies with some factors of which species of fish is an important factor.

TVB-N of the sun-dried fish samples was comparable among the species under present study. The range of TVB-N in the sundried fishes was from 7.54 to 8.32 mg/100g (Table 2). TVB-N express the degree of bacterial spoilage during processing in other word the degree of freshness. In the present study, the TVB-N of the sun-dried fish was within the acceptable level. The highest TVB-N was detected in the dried fish produced from *W. attu* followed by *L. rohita* and *C. striatus*. Despite this minor difference in TVB-N content in the sun-dried fishes, the overall result on TVB-N is comparable among the sun-dried fishes and the result is within allowable limit that is, acceptable.

TBC of the sun-dried fishes was also conducted during the present research study. The TBC of the sun-dried fish samples was also highly varied among the species probably indicating the different degrees of spoilage by bacteria. The result of TBC in the sun-dried fishes were 1.84×10^4 per g in *C. striatus*, 2.32×10^4 per g in *L. rohita* and 5.3×10^6 per g in *W. attu*. A correlation was found between TBC and TVB-N. The sample with high TVB-N content showed maximum bacterial load while comparatively low TVB-N content had minimum bacterial

Table 3. Detection of heavy metal in the sun-dried fishes.

Species	Arsenic (µg/g)	Cadmium (µg/g)	Chromium (µg/g)
C. striatus	0.003 ± 0.001	0.089 ± 0.004	0.045 ± 0.003
L. rohita	0.001 ± 0.001	0.053 ± 0.003	0.025 ±0.002
W. attu	0.004 ± 0.002	0.097 ± 0.003	0.068 ± 0.002

count. On the other hand, higher moisture content promoted the growth of bacteria and accelerated the TVB-N content irrespective of the samples analyzed. Most of the samples with increased TVB-N and TBC contained around 20% moisture. Although there is a close relationship observed between the higher TBC and the corresponding high level of TVB-N content, when initial moisture level was close to 20%, the samples picked up sufficient moisture. Sen et al. (1961) reported that, when water content of the fish fall below 25% of the wet weight, bacterial activity stops; when the water content is further reduced to 15% mould ceases to grow. This indicates that, moisture level of 20% was quite unsuitable for the growth of bacteria. It is of little use of insisting production of sun-dried fish with water content below 20% when there is no option but to store it even in a climate of 90% humidity.

Result of heavy metal concentration (pollution by metal and element) in the sun-dried fishes of this research study is shown in Table 3. The sun-dried fishes contained very low concentration of As, Cd and Cr. The range of the heavy metal concentration was within the acceptable limit for human consumption. Even this concentration of heavy metal accumulated in dried fish flesh was less than that of acceptable range recommended for drinking water. A large number of potentially harmful metals and elements are known as pollutants despite Hg has been implicated in disease to man caused by eating fish and fish product. Pollution from any metal or element may cause unsuspected hazards to man. The elements of most concern are cumulative poisons, that is, those that cause injury to health through progressive and irreversible accumulation in the body as a result of ingestion of repeated small amounts. These include Hg, Cd, Pb, Se, and As (Connell, 1975).

Many countries are now taking voluntary or mandatory action to reduce pollution of the aquatic environment with heavy metal for the food safety of aquatic food particularly fish. Considering the affect of heavy metal on fish quality and safety the food regulatory and health authorities in some developed countries have taken serious view and adopted maximum allowable limit of harmful metals and elements. Usually pollutants of metal and element category contaminate the raw fish. The concentration of harmful metal and element is much higher (calculated value) in the processed fish as moisture percentage is reduced considerably. As a result

of this, the concentration of pollutants in per unit weight/mass is increased remarkably. But in the present study, the concentration of As, Cd, and Cr in the final product that is, the sun-dried fish was very low. The range was from 0.001 to 0.09 µg/g. It clearly indicates that, the raw fish was caught from unpolluted water and the sun-dried fish quality was good, it is safe to eat when pollution by metal and elements is considered.

Water reconstitution behaviour of the dried fishes soaked in water at room temperature (30°C) and in hot water (80°C) are presented in Figures 1 and 2, respectively. Among the three sun-dried fishes, reconstitution rate was faster in C. striatus and L. rohita but slow in the other dried fish W. attu. A close relationship was observed between reconstitution behaviour and physical properties of the samples. The overall reconstitution power of the sun-dried fish samples was comparatively slow with poor texture such as tough, rubbery and compact structure with few interfibrillar spaces. This was especially true for C. straitus and L. rohita. The reasons for the failure of these dried products to attain perfection are the irreversible changes which took place during sun drying, and severe damage suffered by the cellular structure. The real reconstitution of sun-dried fish is impossible.

The best way of reconstitution is to conserve a porous structure by a suitable method which absorbs and retain sufficient water. Compressed products absorb slowly and this is usually incomplete. The fibres of these sample muscles appeared to be cemented together and suffered severely. The large difference in rehydration rate which existed among the sun-dried fishes is due to their micro-structural differences. The C. straitus and L. rohita exhibited an enormously rapid initial rate of rehydration. This is due to water being carried into the deep part of the piece by a porous structure which absorbed and retained sufficient water by capillary. With a tough and rubbery tissue water penetrates mostly to the centre of large pieces by diffusion through the protein of the fibre itself and the process is very slow (Sen et al., 1961; Connell, 1957; Lahiry et al., 1961).

On the basis of the results of the present study, it can be concluded that, the sun-dried C. striatus and L. rohita is better than that of W. attu, when quality and safety is considered. It can also be concluded that, the heavy metal concentration in the sun-dried fishes was within the safe level.

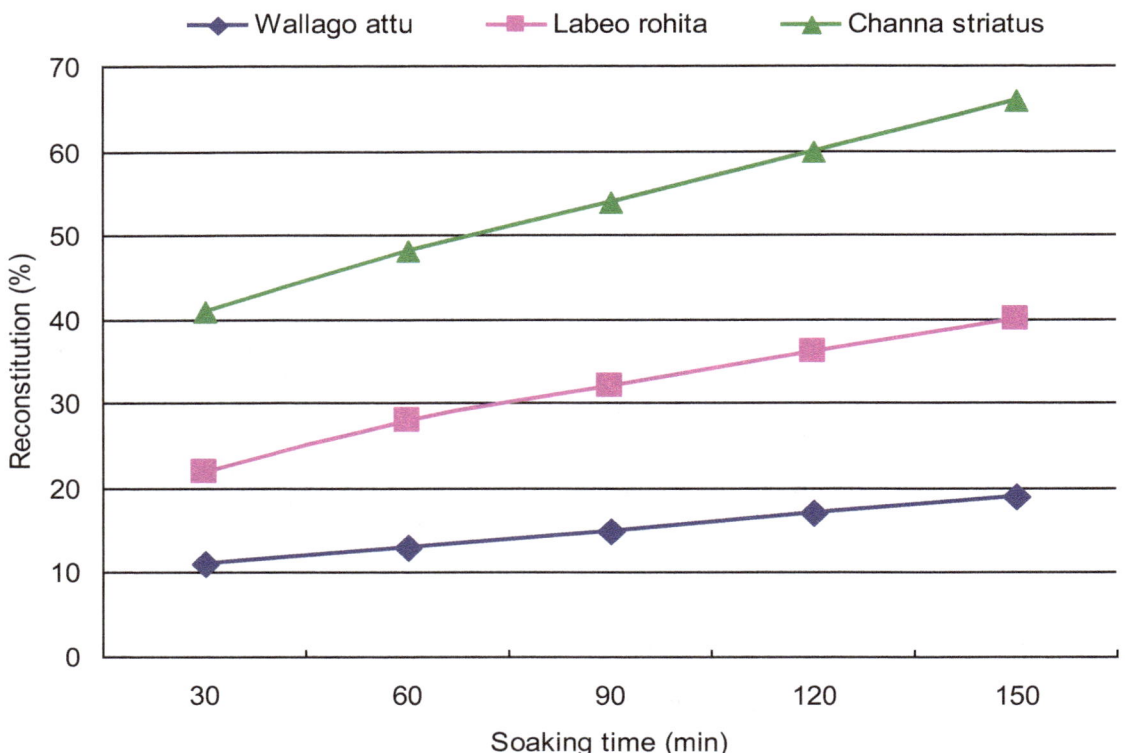

Figure 1. Reconstitution behaviour of the sun-dried fishes soaked in water at room temperature for 150 min.

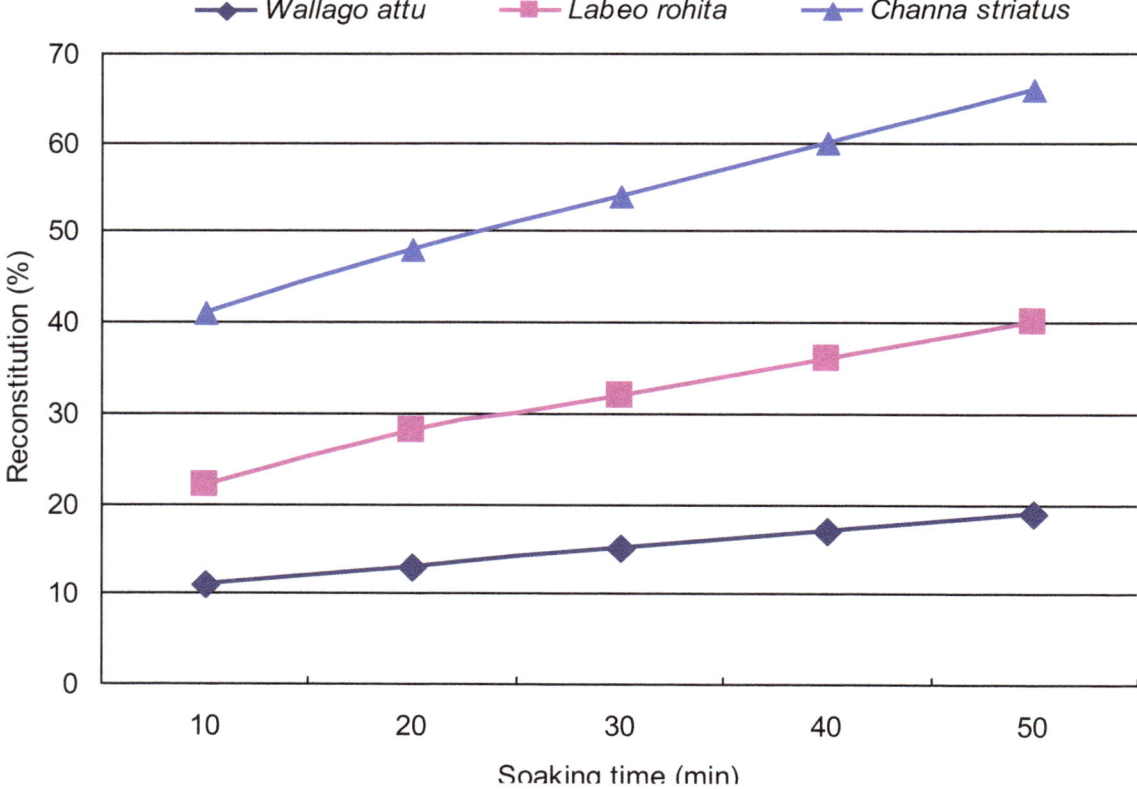

Figure 2. Reconstitution behaviour of the sun-dried fishes soaked in hot water at 80°C for 50 min.

REFERENCES

Ahmed M, Bhuiyan AD, Alam AMS, Huda SMS (1978). Radiation distribution studies on sun-dried fish. IPFC Syn. Fish Utilization and marketing in the IPFC region, Manila. March 8-11.

Analytical Methods Committee (1979). Recommended General Methods for the Examination of Fish and Fish Products. Analyst 104:434-450.

Bhattacharya SK, Bandyopadhyay JK, Chattopadhyay AK (1985). Improved dried product on Blanching of *Gudusia chapra* prior to sun-drying. In: "Harvest and Post-Harvest Technology of Fish. Society Fish. Technol. India P. 531.

Carpenter KJ, Booth VH (1973). Damage to lysine in food processing: its measurement and its significance. Nutr. Abstr. Rev. 43(6):423-451.

Clesceri LS, Greenberg AE, Trussed RR (1989). Standard method for the examination of water and waste water. 17th ed. American Public Health Assoc. Washington D.C. pp. 1-30, 40-175.

Connell JJ (1957). Some aspects of the texture of dehydrated fish. J. Sci. Food Agric. 8(9):326-353.

Connell JJ (1975). Intrinsic Quality. In: "Control of Fish Quality". Fishing News Books Ltd. Farnham, Surrey, England. pp. 4-30.

Connell JJ (1980). Intrinsic Quality. In: "Control of Fish Quality". (Second Edition) Fishing News Books Ltd. Farnham, Surrey, England. pp. 4-30.

Cutting CL (1962). The influence of drying, salting and smoking on nutritive value of fish. In: "Fish in Nutrition". E. Heen and R. Kreuzer (Editors). Fishing News Books Ltd. London.

De HN (1967). Processing of fish protein concentrate in East Pakistan. Transaction of the fish protein concentrate seminar, Dhaka, November 7-8, pp. 15-23.

DoF (Department of Fisheries) (2008). Fish Fortnight Publication-2008. Department of Fisheries, Ministry of Fisheries and Livestock, Government of the People's Republic of Bangladesh, Dhaka, Bangladesh.

Doe, PE, Ahmed M, Muslemuddin M, Sachithananthan K (1977). A polythene tent drier for improved sun-drying of fish. Food Technol. Aust. 29(11):437-441.

Eboh L, Mepba HD, Ekpo MB (2006). Heavy metal contamination and processing effects on the composition, storage stability and fatty acid profiles of 5 common commercially available fish species in Oron/Local Govt. Nog. Food Chem. 97:490-497.

Fazal MA, Kawachi T, Ichio E (2001). Validity of the latest research findings on causes of ground water arsenic contamination in Bangladesh. Water Int. 26(2):380-389.

Hadson PV (1998). The effect of metabolism on uptake, disposition and toxicity in fish. Aquat. Toxicol. 11:3-18.

Hoffman A, Barranco A, Francis BJ, Disney JG (1977). The effect of processing and storage upon the nutritive value of smoked fish from Africa. Trop Sci. 19(1):41-53.

Horner WFA (1992). Preservation of fish by curing (drying, salting and smoking). In: "Fish Processing Technology". Hall GM (editor). Blackie Academic and Professional. pp. 31-71.

Howgate PA, Johnson P, Whittle KJ (1992). Multilingual guide to EC freshness grades for fishery products. Aberdeen: Torry Research Station, Food Safety Directorate, Ministry of Agriculture, Fisheries and Food; Aberdeen, Scotland, UK.

Humayun NM (1985). Studies on the improvement of traditional preservation method of fish drying augment the quality and the shelf-life of the product. M.Sc. Thesis. Department of Fisheries Technology, Bangladesh Agricultural University, Mymensingh, P. 65.

Jason AC (1965). Drying and dehydration. In: "Fish As Food" Vol III. G. Borgstrom (Editor). Academic Press Inc., New York and London.

Lahiry NL, Sen DP, Visweswariah (1961). Effect of varying proportions of salt to fish on the quality of sun drying mackerel. Food Sci. 10(5):139-143.

Poulter RG, Ames GR, Walker DJ (1988). Post-harvest losses in traditionally processed fish products. In: M. Mohan Joseph (Ed.) The First Indian Fisheries Forum, Proceedings. Asian Fisheries Society, Indian Branch, Mangalore. pp. 409-412.

Quadrat-I-Khuda M, De HN, Khan NM, Debnath JC (1962). Biochemical and nutritional studies on East Pakistan Fish, Part-VII. Chemical Composition and Quality of the traditionally Processed fish. Pak. J. Sci. Ind. Res. 5(20):70-73.

Rao SVR, Valsan AP, Kandoran MK, Nair MR (1962). Storage behaviour of salted and dried fish in relation to humidity conditions. Indian J. Fish (B) 9(1):156-161.

Sen DP, Anandaswamy B, Iyenger NVR, Lahiry NL (1961). Studies on the storage characteristics and packaging of the sun-dried salted mackerel. Food Sci. 10(5):148-156.

Tarrio J, Jafor M, Ashra M (1991). Levels of selected heavy metals in commercial fish from freshwater Lake of Pakistan. Toxicol. Environ. Chem. 33:133-140.

22

Utilization of maggot meal in the nutrition of African cat fish

Olaniyi C. O and Salau B. R.

Department of Animal Production and Health, Ladoke Akintola University of Technology, Ogbomosho, Oyo State, Nigeria.

A twelve week feeding trial was carried out to evaluate the growth and nutrient utilization of African catfish (*Clarias gariepinus*) fingerlings fed varying levels of maggot meal. Maggot meal was produced from a mixture of 25 kg of cattle blood, 70% wheat bran, and 30% saw dust and substituted for fish meal in the fish diet to produce five varying levels of maggot meals- Treatment 1 (Trt 1) contained 0% of maggot meal (control), Trt 2 (50% maggot meal), Trt 3 (33.3% maggot meal), Trt 4 (66.6% maggot meal), and Trt 5 (75% maggot meal). A total of 150 fingerlings (two weeks old ±21.74 g) were allotted to five dietary treatments (fifteen fingerlings per tank) in a randomized design. Highest mean weight gain (MWG), percentage weight gain (PWG) and specific growth rate (SGR|) (35.47 g, 63.23%, 1.30%/day) were observed in Trt5 (75% maggot inclusion) respectively while the least values, 13.545 g, 62.29%, 0.69%/day were observed in Trt1 (control) respectively. In the same vein highest value of feed intake (116.4 kg) was recorded for Trt5 and least value (71.8 kg) was found with Trt3. However, FCR were significantly ($P < 0.05$) higher, Trt5 had the least FCR (3.13) and the highest value of 5.07 was found with fish fed control diet. Trt5 had the highest values of protein intake (PI) 47.60 and protein efficiency ratio (PER) 0.75, respectively. The results observed for FI showed that the feed were highly acceptable by the fish and the best growth response was achieved at higher level of maggot inclusion (75% maggot meal inclusion). Highest PER is also an indication that maximum utilisation of nutrients was obtained at the higher levels of maggot meal in the diet. Also, best of feed to flesh gain ratio was attained at the higher level of maggot inclusion. Therefore, it can be concluded that maggot meal can replace fish meal up to 75% without any adverse effect on the growth of the fish.

Key words: Replacement value, maggot meal, growth performance, weight gain.

INTRODUCTION

Recently, the increasing cost of feed has been at alarming rate and this has been affecting the development and expansion of aquaculture in African countries particularly Nigeria (Sogbesan et al., 2006). Fish meal been the major protein source in the fish diet constitutes the highest cost thereby making the price of the feed to rise exponentially. Therefore, several attempts have been made to substitute fish meal with other animal protein sources such as earth worm, shrimp waste, insect and plant protein sources such as sun flower, rape seed, soy bean meal and cottonseed meal. However, they cannot replace fish meal wholly but partially due to the presence of chitin in their exoskeleton. Ng et al. (2001) reported that chitin found in the exoskeletons is a polymer of glucosamine insoluble in common solvents and the presence of chitin do leads to depression of growth performance and protein utilization in catfish fed high levels. In the same vein problem of anti- nutritional factor in tropical legumes have limited their usage and direct incorporation into animal feeds (Ogunji and Writh, 2001) Maggot meal is an animal protein source produced from waste, it has been reported to be highly nutritive with crude protein ranging between 43.9 and 62.4%, lipid 12.5 and 21%, and crude fbre 5.8 and 8.2% (Awoniyi et al., 2003; Fasakin et al., 2003a,b; Ajani et al., 2004).Maggot meal is also rich in phosphorus, trace elements and B complex, vitamins (Teotis and Milles, 1973). According to Fashina– Bombata and Balogun (1997), the cost of harvesting and processing one kilogram of maggot meal is smaller compared to the cost of 1kg of fish meal, thereby showing the cost effectiveness of using maggot meal in the diet of African cat fish. This study therefore evaluates the nutritive value and growth performance of Clarias gariepinus fed with maggot meal.

MATERIALS AND METHODS

Experimental site

The experiment was carried out at the fisheries unit Ladoke Akintola University of Technology Teaching and Research Farm, Ogbomoso, Oyo State.

Maggot production

25 kg of fresh cattle blood, 5 kg of wheat bran and saw dust (70 and 30% respectively) were mixed together to a thickness of 5 cm to constitute substrate. The mixture was spread in a box of 1.4 m by 1.4 m screened with plastic net with one end open. The blood product were left to ferment for 48 h and the odour released attracted flies to perch on the mixture, then the open side was closed the second day. The eggs laid by the female flies on the substrate hatched into larvae within two days, after which the larval matured and was harvested. The whole process took five days. The larvae were harvested and oven dried to a constant weight, after which the dried maggots were milled into a meal and kept in an air tight container at room temperature.

Experimental diets

Feed ingredients were purchased from a feed ingredient store in Ogbomoso. The major feed ingredients used were yellow maize, wheat bran, groundnut cake (GNC), fishmeal, vitamins, mineral premix, oyster shell (Table 1).
Five isonitrogenous diets of 40% crude protein were formulated: Treatment 1 (Trt 1) contained 0% of maggot meal (control), Trt 2 (50%maggot meal),Trt 3 (33.3% maggot meal), Tit 4 (66.6% maggot meal),.Trt5 (75% maggot meal), with these ratios of maggot meal to fish meal 0, 1:1, 1:2. 2:1. 3:1 respectively.
The feed were made into pellets with the use of pelletizing machine; pellets were sun dried for three days to reduce the moisture content and to prevent deterioration. The feeds were packed and stored in air tight container at room temperature.

Experimental procedure

One hundred and fifty catfish fingerlings (Clarias gariepinus) of the same age (two weeks old) and uniform size (± 21.64 g) were procured and allotted to ten circular bowls of 50 L capacity at the rate of fifteen juveniles per bowl. The fish were acclimatized for two weeks during which the fish were fed floating feed to empty their gut in preparation for the experiment. After the acclimatization, the fish were fed 3% body weight subject to change every two weeks during which the fish were weighed

Data collection

The parameters measured were weight gain, average daily weight gain and feed intake. Each of these parameters was measured at 2 weeks interval. Performance characteristics were evaluated according to the method of Olvera- Novoa et al. (1990) as follows:

Mean weight gain (MWG) = Final mean weight (g) - Initial mean weight (g);
Average daily weight gain (ADWG) = Mean weight gain (g) / length of feeding trial (days);
Percentage Weight Gain (PWG) = Mean weight gain (g) / Initial mean weight × 100
Specific growth rate (SGR % /day) = $100[(Log_e W_2 – Log_e W_1)/N_O$ of days
Feed conversion ratio (FCR) = total feed fed (g) / net weight gain (g);
Protein Intake (PI) = total feed consumed × %Crude protein in feed
Feed intake (FI) = amount of feed throughout the period of the experiment;
Protein gain (PG) = mean protein intake (g) / length of feeding trial (days);
Protein efficiency ratio (PER) = Net weight gain (g) / Amount of protein fed (g) while
Protein productive value (PPV) = protein gain in fish (g) / protein in food (g) × 100.

Chemical analysis

The proximate analysis of maggot meal diet and the fish carcass were done, using the procedure outlined by AOAC (1990).

Statistical analysis

The data collected were subjected to analysis of variance (ANOVA) using completely randomized design (CRD) SPSS computer package (Field, 2000) means were separated by Duncan's option of the same statistical package.

RESULTS

The proximate compositions of diets are presented in Table 2. The ether extract, crude fiber and ash content showed a progressive increase in their content as maggot meat increased in the diets, while the crude protein values followed no definite order.

Table 3 showed the growth performance and nutrient utilizationof Clarias gariepinus fingerlings fed maggot meal, The mean weight gain (MWG), Percentage mean weight Gain (PWG), mean weight gain /day(MWG/Day), and specific growth rate were significantly different at p>005 in all the treatment. Treatment 5 (7 5% maggot meal) had the highest values in MWG, %MWG,MWG/day

Table 1. Gross composition of the experimental diets (%).

Parameter	Trt$_1$ (0%): Control	Trt$_2$ (50%)	Trt$_3$ (33%)	Trt$_4$ (66%)	Trt$_5$ (75%)
Yellow maize	22.36	19.13	22.15	17.32	16.17
Wheat bran	11.18	9.57	11.07	8.66	8.80
Groundnut cake	31.73	22.77	15.95	17.76	14.55
Fishmeal	31.73	22.77	31.89	17.76	14.55
Maggot meal	-	22.77	15.95	35.57	43.65
Mineral premix	1.50	1.50	1.50	1.50	1.50
Vitamin premix	0.50	0.50	0.50	0.50	0.50
Salt	0.25	0.25	0.25	0.25	0.25
Vegetable oil	0.25	0.25	0.25	0.25	0.25
Total	100.00	100.01	100.01	100.01	100.01
Calculated CP (%)	39.94	39.71	39.51	39.57	39.55

Trt$_1$, Trt$_2$, Trt$_3$, Trt$_4$, and Trt$_5$ represent control, 1:1, 1:2, 2:1 and 3:1 replacement level of maggot meal:fish meal in the diet fed to African catfish (*Clarias gariepinus*) fingerlings.

Table 2. Proximate composition of experimental diets.

Parameter	Trt$_1$ (0%) control	Trt$_2$ (50%)	Trt$_3$ (66%)	Ttr$_4$ (33%)	Trt$_5$ 75%)
CP	44.20^a	43.50^b	44.23^a	43.48^b	44.03^{ab}
CF	3.74^b	3.76^b	3.74^b	3.85^b	4.10^a
EE	4.76^a	4.60^a	4.50^a	4.76^a	4.85^a
ASH	13.37^{ab}	12.77^c	13.11^b	13.16^b	13.60^a
DM	90.04^b	90.10^b	90.19^b	90.06^b	95.11^a
NFE	23.98^d	25.47^b	24.52^d	24.80^c	28.54^a

Means within the row with different superscript are significantly different (p<0.055).

and SGR (35.47 g, 163.23%, 506.72 mg,1.37 g respectively), while the least values of MWG, %MWG, MWG/day, and SGR were recorded in Treatment 1 (control) (13.54g, 62.39 %,193.36mg, and 0.69%/day respectively). Feed intake was very high in Treatment 5 (75% maggot meal) followed by Treatment 3 (33%maggotmeall) while the fish fed control (diet T1) consume less compare to the fish in all other treatments.The FCR were significantly different (p<0.05), fish fed at 75% inclusion level (Treatments 5) had the best FCR (3.1 3 g) followed by Treatments 3 (33% inclusion) (3.57 g) while Treatments 1 (control) has the highest FCR which makes it to be less efficient in terms of feed conversion (5.07 g). In the same vein Treatment 5 also have the highest gross feed conversion efficient (GFCE) while Treatment 1 (control) had the lowest GFCE.

DISCUSSION

All the five diets were accepted and utilized for growth by the fingerlings. The proximate composition of five experimental diets was as follows: Crude protein ranging from 43.48 to 44.20%, crude fibre 3.74 to 4.10%; ash

12.77%, to 13.60%, dry matter 90.04 to 95.11%, ether extract 4.50 to 4.88% (Table 2). The result of this study, compare favourably with Aniebo et al. (2009) who reported a CP of 40.59 to 40.74%. The crude fiber in this study ranging from 3.74 to 4.10% agreed with Aniebo et al. (2009) who reported crude fiber ranging from 4.1 to5.2%.

Growth response by catfish fed maggot meal based diets in Table 3 showed that an increase in the maggot meal in the diets gave an improvement in growth and feed conversion efficiency. Progressive increment in MWG, PWG and SGR were observed with increasing levels of maggot meal. The best growth for all treatments was observed in catfish fed 75% maggot meal based diet (Trt 5). The improving growth response observed with increasing levels of maggot, may be caused by the high level of crude protein in maggot meal. This agreed with Ogunji et al.(,2009) who observed a better performance of diets containing maggot meal over those fed 100% fish meal. Fish fed 75% maggot meal inclusion level (Trt5) recorded the least FCR (3.13) this is an indication that it has lower feed to flesh conversion.

The lowest growth rate was noticed in the control diet, this may be attributed to the low feed conversion efficiency. A11 the five treatment were accepted and

Table 3. Growth performance and feed utilization of *C. gariepinus* fingerlings fed diet containing varying inclusion of maggot meal.

Parameter	Trt$_1$- (0%) Control	Trt$_2$ (50%)	Trt$_3$ (66%)	Tr$_{t4}$ (33%)	Trt$_5$ (75%)	SEM
Initial MW(g)	21.73[b]	21.80[a]	21.67	21.80[a]	21.73[b]	0.00
Final MW(g)	35.26	42.67	47.36[b]	40.93	57.20[a]	0.38
MWG(g)	13.54	20.87	25.70[b]	19.14	35.47	0.23
MWG/day (mg)	193.26	298.16	367.08	273.35	506.72[a]	5.39
%MWG (%)	62.39	95.73	118.57[b]	87.77	163.23[a]	1.73
SGR(%/ day)	0.69[b]	0.96[ab]	1.12[a]	0.90[ab]	1.37[a]	0.32
TFI(g)	71764.14	86822.40	96386.43	83295.24	116396.28	767.21
AFI(g)	68.35	82.69	91.80	79.33	110.85[a]	0.73
FCR	5.07	3.96	3.57	4.16	3.13	0.05
GCFE	19.79	25.25	27.98	24.08	32.01[a]	0.23 2.23
PI	29.35[d]	35.68[c]	41.4[b]	33.35[c]	47.60[a]	0.12
PER	0.46[c]	0.58[b]	0.62[a]	0.57[c]	0.75[a]	

Means within the row with difference superscripts are significantly different (p<0.05); PMWG, Percentage means weight gain; MWG/day, mean weight gain/day; SGR, specific growth rate; TFI, average feed intake; FCR, feed conversion ratio; PI, protein intake; PER, protein efficiency ratio.

utilized by the fingerlings with the highest total feed intake observed in Trt 5 (75% maggot meal) while control (Trt 1) showed the lowest value, this is in agreement with Alegbeleye et al. (1991) and Idowu et al. (2003) that maggot meal like other animal protein sources was well accepted and utilized by the fish. The significantly (P > 0.05) higher protein efficiency ratio (PER) of fish fed 50, 66 and 75% maggot meal based diets compare to others, attests to the fact that maximum utilisation of nutrients was obtained at the higher levels of maggots in the diet.

CONCLUSION AND RECOMMENDATION

From the results obtained, diet 5 (75% maggot meal) has the best performance in term of F'CR, MWG, SGR, PER therefore it can be concluded that maggot meal can be included in fish diet up to 75% in the diet of catfish fingerlings.Based on the result obtained from the experiment, it is hereby recommended that 75% maggot meal can be included in the diet of *C. gariepinus* to reduce cost and maximize profit.

REFERENCES

Ajani EK, Nwanna LC, Musa BO (2004). Replacement of fishmeal with maggot meal in the diets of Nile tilapia, Oreochromis niloticus. World Aqua. 35:52-54.

Alegbeleye WO, Anyanwu DF. Akeem AM (1991). Effect of varying dietary protein levels on the growth and utilization performance of catfish, *Clarias gariepinus*. Proceedings of the 4th Annual Conference of Nigerian Nigeria, pp. 51-53.

Aniebo AO, Erondu ES, Owen OJ (2009) Replacement of fish meal with maggot meal in African Catfish (*Clarias gariepinus*) diets Revista UDO Agricola 9(3):666-671.

AOAC (1990). Official methods of Analysis, 13th edition. Association of Official Analytical Chemists, Arlington, Virginia.

Awoniyi TAM, Aletor VA, Aina JM (2003). Performance of Broiler Feel on Maggot Meal in Place of Fishmeal. Int. J. Poult. Sci. 2 (4):271-274.

Fasakin EA, Balogun, AM, Ajayi OO (2003a). Nutrition implication of processed maggot meals; hydrolysed, defatted, full-fat, sun-dried and oven dried in the diets of Clarias gariepinus fingerlings. Aquac. Res. 9(34):733-738.

Fasakin EA, Balogun AM, Ajayi OO (2003b). Evaluation of full-fat and defatted maggot meals in the feeding of Clariid catfish Clarias gariepinus fingerlings. Aqua. Res. 34:733-738.

Fashina-Bombata HA, Balogun O (1997). The effect of partial or to tal replacement of fish meal with maggot meal in the diet of (Oreochromis niloticus) fry. J. Prospects Sci. 1:178-181.

Field A (2000) Discovering statistics using SPSS for windows, sage Publication. p.496.

Idowu AB, Amusan AAS, Oyediran AG (2003). The response of C. griepinus (Burchell 1822) to the diet containing housefly maggot (Musa domestica) Nigerian J. Anim. Prod. 30(1):139-144.

Ng WK., Liew FL, Ang LP, Wong LX (2001). Potential of mealworm (*Tenebriomolitor*) as an alternative protein source in practical diets for African Catfish, *Clarias gariepinus*. Agric. Res. 32(1):273–280.

Ogunji, Slawski H, Schulz C, Wener C, Wirth M (2006a). Preliminary evaluation of housefly maggot meal as an alternative protein source in diet of carp (Cyprinus carpio L.) World Aquaculture Society Abstract Data Aqua 2006 – Meeting, Abstract 277.

Ogunji JO, Kloas W, Wirth M, Schulz C, Rennert B (2006b). Housefly maggot meal (Magmeal): An emerging substitute of fishmeal in tilapia diets. http://www.tropentag.de/2006/abstracts/full/76.pdf

Ogunji JO, Wirth M (2001). Alternative protein sources as substitutes for fish meal in the diet of young Tilapia Oreochromis niloticus (Linn.). Israeli J. Aqua – Bamidgeh 53:34-43.

Olvera – Novoa MA, Campos GS, Sabido GM, Martinez – Palacios CA (1990). The use of alfalfa leaf protein concentrate as protein source in diets for tilapia (Oreoclronics mossambicus). Aquaculture 90:291-302.

Sogbesan AO, NAjuonu N, Musa BO, Adewole AM (2006). Harvesting techniques and evaluation of maggot meal as animal dietary protein source for "Heteroclarias" in outdoor concrete tanks. World J. Agric. Sci. 2(4):394-402.

Teotis JS, Milles BF (1973). Fly pupae as a dietary ingredient for starting chicks. Poult. Sci. 52:1830-1835.

23

Impacts of flooding on coastal fishing folks and risk adaptation behaviours in Epe, Lagos State

A. S. Oyekale[1], O. I. Oladele[1] and F. Mukela[2]

[1]Department of Agricultural Economics and Extension, North-West University Mafikeng Campus, Mmabatho, 2735 South Africa.
[2]Department of Agricultural Economics, University of Ibadan, Ibadan, Oyo State, Nigeria.

Climatic changes have made flooding a major environmental hazard in the coastal areas of Nigeria. This study assessed the impacts and households' adaptation mechanisms in Epe Division of Lagos State. Data obtained from some fishing folks were subjected to descriptive, Probit regression and Tobit regressions analytical methods. The results show that the adverse impacts of flooding reduce with ability to migrate, monthly income and possession of other secondary occupations. The females, educated and rich among the fishing folks are willing to pay significantly higher amounts on insurance against flooding. The study recommended that efforts to address flooding should include alternative skill development, migration, offering of assistances, provision of affordable and quality health services for the treatment of malaria, cholera and dysentery and proper development of early warning signal, among other.

Key words: Flooding, climate change, impact, vulnerability, fishing folks.

INTRODUCTION

In 1976, international communities ratified the right to adequate housing and continuous improvement of living conditions as contained in the International Covenant on Economic, Social and Cultural Rights (The Office of the High Commission for Human Right). This pledge was reactivated at the UN Millennium Summit where, in line with the Millennium Development Goals (MDGs), some specific targets for realizing a significant improvement in the lives of at least 100 million slum dwellers' by 2020 was set. In recent years, however, climate change presents significant threats to the achievement of those laudable Millennium Development Goals (MDGs), especially those related to eliminating poverty and hunger and promoting environmental sustainability (United Nations, 2007). This is because, variations in rainfall and extreme weather events are likely to place additional strains on poorer countries already facing serious challenges due to food insecurity, indebtedness, HIV/AIDS, environmental degradation, armed conflicts, economic shocks and the negative effects of globalization (UNDP, no date). Therefore, attentions have been given to extreme weather events and climatic patterns resulting from global warming, which is caused by emission of green house gases (GHGs) like carbon dioxide (CO_2), methane (CH_4) and nitrous oxide (N_2O). UN (2007) submitted that the primary contributor to climate change is carbon dioxide (CO_2), released by the burning of fossil fuels, whose concentration had reached 29 billion metric tons in 2004 and continue to rise, as evidenced by increasing concentrations of CO_2 in the atmosphere.

The Intergovernmental Panel on Climate Change (IPCC) has linked the rise in sea level to climate change. Between 1960 and 1970, a mean sea level rise of 0.462 m was recorded along the Nigerian coastal water (Udofa and Fajemirokun, 1978). Flooding of low-lying areas in the Lagos and the coastal State of the Niger Delta region has been observed. Settlements in the coastal region have been uprooted by coastal erosion. The inundation arising from the rise in sea level will increase problems of floods, intrusion of sea-water into fresh water sources and ecosystems, destroying such stabilizing systems as mangroves, and affecting agriculture, fisheries and general livelihoods (Okali and Eleri, 2004). Occupants of some of the affected houses, who are unable to relocate for financial reasons, will have to cope with the situation. This makes them vulnerable to different kinds of water-related disease such as malaria, dysentery, cholera, and diarrhea. Trauma resulting from the problem can lead to nonpathogenic diseases such as hypertension and diabetes (Uyigue, 2007).

Murty (1984) defined storm surges as "oscillation of the water level in a coastal or inland water body in period range of few a minutes to a few days, resulting from forcing from weather system". Over the years, the combination of strong onshore winds, low atmospheric pressure and astronomical tides has often resulted in ocean surges and associated coastal flooding in Lagos State. The threat imposed by these surges and associated flood over the low-lying coast of Nigeria is becoming unbearable (Olaniyan and Afiesimama, 2002). The most serious flooding often results when an extreme storm surge event occurs concomitantly with a tidal maximum (Lowe et al., 2001). Water level during some ocean surges in Nigeria can rise up to about 5 m (Awosika et al., 1992). Surface waves and other coastal circulation changes can modify or amplify the impact of a surge in the coastal region.

Flood hazard is one of the most frequent and disastrous phenomena in the world (Marfai and King, 2007). Sivakumar (2005) noted that between 1993 and 2002, there were 2,654 hazard events where floods and windstorms accounted for about 70% of the hazards while the remaining 30% of the disasters were brought about by droughts, landslides, forest fires, heat waves and others. The problem has also led to contamination of coastal water resources, decimation of coastal agricultural and recreational area, destruction of settlement and major roads, dislodgement of oil producing and export handling facilities, loss of properties, income and sometimes lives. The anomalous forcing is generally sustained for several days with largest disturbances arising from storm tracks (Komen and Smith, 1999).

Nigeria's high vulnerability to climate change mainly stems from its geographical location – in the tropics and with a long coastline. With climate change predicted to cause global sea level rise to which 29,000 km^2 of Nigerian coastline is highly vulnerable, the frequency of incidence of extreme weather condition in the form of flooding resulting from ocean surge cannot be prevented (IPCC, 2001). Flooding in one form or other affects at least 20% of the nation's population. Flooding and its twin problem of

erosion cause losses that run into billions of Naira per year (Ayo-Lawal, 2007). It should be noted that the types of flooding known to happen in Nigeria include flash floods (river flooding) after torrential rains, dam bursts which follow flash floods and urban flooding of low-lying areas with poor surface drainage as well as coastal flooding. The rainy season is marked by widespread phenomenon of flooding experienced due to heavy torrential rains or storm surge exacerbated by the low lying (coastal) topography, vulnerable soil characteristics, and intense wave and tidal action (UNESCO/CSI, 2000). Also, urbanization and population pressure with its attendant problems of poor land use leading to alterations and blockage of channels constitute the human induced causes of flooding like dumping of refuse in or erecting buildings on drainage routes.

It is pathetic to note that rural dwellers that are most vulnerable to climate change have the least capacity to adapt. Flooding becomes more severe when fishing communities that are located along the coasts are involved. This is due to high level of and their low adaptation mechanism. This study therefore assessed the impact of flooding among coastal fishing households in Epe Division of Lagos State and evaluated their adaptation mechanisms. It is expected that the findings from the study will pave ways for providing assistance to these vulnerable households.

Conceptual framework

This study adopted the livelihood strategies framework presented in Figure 1. It is the basis of most frameworks used in impact and adaptation studies such as the one utilized by Leary and Beresford (2007). This framework shows how the interactions between human induced factors and natural factors result in environmental hazards such as flooding. The exposure of vulnerable people, places and systems determines the amount of impact the hazards will have or the level of risk experienced.

Therefore, vulnerability to climatic hazards is a function of the interactions between sensitivities, capacities and resilience, which are all determined by human-environmental conditions and processes. Resilience is the amount of change a system can undergo without changing state (IPCC, 2001). These conditions and processes consist of demographic, social, economic, political, biophysical and ecological. Adaptation options of stakeholders are shown to be a function of information. The stakeholders include decision makers (or policy makers), vulnerable groups like fishermen, and the general public. From the figure, information on assessment of changes, vulnerabilities, exposures, risks/impacts, adaptation responses and driving forces must flow to and from the stakeholders.

Climate change is expected to manifest itself through change in frequency and magnitude of extreme events

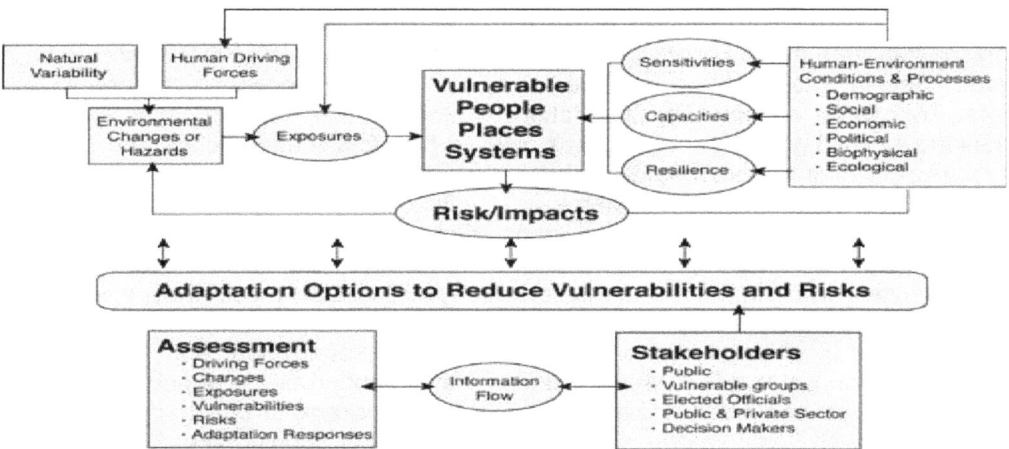

Figure 1. Vulnerability, adaptation and responses assessments (Leary and Beresford, 2007).

like flooding, drought, storm etc. (IPCC, 2001). Although there is dearth of data on the frequency of disasters due to climate change in many countries, Jepma and Munasinge (1998) estimate that with a 70 cm sea level rise, the number of people at risk of annual flooding could increase from 46 to 90 million. These changes (floods and storms) are expected to bring about, in addition to loss of biodiversity, life and livelihood, the spread of diseases such as cholera, dysentery, malaria and yellow fever and increase the chances of famine in areas with inadequate coping systems (Blaikie, 1994).

Literature maintains that areas where malaria is currently endemic could experience intensified transmission because flooding provides breeding ground while higher temperatures worsen the situation. Increases in non-vector-borne infectious diseases, such as salmonellosis, cholera and giardiasis, also could occur as a result of elevated temperatures and increased flooding. This means that a great obstacle stands in the way of achievement of the Millennium Development Goals because malaria is endemic in this part of Africa and translates into drastically reduced food production leading to food insecurity and all its attendant problems.

There are varieties of underutilized options for reducing flood, erosion and drought (Osman-Elasha et al., 2007). However, application of these measures depends to a large extent on perception of risk. Maddison (2006) reported that perception results on climate change showed that a significant number of farmers believe that temperature has already increased and that precipitation has declined. The study also reported that experience, access to free extension services and markets are important determinants of adaptation.

Since perception is the fundamental prerequisite of long-term livelihood adaptation facilitated by coordination of agency planning, communication and field operations activities, as well as the activities of government line agencies and departments, NGOs, and the rural people themselves; it follows that adaptation is to have a

multisectoral approach. It should be participatory in nature to ensure sustainability and faster spread as well as effective use. It is also important to recognize that no single adaptation strategy can be encouraged, given the uniqueness of each situation. This is clearly recognized by UNFCCC (2002) which emphasizes the lack of a "one size fits all formula". Therefore, there is a need to identify existing knowledge and coping strategies of the poor as well as their risk perception which would determine their willingness to pay or adapt as well as the effectiveness of adaptation.

MATERIALS AND METHODS

Area of study

Lagos State, the city of aquatic splendour, was created on 27th May, 1967 and lies to the South Western part of Nigeria. It has boundaries with Ogun State in the North and East, Republic of Benin in the West and the Atlantic Ocean in the South with a shore line stretch of 180 km. It comprises five geographical divisions which are Epe, Ikorodu, Ikeja, Lagos and Badagry. Lagos has a population of 17 million out of the national estimate of 140 million in 2006 and an area of 356,861 ha of which 75,755 ha are wetlands. It has a maze of lagoons and waterways make up about 22% of the total landmass.

This study was carried out in Epe, which is a coastal town demarcated by a long range of hills into equal parts. Epe local government area has a total population of 323,634 people of which 153,360 are males (2006 census). Most of the inhabitants of the division engage in fishing and farming activities for their livelihood though the fishing activities is seasonal and synchronized to the lifecycle of the fish. The land is highly endowed with forest wood and the types of crops grown are rice, coconut, pineapple, cassava, cocoa, palm tree, banana or plantain, maize, vegetables and ginger favoring conditions. Silica sands, fish, reptiles, shrimps and bitumen are also extractable in the area.

Sources and method of data collection

This study used primary data that were collected through personal interviews that were aided by structured questionnaires. Simple random sampling technique was applied in this study. The respondents, which are fishermen were randomly selected. The selection of communities in Epe was based on degree of fishing activities and proximity to water source. Out of the 120 questionnaires distributed only 100 were returned of which only 94 were completely and properly filled by the respondents. The communities visited included Owode and Olowo market. These were chosen because of high degree of fishing activities.

Method of data analysis

The data collected from the field were processed using various statistical and analytical tools. These include simple statistical tools like frequencies, percentages, and descriptive statistics and cross tabulations.

Probit regression

In order to identify the factors explaining experiences of some forms of problems resulting from flooding, we used the Probit regression. We intended to estimate the factors explaining some probability values ranging between 0 and 1. The dependent variable (Yi) will be binary with values of 1 if respondents answered yes to some questions of whether flooding is the major environmental problem, experienced activity disruption by flooding, experienced of property losses through flooding, experienced of death through flooding, experienced water contamination due to flooding and spend more on health immediately after flooding, and 0 otherwise. The model is stated as:

$$Y_{ij} = \alpha_j + \beta_j \sum_{j=1}^{n} X_{ij} + e_i$$

X_1 = sex (1 = male, 0 otherwise), X_2 = age of respondents in years, X_3 = marital status (married = 1, 0 otherwise), X_4 = household size, X_5 = years of education, X_6 = monthly income in Naira, X_7 = fishing experience in years, X_8 = association membership (yes = 1, 0 otherwise), X_8 = malaria incidence increased during flooding, X_9 = typhoid incidence increased during flooding (yes = 1, 0 otherwise), X_{10} = cholera incidence increased during flooding (yes = 1, 0 otherwise), X_{11} = diarrhea incidence increased during flooding (yes = 1, 0 otherwise), X_{12} = dysentery incidence increased during flooding (yes = 1, 0 otherwise), X_{13} = influenza incidence increased during flooding i(yes = 1, 0 otherwise), X_{14} = tuberculosis incidence increased during flooding (yes = 1, 0 otherwise), X_{15} = more labour required during flooding (yes =1, 0 otherwise), X_{16} = fish species change during flooding (yes =1, 0 otherwise), X_{17} = early warning received before flooding (yes =1, 0 otherwise), X_{18} = other occupation (yes = 1, 0 otherwise), X_{19} = received help during flooding (yes = 1, 0 otherwise); ei = error disturbance term.

Tobit regression for risk insurance

The fishing folks were asked for their willingness to participate in a functioning insurance program against flooding. We found this to be relevant given increasing attentions that are being given to issues of insurance companies and implementation of the National Heath Insurance Scheme (NHIS). The households were asked for the among they would be willing to pay monthly in order to secure a kind of insurance for their properties, in case of flooding. Because we have zero amounts, a Tobit model was estimated. The equation is given as:

$$Z_i = \phi + \lambda_j \sum_{1=j}^{n} X_j + v \tag{2}$$

Where Z_j is the amount willing to be paid and Xs are as defined for Equation (1).

RESULTS AND DISCUSSION

Experience of flooding across socio-economic groups

Table 1a shows that 88.30% of the fishing folks experienced one form of flooding or the other. Among these, 77.66% are males. The table also reveals that males have higher involvement in fishing as an occupation. This confirms earlier findings by various researchers such as Haakonsen (1992) and Jinadu (2003) that fishing is primarily done by males, although women are predominantly involved in the processing. Average age of the fishing folks is 36.42 years with 27.32% as the coefficient of variation. In Table 1a, those that experienced flooding among those that are less than 30 years are 35.11% while 7.45% did not.

Average fishing experience is 9.64% with its variability index being 116.51%. Based on the groupings of the fishing folks' experiences, among those that are less than 5 years in fishing, 39.36% experienced flooding, while only 8.51% have not experienced it. It should be noted that all those who indicated to have spent more than 16 years in fishing experienced flooding.

The table shows that 55.32% of the respondents are married and experienced flooding, while only 4.26% did not experience it. Average household size is 6.14 with standard deviation of 2.91. In the groupings contained in Table 1a, most of the respondents have household sizes of 5 to 6, while the least have household size of more than 10 members. From this, it can be inferred that respondents have a larger human capital asset base which can be translated to enhanced capacity to cope especially if household members are in their productive phase of life. On the other hand, it may mean a reduced capacity to cope with stresses and shocks if household members are all dependent upon the household head. The table also reveals that most of the fisher folk lack formal education and only a small fraction managed to attend tertiary institutions. Also, 73.40% of the respondents primarily engage in fishing as means of livelihood. The results show that artisans, traders and farmers are among the most prominent occupations, besides fishing.

Impact of flooding on fishing households

We estimated some equations in order to assess the impacts of flooding on the fishing folks. These equations

Table 1. Respondents' experience of flooding across some socio-economic groups.

Socio-economic groups	Yes	Percent	No	Percent	Total frequency	Percent
Sex						
Male	146	77.66	16	8.51	162	86.17
Female	20	10.64	6	3.19	26	13.83
Age						
less 30	66	35.11	14	7.45	80	42.55
30-39	46	24.47	4	2.13	50	26.60
40-49	26	13.83	2	1.06	28	14.89
50-59	24	12.77	2	1.06	26	13.83
More than 60	4	2.13	0	0.00	4	2.13
Fishing experience						
0-5	74	39.36	16	8.51	90	47.87
6-10	29	15.43	4	2.13	33	17.55
11-15	18	9.57	2	1.06	20	10.64
16-20	26	13.83	0	0.00	26	13.83
More than 20	19	10.11	0	0.00	19	10.11
Household size						
2-4	66	35.11	14	7.45	80	42.55
5-6	46	24.47	4	2.13	50	26.60
7-8	26	13.83	2	1.06	28	14.89
8-10	24	12.77	2	1.06	26	13.83
More than 10	4	2.13	0	0.00	4	2.13
Marital status						
Single	36	19.15	10	5.32	46	24.47
Married	104	55.32	8	4.26	112	59.57
Separated	8	4.26	0	0.00	8	4.26
Widowed	14	7.45	4	2.13	18	9.57
Divorced	4	2.13	0	0.00	4	2.13
Education						
None	44	23.40	2	1.06	46	24.47
Primary	32	17.02	8	4.26	40	21.28
Secondary	44	23.40	6	3.19	50	26.60
Tertiary	12	6.38	4	2.13	16	8.51
Adult education	8	4.26	0	0.00	8	4.26
Other non-formal	26	13.83	2	1.06	28	14.89
Primary occupation						
Fisherman	122	64.89	16	8.51	138	73.40
Teaching	0	0.00	2	1.06	2	1.06
Trader	12	6.38	2	1.06	14	7.45
Okada rider	4	2.13	0	0.00	4	2.13
Artisan	12	6.38	0	0.00	12	6.38
Student	4	2.13	0	0.00	4	2.13
Clerk	2	1.06	0	0.00	2	1.06
Farmer	6	3.19	0	0.00	6	3.19
Civil servant	2	1.06	2	1.06	4	2.13
Sand loader	2	1.06	0	0.00	2	1.06

Table 1b. Description of variables used for analysis.

Variable	Mean	Std. deviation
Problems associated with flooding		
Flooding as environmental problem	0.6809	0.4674
Flooding disrupt economic activities	0.6596	0.4751
Flooding leads to property loss	0.6489	0.4786
Flooding leads to water contamination	0.3723	0.4847
Flooding lead to human and livestock death	0.2766	0.4485
Spend more on health during flooding	0.1489	0.3570
Socio-economic characteristics		
Sex	0.8617	0.3461
Age	33.8298	11.4990
Marital status	0.7021	0.7988
Household size	5.4043	3.4403
Education of house head	6.6596	5.2378
Monthly income	32638.3000	63213.6900
Fishing experience	9.6383	11.2291
Membership of association	0.6383	0.4818
Disease incidence		
Malaria incidence during flooding	0.2340	0.4245
Typhoid incidence during flooding	0.2553	0.4372
Cholera incidence during flooding	0.2340	0.4245
Diarrhea incidence during flooding	0.2766	0.4485
Dysentery incidence during flooding	0.1702	0.3768
Influenza incidence during flooding	0.3191	0.4674
Tuberculosis incidence during flooding	0.1383	0.3461
Other impacts and coping methods		
More labour required during flooding	0.7553	0.4310
Received warning signals	0.1702	0.3768
Have other occupation	0.3085	0.4631
Received help during flooding	0.9149	0.2798
Migration	0.5160	0.5011
Fish composition changed during flooding	0.1277	0.3346

are based on whether flooding is constituting a major environmental problem, disrupting economic activities, leading to lose of properties, leading to death, resulting into water contamination and households' spend more on health immediately after flooding. Table 1b shows the description of the variables that were used for the analysis.

Flooding as environmental problem

Table 2 shows the Probit results for analyzing the factors predisposing flooding to constitute a major environmental problem to the fishing folks. The pseudo adjusted coefficient of determination shows that the model explained 26.88% of the variations in the probability. The computed Chi Square value for the likelihood ratio is statistically significant (p<0.01). This shows that the model fits the data very appropriately.

Out of the included variables, fishing experience, higher incidence of influenza during flooding, tuberculosis during flooding, more labour required during flooding and having other occupation show some statistical significance (p<0.10) and have varied signs. The results show that flooding is less likely to be an environmental problem to those with higher fishing experience, with reported cases of influenza and those with other occupations. Precisely, flooding constitutes some environmental problems as homes are flooded and debris of refuses liter the whole compound and rooms. Those with higher fishing experience are expected to have devised a better way of coping so that they are not always affected by flooding.

Table 2. Probit regression of factors explaining flooding as the major environmental problem.

Variable	Coefficients	Standard error	z value
Socio-economic characteristics			
Sex	0.2948	0.4058	0.73
Age	-0.0005	0.0161	-0.03
Marital status	0.1384	0.1850	0.75
Household size	-0.0242	0.0461	-0.53
Education of house head	-0.0135	0.0244	-0.55
Monthly income	0.0000	0.0000	-0.89
Fishing experience	-0.0230*	0.0121	-1.89
Membership of association	0.3981	0.3159	1.26
Disease Incidence			
Malaria incidence during flooding	-0.2940	0.3541	-0.83
Typhoid incidence during flooding	0.3762	0.3312	1.14
Cholera incidence during flooding	-0.3113	0.3547	-0.88
Diarrhea incidence during flooding	-0.0722	0.3609	-0.2
Dysentery incidence during flooding	0.6615	0.4444	1.49
Influenza incidence during flooding	-0.9214***	0.3138	-2.94
Tuberculosis incidence during flooding	0.8988*	0.5098	1.76
Other impacts and coping methods			
More labour required during flooding	1.3554***	0.3512	3.86
Fish composition changed during flooding	0.3072	0.4503	0.68
Received warning signals	0.4606	0.3700	1.24
Have other occupation	-0.6159**	0.2831	-2.18
Received help during flooding	-0.9359	0.5905	-1.58
Migration	-0.3841	0.2658	-1.45
Constant term	0.7939	0.8032	0.99

LR Chi^2 (21) = 63.30; Prob > Chi^2 = 0.0000; Pseudo R^2 = 0.2688; ***Significant at 1%; **Significant at 5%, *Significant at 10%.

Orlove et al. (2004) reported that some workers did switch to other occupations in the event of flooding (or reallocated their effort to secondary employment), but the opportunities in the labor market were extremely limited, due to the general economic downturn of the affected communities. However, those people with reported cases of tuberculosis after flooding and who would require more labour more their fishing activities during flooding have higher probability of seeing flooding as a major environmental problem. During flooding, Orlove et al. (2004) reported that transportation networks are often disrupted thus constituting serious environmental problem.

Flooding disrupts economic activities

Table 3 contains the result for those whose economic activities are disrupted by flooding. The Chi square of the likelihood ratio is also statistically significant (p<0.01), showing that the model fits the data very well. Out of the variables that were included, age, household size, fishing experience, typhoid fever, cholera, dysentery, influenza, fish composition change, received warning, received assistance during flooding and migration are statistically significant (p<0.10).

The results further show that those with more fishing experience have significantly lower probability of having their economic activities disrupted by flooding. However, those with large household size have significantly higher probability of having their economic activities affected by flooding. This may result from higher pressure to meet their needs. The health variables reveal that those that reported more cases of typhoid fever, cholera, and dysentery have higher probability of having their economic activities disrupted by flooding. Douglas et al. (2008) reported that in Sierra Leone, cholera and dysentery were largely associated with flooding with serious economic consequences. Those who received warning, and indicated that fish composition changes during flooding have higher probability of having their economic activities disrupted by flooding. The result for

Table 3. Probit regression of factors explaining experience of activity disruption by flooding.

Variable	Coefficients	Standard error	z value
Socio-economic characteristics			
Sex	0.2804	0.4074	0.69
Age	-0.0418**	0.0171	-2.45
Marital status	0.0397	0.1814	0.22
Household size	0.1583***	0.0547	2.90
Education of house head	0.0385	0.0258	1.49
Monthly income	0.0000	0.0000	0.86
Fishing experience	-0.0296**	0.0120	-2.46
Membership of association	0.3218	0.3200	1.01
Disease Incidence			
Malaria incidence during flooding	0.2568	0.3640	0.71
Typhoid incidence during flooding	0.7011*	0.3614	1.94
Cholera incidence during flooding	0.7894*	0.4178	1.89
Diarrhea incidence during flooding	-0.4962	0.3650	-1.36
Dysentery incidence during flooding	0.8062*	0.4622	1.74
Influenza incidence during flooding	-1.4635***	0.3543	-4.13
Tuberculosis incidence during flooding	0.1673	0.4237	0.39
Other impacts and coping methods			
More labour required during flooding	0.5841	0.3597	1.62
Fish composition changed during flooding	1.0106**	0.4450	2.27
Received warning signals	0.7401*	0.3886	1.90
Have other occupation	-0.4234	0.2918	-1.45
Received help during flooding	-1.4892**	0.6335	-2.35
Migration	-0.5305**	0.2630	-2.02
Constant term	1.7801**	0.8322	2.14

LR chi^2(21) = 68.02; Prob > chi^2 = 0.0000; Pseudo R^2 = 0.2821; ***Significant at 1%; **Significant at 5%; *Significant at 10%.

warning is not what is expected. However, Orlove et al (2004) found that it is not in all cases that fishing folks take to warnings. However, those who received assistance during flooding have significantly lower probability of having their economic activities disrupted by flooding. Also, those that are able to migrate during flooding have significantly lower probability of having their economic activities disrupted by flooding. Douglas et al. (2008) identified migration as a viable form of adaptation being used by some flood affected households in Accra.

Experience of property losses

Table 4 shows the results of the analysis for those who indicated that they lose their properties during flooding. The Pseudo Adjusted R square implies that 29.64% of the variations in the values of the probabilities have been explained by the included variables. The likelihood ratio of the Chi Square is statistically significant (p<0.01) and implies that the model produced a good fit for the data.

Out of the variables that were included, monthly income, diarrhea, more labour required, received warning signals, have other occupation and migration are statistically significant (p<0.10).

Also, as monthly income increases, the probability of losing properties due to flooding significantly decreases. Those that required more labour and received initial warnings have significantly higher probability of losing their properties during flooding. Possession of other occupation significantly reduced the probability of losing properties due to flooding. Those that are able to migrate during flooding have significantly lower probability of losing their properties during flooding.

Flooding leading to loss of life of human beings and livestock

Table 5 estimated the equation for the likelihood of flooding resulting into death of human and animals. It should be noted that occasionally, flooding resulted into

Table 4. Probit regression of factors explaining experience of property losses through flooding.

Variable	Coefficients	Standard error	z value
Socio-economic characteristics			
Sex	0.1805	0.4012	0.45
Age	-0.0055	0.0163	-0.34
Marital status	0.3383	0.2242	1.51
Household size	0.0719	0.0495	1.45
Education of house head	-0.0295	0.0243	-1.22
Monthly income	0.0000**	0.0000	-2.06
Fishing experience	-0.0181	0.0122	-1.48
Membership of association	-0.1862	0.3116	-0.6
Disease incidence			
Malaria incidence during flooding	0.0483	0.3410	0.14
Typhoid incidence during flooding	0.3453	0.3136	1.1
Cholera incidence during flooding	-0.0461	0.3811	-0.12
Diarrhoe incidence during flooding	-0.7107**	0.3358	-2.12
Dysentery incidence during flooding	0.5248	0.4258	1.23
Influenza incidence during flooding	0.4362	0.3450	1.26
Tuberculosis incidence during flooding	0.5398	0.4947	1.09
Other impacts and coping methods			
More labour required during flooding	0.8018**	0.3502	2.29
Fish composition changed during flooding	0.7865	0.4943	1.59
Received warning signals	1.2896***	0.4564	2.83
Have other occupation	-0.4675*	0.2814	-1.66
Received help during flooding	-0.2510	0.4640	-0.54
Migration	-0.4687*	0.2608	-1.8
Constant term	0.1179	0.7381	0.16

LR chi^2 (21) = 72.23, Prob > chi^2 = 0.0000, Pseudo R^2 = 0.2964; ***Significant at 1%; **Significant at 5%; *Significant at 10%.

Table 5. Probit regression of factors explaining experience of death through flooding.

Variable	Coefficients	Standard error	z value
Socio-economic characteristics			
Sex	0.1955	0.5551	0.35
Age	0.0001	0.0182	0
Marital status	-0.0540	0.1887	-0.29
Household size	0.0155	0.0529	0.29
Education of house head	0.0015	0.0279	0.05
Monthly income	0.0000**	0.0000	2.11
Fishing experience	-0.0262**	0.0131	-2.00
Membership of association	0.0061	0.3443	0.02
Disease incidence			
Malaria incidence during flooding	-0.0132	0.3597	-0.04
Typhoid incidence during flooding	0.5640	0.3513	1.61
Cholera incidence during flooding	-0.2110	0.3898	-0.54
Diarrhoe incidence during flooding	-0.1117	0.3572	-0.31
Dysentery incidence during flooding	-2.4278***	0.6407	-3.79
Influenza incidence during flooding	1.4761***	0.3886	3.8

Table 5. Contd.

Tuberculosis incidence during flooding	2.4110***	0.6080	3.97
Other impacts and coping methods			
More labour required during flooding	0.6130	0.4351	1.41
Fish composition changed during flooding	0.2337	0.4421	0.53
Received warning signals	-0.8129	0.3874	-2.1
Have other occupation	0.6360**	0.2844	2.24
Received help during flooding	-0.0041	0.5792	-0.01
Migration	1.0947***	0.3173	3.45
Constant term	-2.2677**	0.9288	-2.44

LR chi^2 (21) = 95.22; Prob > chi^2 = 0.0000; Pseudo R^2 = 0.3836; ***Significant at 1%; **Significant at 5%; *Significant at 10%.

death of human beings in the study area. However, the majority of those who indicated to be affected in this category of analysis only lost their livestock. The pseudo R square indicated that 38.36% of the variations in the values of the probabilities have been explained by the included variables.

As the monthly income increases, the probability of suffering loses of livestock due to flooding increases. Those who reported higher incidence of influenza and tuberculosis also have significantly higher probability of experiencing death of livestock/human due to flooding. Those that received warning signals have significantly lower probability of suffering loss of livestock during flooding. Also, those that have other occupations have higher probability of suffering losses of livestock due to flooding. Those that indicated to be migrating due to flooding have significantly higher probability of suffering death of humans and livestock due to flooding.

Water contamination through flooding

Table 6 shows the results of Probit regression explaining the factors that are associated with those that indicated to be suffering water contamination due to flooding. The Pseudo Adjusted R Square indicates that 53.57% of the variations in the probability of suffering water contamination due to flooding. The chi square of the likelihood ratio is also statistically significant (p<0.01). Those variables that are statistically significant (p<0.10) are sex, education, fishing experience, membership of association, malaria, typhoid, cholera, dysentery, and fish composition changed due to flooding.

Being a male respondent significantly reduces the probability of having water contaminated due to flooding. Also, as the years of education and fishing experience increase, the probability of having water contaminated through flooding increases. Membership of associations significantly reduces the probability of having water contaminated by flooding. Those that indicated to suffer more incidences of malaria, cholera, and dysentery have

significantly probability of having their water contaminated by flooding. However, contrary to expectation, those that indicated more incidences of typhoid fever and diarrhea during flooding have significantly lower probability of having their water contaminated. Also, those that indicated that the composition of fish changes after flooding have significantly higher probability of having their waters contaminated.

Health expenditures increases during flooding

Table 7 shows the Probit regression results of the factors explaining spending more on health during flooding. The pseudo coefficient of determination shows that 61.22% of the variations in the values of probability are explained by the included variables. The Likelihood ratio chi square is also statistically significant (p<0.01). Out of the variables that were included, marital status, household size, education, fishing experience, malaria, cholera, diarrhea, influenza, tuberculosis and migration are statistically significant (p<0.10).

The results show that those that are married have significantly higher probability of spending more on health during flooding. However, contrary to expectation, household size variable has negative sign, implying that those that have higher family size have significantly lower probability of spending more on health during flooding. Also, as households' education increases, the probability of spending more on health during flooding increases. As fishing experience increases, the probability of spending more on health during flooding increases. The included health variables shows that those that indicated higher incidence of malaria and diarrhea have significantly lower probability of spending more on health during flooding. However, those that indicated higher incidence of influenza and tuberculosis during flooding have significantly higher probability of spending more on health during flooding. Those that indicated to be migrating due to flooding have significantly higher probability of spending more on health during flooding.

Table 6. Probit regression results of factors explaining suffering water contamination due to flooding.

Variable	Coefficients	Standard error	z value
Socio-economic characteristics			
Sex	-2.2774**	0.8911	-2.56
Age	0.0179	0.0270	0.66
Marital status	0.2645	0.1733	1.53
Household size	-0.1062	0.0723	-1.47
Education of house head	0.0613*	0.0315	1.95
Monthly income	0.0000	0.0000	0.15
Fishing experience	0.0530***	0.0187	2.84
Membership of association	-0.7084*	0.4083	-1.74
Disease incidence			
Malaria incidence during flooding	1.8704***	0.4969	3.76
Typhoid incidence during flooding	-2.3220***	0.6945	-3.34
Cholera incidence during flooding	2.1653***	0.7618	2.84
Diarrhoe incidence during flooding	-2.2979***	0.8659	-2.65
Dysentery incidence during flooding	2.6946***	0.6472	4.16
Influenza incidence during flooding	-0.2384	0.4213	-0.57
Tuberculosis incidence during flooding	-0.5632	0.6192	-0.91
Other impacts and coping options			
More labour required during flooding	1.8708**	0.7781	2.4
Fish composition changed during flooding	1.9174**	0.5232	3.66
Received warning signals	-0.2859	0.4483	-0.64
Have other occupation	-0.5244	0.4127	-1.27
Received help during flooding	0.0978	0.6659	0.15
Migration	0.2299	0.3624	0.63
Constant term	-1.7186	1.0788	-1.59

LR chi^2 (21) = 118.78; Prob > chi^2 = 0.0000; Pseudo R^2 = 0.5357; ***Significant at 1%; **Significant at 5%; *Significant at 10%.

Table 7. Probit regression results of factors explaining spending more on health due to flooding.

Variable	Coefficients	Standard error	z value
Socio-economic characteristics			
Sex	-6.7524	33.3082	-0.2
Age	0.0119	0.0437	0.27
Marital status	0.9249**	0.4115	2.25
Household size	-0.2652*	0.1484	-1.79
Education of house head	0.2917***	0.1108	2.63
Monthly income	0.0000	0.0000	0.09
Fishing experience	0.0819**	0.0345	2.37
Membership of association	-0.6755	0.8064	-0.84
Disease incidence			
Malaria incidence during flooding	-1.1217*	0.6039	-1.86
Typhoid incidence during flooding	0.4943	0.6735	0.73
Cholera incidence during flooding	3.4892***	1.3218	2.64
Diarrhoe incidence during flooding	-1.5441*	0.9242	-1.67
Dysentery incidence during flooding	-0.5682	0.7670	-0.74
Influenza incidence during flooding	2.1866**	0.9210	2.37

Table 7. Contd.

Tuberculosis incidence during flooding	3.2557***	0.9758	3.34
Other impacts and coping methods			
More labour required during flooding	4.0276	33.2485	0.12
Fish composition changed during flooding	1.0855	0.8386	1.29
Received warning signals	-2.0572	1.0311	-2
Have other occupation	-0.4889	0.6835	-0.72
Received help during flooding	1.6289	1.1271	1.45
Migration	2.0575**	0.8811	2.34
Constant term	-6.3282**	2.5501	-2.48

LR chi^2 (21) = 96.88; Prob > chi^2 = 0.0000; Pseudo R^2 = 0.6122; ***Significant at 1%; **Significant at 5%; *Significant at 10%.

Table 8. Tobit regression results of factors influence amounts willing to pay for insurance against flooding.

Variable	Coefficients	Standard error	z value
Socio-economic characteristics			
Sex	-1281.3100***	442.1992	-2.9
Age	-3.9920	16.3067	-0.24
Marital status	124.3315	178.7979	0.7
Household size	-32.3062	46.9974	-0.69
Education of house head	41.5010*	23.8260	1.74
Monthly income	0.0040	0.0017	2.3
Fishing experience	21.3371*	11.2567	1.9
Membership of association	-216.2850	308.3372	-0.7
Malaria incidence during flooding	489.5611*	298.4399	1.64
Disease incidence			
Typhoid incidence during flooding	417.9736	299.2814	1.4
Cholera incidence during flooding	403.3123	338.4940	1.19
Diarrhoe incidence during flooding	-94.8162	328.6289	-0.29
Dysentery incidence during flooding	-5.4967	376.6810	-0.01
Influenza incidence during flooding	101.5781	310.8857	0.33
Tuberculosis incidence during flooding	-214.2830	375.4206	-0.57
Other impacts and coping methods			
More labour required during flooding	1047.5220**	416.7032	2.51
Fish composition changed during flooding	206.1422	380.6176	0.54
Received warning signals	-678.8190*	382.9870	-1.77
Have other occupation	269.5328	250.9725	1.07
Received help during flooding	-1212.2300***	387.5425	-3.13
Migration	-375.4150	262.0193	-1.43
Constant term	444.2584	672.3573	0.66
Sigma	1170.3100	113.7827	10.2854828

***Significant at 1%; **Significant at 5%; *Significant at 10%.

Tobit regression results of amounts willing to pay for insurance against flooding

Table 8 shows the results of the factors influencing the amounts that fishing folks are willing to pay for insurance against flooding. The pseudo coefficient of determination shows that 4.11% of the variations in the amounts that are willing to pay have been explained by the included variables. The likelihood ratio chi square is statistically significant (p<0.01). The sigma parameter is also statistically significant. This shows that the model produced a good fit for the data.

The results show that the parameters of sex, education, income, fishing experience, malaria, more labour required during flooding, received warning signals and received help during flooding are statistically significant ($p < 0.01$). The amounts that the males fishing folks are willing to pay are significantly lower than that of the female. Also, as education and monthly income increase, the amounts that are willing to be paid significantly increase. As the fishing experience increases, the amounts being willing to pay significantly increases. Those that indicated that more labour are always required for fishing during flooding are willing to pay significantly higher amount for insurance. However, those that indicated to have received warning signals and assistance during flooding are willing to pay significantly lower amounts.

Conclusions

This study was borne out of a desire to study climate change impacts in Lagos with a focus on naturally caused flooding and how the artisannal fishermen are affected because of the pertinent role they play in ensuring food security of the nation. It is obvious that there are many varied impacts on the economy and the need for adaptation as well as mitigation studies and interventions cannot be overemphasized. Given the results of our analyses, the following policy issues clearly emerge from this study:

1. Flooding is an environmental problem that affects majority of the coastal fishing folks, and efforts to mitigate its impact should be earnestly devised. This becomes important due to losses of lives and properties. There is need for government's interventions in ensuring alternative skill development for those along the coast. This is important because possession of alternative occupation reduces vulnerability risks in most of our results.
2. We found that migration and offering of assistances are important factors for reducing vulnerability to some adverse consequences of flooding. There is the need for stakeholders' interventions in providing permanent refugee camps for those that are likely to be affected by flooding. This shows that resettlement may be a workable option. Although this may be abused in a kind of mega city like Lagos, development of appropriate modality for resettlement will assist in averting future problems.
3. Government's interventions in providing affordable health services to those affected by flood are also promising. This is particularly welcomed for treatment of disease outbreaks like malaria, cholera and dysentery that are highly associated with flooding.
4. Proper development of early warning signal is also essential for reducing vulnerability. Such information may

be passed gathered via monitoring by appropriate agencies in Lagos and passed to the people through such media as radio and community leaders.

REFERENCES

Awosika LF, French GT, Nicholls RJ, Ibe CE (1992). The Impact of Sea level Rise on the Coastline of Nigeria. In: Proceedings of IPCC Symposium on the Rising Challenges of the Sea. Magaritta, Venezuela. 14-19 March, 1992.

Ayo-Lawal G (2007). Showers of Agony, Ruin and Death" Nigerian Tribune Wed.28th Nov. 2007. http://www.tribune.com.ng

Blaikie P (1994). At Risk: Natural Hazards, People's Vulnerability and Disasters, Routledge, LondonBotkin

Douglas I, Alam K, Maghenda M, Mcdonnell Y, Mclean L, Campbell J. (2008). Unjust Waters: Climate Change, Flooding and the Urban Poor in Africa. Environ. Urbanization 20(1):187-205.

Haakonsen MJ (1992). West African Artisanal Fisheries, In: Fishing For Development: Small–Scale Fisheries in Africa, Tredten I, Hersoug B (eds.) The Scandinavian Institute of African Studies, Uppsala, Motala, Grafisca AB Motala, (1992).

IPCC (2001). Climate Change 2001: Impacts, Adaptation and Vulnerability, contribution of Working Group II to the Third Assessment Report of the IPCC, Cambridge University Press, New York.

Jepma C, Munasinghe M (1998). Climate Change Policy: Facts, Issues and Analyses, Cambridge University Press, Cambridge

Jinadu OO (2003). Small-Scale Fisheries In Lagos State, Nigeria: Economic Sustainable Yield Determination." Federal College of Fisheries and Marine Technology Wilmot Point, Victoria Island, Lagos, Nigeria (2003).

Komen G, Smith N (1999). Wave and Sea Level Monitoring and Prediction in the Services Module of the Global Ocean Observing System (GOOS). J. Mar. Syst19:35-250.

Leary N, Beresford S (2007). Vulnerability of people, places, and systems to environmental change. In: Knight, G. and Jaeger, J. (eds): Integrated regional assessment. Cambridge University Press, Cambridge, UK.

Lowe JA, Gregory JM, Flather RA. (2001): Changes in the occurrence of storm surges around the United Kingdom under a future climate scenario using a dynamic storm surge model driven by Hadley Center climate models. Climate Dynamics. 18:179– 188.

Maddison D (2006). The Perception of and Adaptation to Climate Change in Africa. CEEPA. Discussion Paper No. 10. Centre for Environmental Economics and Policy in Africa. Pretoria, South Africa: University of Pretoria.

Marfai MA, King L (2007). Coastal Flood Management in Semarang, Indonesia. Environ Geol

Murty TS (1984). Storm Surges – Meteorological Ocean Tides. Canadian Bulletin of Fisheries and Aquatic Science. P. 212.

Okali D, Eleri EO (2004). Climate Change and Nigeria: A Guide for Policy Makers. The Publication of the Nigerian Environmental Study Action Team (NEST).

Olaniyan E, Afiesimama EA (2002). On Marine Winds Waves and Swells Over the West African Coast for Effective Coastal Management: A Case Study of the Victoria Island Beach. Proceedings of Ocean 2002 MTS/IEEE Conference Oct 29-31 Biloxi, Mississippi. pp. 561-568.

Orlove BS, Broad K, Petty AM (2004). Factors That Influence the Use of Climate Forecasts: Evidence from the 1997/98 El Niño Event in Peru.

Osman-Elasha B, Spanger-Siegfried E, Goutbi N, Zakieldin S, Hanafi A (2007). Sustainable Livelihood Measures: Lessons for Climate Change Adaptation in Arid Regions of Africa. Annals of Arid Zone 44(3&4):403-419.

Sivakumar MVK (2005). Impacts of Natural Disasters in Agriculture, Rangeland and Forestry: An Overview. In: Sivakumar MVK, Motha RP, Das HP (eds) Natural Disasters and Extreme Events in Agriculture, Impacts and Mitigation. World Meteorological Organization, Geneva

The Office of the High Commission for Human Right (1966).

International Covenant on Economic, Social and Cultural Rights Adopted and opened for signature, ratification and accession by General Assembly resolution 2200A (XXI) of 16 December 1966, Internet file downloaded from http://www.unhchr.ch/html/menu3/b/a_cescr.htm on 9th August 2008.

Udofa IM, Fajemirokun FA (1978). On A Height Datum for Nigeria. In Proceedings: International Symposium on Geodetic Measurements and Computations. Ahmadu Bello University, Zaria, Nigeria.

United Nations (2007). The Millennium Development Goals Report 2007

United Nations Development Programme . Poverty Eradication, MDGs and Climate Change. Internet file downloaded from http://www.undp.org/climatechange/adap01.htm on 7th August 2008.

UNESCO/CSI (2000). Special Project: Study of Main Drainage Channels of Victoria and Ikoyi Islands in Lagos Nigeria and their Response to Tidal and Sea Level Changes (2000). http://www.unesco.org/csi

UNFCCC (2002). Report of the Conference of Parties on its Seventh Session held at Marrakesh from 29th October to 10th November 2001: Part II – Action Taken by the Conference of Parties.

Uyigue E (2007) Climate Change in the Niger Delta. Internet file downloaded from http://www.ciel.org/Publications/Climate/CaseStudy_Nigeria_Dec07.pdf On 5th August 2008.

24

Growth and economic performance of fingerlings of *Oreochromis niloticus* fed on different non-conventional feeds in out-door hapas at Akosombo in Ghana

Emmanuel D. Abarike[1], Edward A. Obodai[2] and Felix Y. K. Attipoe[3]

[1]Department of Fisheries and Aquatic Resources Management, University for Development Studies, P. O. Box TL 1882, Tamale-Ghana.
[2]Department of Fisheries and Aquatic Sciences, University of Cape Coast, Ghana.
[3]Aquaculture Research and Development Centre (ARDEC), Council for Scientific and Industrial Research (CSIR), Akosombo, Ghana.

The study was conducted at the Aquaculture Research and Development Centre at Akosombo in Ghana to observe the growth and economic performance of the fingerlings of *Oreochromis niloticus* fed on diets prepared using different agro-industrial by-products. Four isonitrogenous (30% CP) and isoenergetic (GE 18 MJ/kg) diets were formulated: Wheat bran (WB) (Diet 1); Pito mash (PM) (Diet 2); Rice bran (RB) (diet 3); and groundnut bran (GB) (Diet 4). They were fed to fingerlings of *O. niloticus* of average weight 7.0 ± 0.23 g stocked at 20 fish m^{-3} in out-door hapas for a period of 24 weeks. The study revealed that fish which were fed on Diet 1 grew significantly ($P < 0.05$) faster than those fed on the other diets. Fish growth was least on those fed on Diet 4. The incidence cost (IC) was highest ($P < 0.05$) in fish fed Diet 4 and lowest ($P < 0.05$) for fish fed Diet 2. The profit index was highest ($P < 0.05$) for fish fed diet 2 and lowest ($P < 0.05$) for fish fed Diet 2. *O. niloticus* fingerlings fed WB based diet produced the fastest growth while fingerlings fed PM based diet was the most cost-effective diet.

Key words: Growth, economic, performance, fingerlings, *Oreochromis niloticus,* non conventional feeds, Akosombo Ghana.

INTRODUCTION

In the development and management of an aquaculture enterprise, fish feed plays a vital role in its growth and expansion (Gabriel et al., 2007). The high cost of feeding has affected the development and expansion of aquaculture enterprises in most African countries contributing in no small way, to the low protein intake in many developing African countries (Abu et al., 2010). Non conventional feed resources are credited for being non competitive in terms of human consumption, very cheap by-products or waste products from agriculture, (Iluyemi et al., 2010). The main source of carbohydrate in many farm-based fish diets in Ghana is Wheat bran (WB) and maize. However, the widespread uses of WB and maize in livestock including poultry feed production is causing a price increase in these feed ingredients. This puts it in a situation that is, subjected to global market shocks and votility should there be shortage of supply of any of the aforementioned commodities. However, a fair proportion of industries in Ghana are agriculture-based and produce a range of by-products which could be rich ingredients for the formulation of fish feed as they already contribute to livestock and poultry feed.

There is the need to look for alternative sources of carbohydrates that are nutritionally rich and locally available. Pito mash (PM), a by-product obtained from sorghum used in the production of a local drink known as "pito", rice bran (RB) and groundnut bran (GB), are agro-industrial by products that are of great importance for developing more nutritive and economical fish diets in developing countries like Ghana. However, information available on the nutritive value of some of these agro-industrial wastes/by-products (that is, PM, RB and GB) in the formulation of fish feed is rather scanty in Ghana. PM, RB and GB are ideal for use as a good plant feed ingredient for the development of fish feed in Ghana. This is because these by-products do not suffer severe competition as human food, livestock (including fish) and poultry feed as it may be with other sources, e.g. WB or maize. Availability of these agro-industrial by-products is a significant step towards more efficient utilization of plant by-products. It will also help to boost the income generated by the local industries involved, hence raising the standard of living of the persons involved. This research therefore closely examined and evaluated the nutritional quality and economic feasibility of PM, RB and GB as ingredients that would be used as feed for fingerling production.

MATERIALS AND METHODS

Site description

The study was conducted at the Aquaculture Research and Development Centre (ARDEC) which lies East of Akosombo in the Eastern Region of Ghana between October 2010 and March 2011.

Procurement of ingredients

PM was purchased from pito brewers at the Akosombo community and RB from a rice milling factory at Akuse all in the Eastern Region of Ghana. GB was obtained from Kumasi in the Ashanti Region of Ghana. Other ingredients such as broiler premix, palm oil, WB and fish meal (FM) were purchased from Ashaiman timber market in the Greater Accra Region of Ghana. The feedstuffs were transported to ARDEC and processed into feed for use during the trial experiment.

Chemical analysis of feed items

Proximate analyses of the feed ingredients for the experiment were carried out at the Animal Nutrition Laboratory of the School of

Agriculture of the University of Cape Coast following the procedures that broadly adhere to Association of Official Analytical Chemists [AOAC] (1990). The protocol was used in determining the percentage (%) dry matter (DM), Crude protein (CP), Ash in percentage, Crude lipids (CL) in percentage also known as the Ether Extract (EE) of fat, Crude fibre (CF) in percentage and moisture in percentage. Nitrogen-free extract (NFE) was computed using the formula: % NFE = (100 - % CP + % CF + % EE + % Ash). The nutritional characteristics of feed ingredients were used in formulating experimental diets. Formulated diets were also analyzed to ascertain their nutritional characteristics and to see if they conform to the desired protein and energy requirements of the species under study based on the formulation.

Diet formulation and preparation

The trial and error method was use to formulate four different is nitrogenous (30% CP) and isoenergetic diets (GE, 18 MJ/kg). Feedstuff were sun dried and finely ground at the corn mill and sieved with a nylon mesh (420 μ/cm^2) to remove stones and larger sized particles (dissimilar sizes results in unstable pellet). The ingredients were weighed using a top pan balance according to the proportion based on the formulation for various treatments and mixed together in a large basin. Broiler (vitamin/mineral) premix, lysine and methionine and palm oil were added and the mixture further mixed thoroughly. In preparing pelleted feed, a little quantity of water was added to moisten the mixed proportions of the prepared feed to enable pelleting using a pelleting machine. Pelleted feed were sun dried, bagged and stored in dry and cool environment for use throughout the study.

Experimental set-up and fish

Twelve (12) fine mesh hapas (size 1 mm) each of capacity 10 m^3 (5 × 2 × 1 m) were installed in a 0.2 ha pond such that three quarters (¾) of the height of the hapas were submerged and one quarter (1/4) above the water surface to prevent the fish from escaping. The hapas were suspended by means of nylon ropes tied to bamboo poles, inserted into the bed of the pond.

Sex reversed fingerlings of O. niloticus of average weight 7 ± 0.23 g were obtained from ARDEC and stocked at 20 fish per meter cube for about 3 days for the fish to get them acclimatized before the start of the actual feeding trial experiment. The weight of fish were measured using a dry celled digital balance (model: Tanita KD 160) to the nearest 1 g and the standard and total lengths measured using a fish measuring board to the nearest 0.1 cm before being transferred to the experimental hapas. The 4 treatments, each in triplicate groups, were randomly assigned to the hapas installed in the pond to reduce biasness. The fish were reared and observed for 24 weeks during which fortnight body measurement (weight and length) were taken to evaluate their growth and response to the feeding on the diets administered.

Feeding regime

Fish were fed by hand-casting twice daily between 0830 to 0930 GMT and between 0300 to 0400 GMT at 10% for 6 weeks, 7% following another 6 weeks, 4% for another 6 weeks and 2% for the last 6 weeks making 24 weeks in all. The fishes were fed for 6 to 7 days a week and the daily ration equally divided between feeding times. Daily feed intakes were recorded and feed adjustments made fortnightly by sampling 25% of the fishes from each replicate of the various treatments and weighed to provide a good significant estimate of the average weight. The lengths and weights of fishes from each replicate were pooled for each treatment and based on these measurements (weights), the ration was adjusted accordingly.

Analysis of experimental data

Experimental data gathered during the growth trial for fingerlings of *O. niloticus* were used to determine various biological parameters such as:

(1) Mean weight gain (MWG) was computed as(Adewolu, 2008):

$$MWG = Final\ mean\ weight\ of\ fish - Initial\ mean\ weight\ of\ fish$$

(2) Specific growth rate was computed (SGR) as:

$$SGR = \frac{\ln W_2 - \ln W_1}{t} \times 100$$

where, W_1 = initial weight (g) at stocking, W_2 = final weight (g) at the end of experiment. $\ln W_2 - \ln W_1$ = Natural logarithms of both the final and initial weight of fish and T = duration (in days) of trial (Adewolu, 2008),

(3) Protein efficiency ratio (PER) as:

$$PER = \frac{total\ weight\ gained\ by\ fish}{total\ protein\ fed\ to\ fish}$$

Where, protein intake per fish is total feed given multiply (x) by the CP percentage in feed (Adewolu, 2008),

(4) Feed conversion ratio (FCR) (Sawhney and Gandotra, 2010) as:

$$FCR = \frac{total\ feed\ given}{total\ weight\ gained\ by\ fish}$$

(5) Survival rate (SR) (Charo-Karisa et al., 2006) as:

$$\%SR = \frac{Initial\ number\ of\ fish\ stocked - mortality}{Initial\ number\ of\ fish\ stocked} \times 100$$

(6) Condition factor (K) computed as:

$$K = \left(\frac{W}{SL^3}\right) \times 100$$

Where: K = condition factor, W = weight of fish in grammes and SL = the standard length of the fish in centimeters (Charo-Karisa et al., 2006).

A simple economic analysis was used to assess the cost effectiveness of diets used in the feed trial. The cost of feed was calculated using market prices taken into consideration the cost of feed and the transport fare with the assumption that all other operating costs remained constant (e.g. cost of constructing hapa, cost of fingerlings and labour). Indices for economic evaluation included:

(1) Incidence cost (IC) (Abu et al., 2010) as:

$$IC = \frac{cost\ of\ feed}{weight\ of\ fish\ produced}$$

(2) Profit index (PI) as:

$$PI = \frac{Weight\ or\ value\ of\ fish\ produced}{cost\ of\ feed}$$

Biological and economic data were subjected to a one-way analysis of variance (ANOVA) using the SPSS version 16 at 5% (P < 0.05) significant level. Variance of data was presented as standard error of means. Where significant differences occurred, treatment means were compared using Fisher's least significant difference (LSD).

RESULTS

Chemical composition of feedstuff

Results of the proximate analysis of the feed ingredients expressed on a dry matter basis (that is, to help standardize information on the ingredients) are shown in Table 1. Among the test by-products, PM recorded the highest CP (28.77%) and RB recorded the lowest CP (6.68%). In terms of EE, GB had the highest (9.00%) and WB had the least (4.59%). The CF content of RB was the highest with 31.47% and lowest in WB with 10.48%. The calculated NFE was and highest in WB (64.29%) and lowest in RB (36.25%).

Inclusive levels and chemical composition of diets for fingerlings of *O. niloticus*

Table 2 shows the composition and chemical analysis of diets for fingerlings of *O. niloticus*. Among the test agro-industrial by-products, Diet 3 had the highest amount of FM (56%) and palm oil (2.58%) and Diet 2 had the lowest amount of FM (16%) and palm oil recorded the least (1.14%). The amount of FM in Diet 1 (45.5%) was higher than in diet 4 (34%). Methionine, lysine and broiler premix were the same for all the diets, because specific quantities were needed in all diets to supplement those naturally occurring in the diets.

All the 4 prepared diets had similar (x^2 < 0.35, P > 0.05) CP levels. Diet 4 had the highest amount of EE (18.74%) and Diet 2 had the lowest EE (8.68%). Crude CF levels in the diets was in the following descending order Diet 2 > Diet 3 > Diet 1 > Diet 4. Ash content was highest in Diet 3 (22.32%) and lowest in diet 2 (13.94%). Diet 2 (41.05%) had the highest NFE and diet 3 (28.64%) had the lowest. Although the gross energy (GE) was highest in Diet 4 (19.98 MJ/kg) and lowest in Diet 3 (17.11 MJ/kg), there was no significant differences (x^2 < 0.35, P > 0.05) among all the diets.

Production data for fingerlings of *O. niloticus*

Data on the growth performance of fingerlings of *O. niloticus* (Table 3) indicate that, the average initial weights (AIW) of the test fish were found to vary slightly but not significantly (P > 0.05) among the four treatments. However, at the end of the trial, the average final weights (AFW) were significantly (P < 0.05) different among the treatments. Fish fed on Diet 1 (88.0 ± 1 1.43 g) were the heaviest. The AFW of other diets took the following descending order Diet 3 > Diet 2 > Diet 4. Among the treatments, Diet 1 supported the best mean weight gain [MWG] (80.80 ± 3.48 g) and the Diet 4 supported the

Table 1. Chemical composition of feed ingredients.

Type of analysis	Fish meal	Pito mash	Rice bran	Groundnut bran	Wheat bran
Dry matter (%)	94.09	92.93	91.78	93.96	92.68
Crude protein (%)	48.95	28.77	6.68	21.69	15.46
Ether extract (%)	12.54	7.81	8.76	9.00	4.59
Crude fibre (%)	0.88	12.77	31.47	17.51	10.48
Ash (%)	27.93	4.42	16.89	4.78	5.18
Nitrogen-free extract (%)	9.70	46.23	36.25	47.02	64.29
GE (MJ/kg)	18.19	17.82	11.26	16.75	16.50

Table 2. Inclusion levels and proximate analysis of diets for fingerlings *O. niloticus*.

Ingredient	Diets			
	1	2	3	4
Fish meal	45.5	16	56	34
Pito mash	-	80.27	-	-
Rice bran	-	-	39	-
Groundnut bran	-	-	-	62
Wheat bran	50.4	-	-	-
Methionine	0.1	0.1	0.1	0.1
Lysine	1.9	1.9	1.9	1.9
Broiler premix	0.5	0.5	0.5	0.5
Palm oil	1.5	1.14	2.58	1.45
Total	100	100	100	100
Proximate analysis				
Dry matter (%)	88.54	91.22	91.52	91.45
Calculated crude protein (%)	30.06	30.22	30.02	30.24
Actual. crude protein (%)	30.42	30.36	30.14	30.25
Ether extract (%)	8.68	9.98	12.84	18.74
Crude fibre (%)	5.86	8.07	6.06	3.54
Ash (%)	14.95	13.94	22.32	16.76
Nitrogen-free extract (%)	40.09	41.05	28.64	30.71
Gross energy (MJ/kg)	17.67	18.15	17.11	19.98

Table 3. Harvest data for fingerlings *O. niloticus* fed on different dietary treatments.

Parameter	Diets (mean ± standard error)			
	1	2	3	4
Average initial weight	7.19[a] ± 1.68	7.18[a] ± 1.66	7.10[a] ± 1.75	7.04[a] ± 1.56
Average final weight	88.01[a] ± 1.43	64.53[c] ± 0.86	75.27[b] ± 1.18	51.35[d] ± 1.07
Mean weight gain	80.80[a] ± 3.48	57.33[b] ± 4.30	67.93[b] ± 4.19	44.30[c] ± 5.41
Specific growth rate	1.49[a] ± 0.02	1.30[bc] ± 0.09	1.38[ab] ± 0.05	1.17[c] ± 0.06
Protein efficiency ratio	0.71[a] ± 0.02	0.67[a] ± 0.14	0.75[a] ± 0.05	0.51[a] ± 0.14
Feed conversion ratio	4.74[b] ± 0.15	5.32[a] ± 1.03	4.46[b] ± 0.28	8.60[a] ± 3.49
Survival rate	68.33[a] ± 0.88	71.50[a] ±7.50	81.00[a] ± 4.48	62.17[a] ±16.85
Condition factor	3.28[a] ± 0.01	3.09[a]± 0.10	3.23[a] ± 0.0.02	2.38[b] ± 0.48

Similar superscript alphabets in the rows denote homogeneous means (LSD, $P > 0.05$).

worse MWG (44.30 ± 5.41 g) The growth curves of fish in response to the test diets over the 24 weeks (6 months) experimental period are shown in Figure 1. Fish fed on Diet 1 maintained the highest growth differentiating clearly

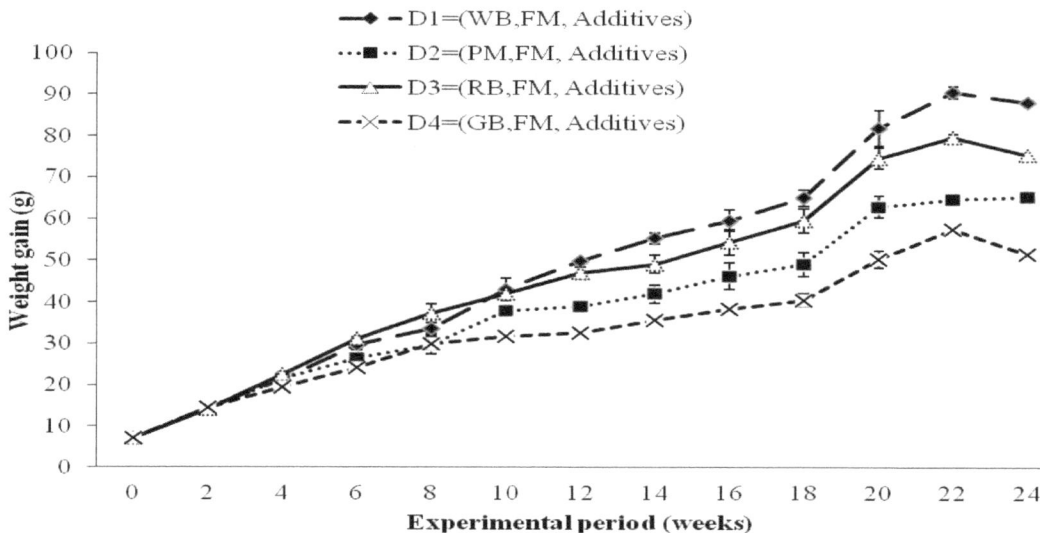

Figure 1. Growth curves for fingerlings of *O. niloticus* fed on different dietary treatments for 24 weeks (Vertical bars represent standard errors).

Table 4. Economic benefits of using different diets.

Diet	Cost /kg. of feed (GH¢)	Total feed input (kg)	Total cost of feed input (GH¢)	Harvested biomass (kg)	Assumed sales of biomass (GH¢)	[1]IC	[2]PI
1	0.71	121.24	86.09	30.02	90.05	8.62[a] ± 0.08	3.14[ab] ± 0.03
2	0.46	107.21	49.32	26.18	78.54	5.96[a] ± 0.33	4.82[a] ± 0.28
3	0.76	126.98	96.50	28.39	85.13	10.40[a] ± 0.23	2.62[b] ± 0.05
4	0.67	104.96	70.32	15.91	47.71	17.64[a] ± 2.43	2.02[b] ± 0.20

Similar alphabets in the columns denote homogenous means (LSD, P > 0.05), Cost price per kilogram of fish (that is. below 150 g) at the farm gate = GH¢ 3. Market price per kilogram of test diets was based on cost per kilogram of feed ingredients used in formulating the test diets. 1. Incidence cost = IC, 2. Profit index = PI.

after the 10th week of culture and Diet 4 maintained the least trend in growth.

Among the treatments, fish fed on diet 1 had the highest specific growth rate [SGR] (1.49 ± 0.02% day^{-1}) and fish fed on Diet 4 had the lowest SGR (1.17 ± 0.06% d^{-1}). However, the SGR were not significantly (P > 0.05) different among fish fed on Diet 1 and those feed on Diet 2. Also the SGR for fed on Diet 2 were not significantly (P > 0.05) different from fish fed on Diet 3. Although, PER as observed in study were not significantly (P > 0.05) different between treatments, PER was highest for fish feed Diet 3 (0.75 ± 0.05) and lowest for fish fed on diet 4 (0.51 ± 0.14). The best performing Diet in terms of feed conversion ratio (FCR) was lowest (that is, better utilization) for fish fed on Diet 3 (4.46 ± 0.28) and highest (that is, poor utilization) was for fish fed on Diet 4 (8.60 ± 3.49). However, the FCR were not significantly (P > 0.05) different for fish fed various diets. The condition factor (K) of fish fed on Diets 1, 2 and 3 were similar (P > 0.05). However, fish fed on Diet 4 was statistically (P < 0.05)

different from the other treatments.

Economic benefits of using the different diets

Shown in Table 4, the cost per kilogram of feed was highest for fish fed on Diet 3 (GH¢0.76) and lowest for fish fed on Diet 2 (GH¢ 0.46). Feed administered was highest for fish fed on Diet 3 (126.98 kg) and lowest for fish fed on Diet 4 (104.96 kg). The total biomass harvested was highest for fish fed on Diet 1 (30.02 kg) and lowest for fish fed on Diet 4 (15.91 kg). Among the treatments, Diet 4 had the highest (that is, highest cost per kilogram of feed) IC of 5.88 and Diet 2 had the lowest (lowest cost per kilogram of feed) IC of 1.99. However, there were no significant (P > 0.05) differences among all the treatments in terms of IC. The profit index (PI) was highest for fish fed on Diet 2 (4.82 ± 0.28) and lowest for fish fed on Diet 4 (2.02 ± 0.02). However, fish fed on Diet 2 were similar (P > 0.05) to fish fed on Diet 1. Also fish fed on the Diets 1, 3 and 4 were similar (P > 0.05).

DISCUSSION

Characteristics of feedstuff

The variability in the composition of agro-industrial by-products and diets formulated and prepared is reflected in growth and development of *O. niloticus*. This is because, growth of fish fed on various diets tended to differ, although not significantly among the tested diets. The ability of *O. niloticus* to utilize various diets could be attributed to wide spectrum of preference for foods. This is in agreement with Gonzalez and Allan (2007) and Audu et al. (2008) who, reported that, *O. niloticus* readily adapts to eating a wide variety of feeds, and that they have very long intestines necessary to digest plant materials.

Acceptance of diets of fingerlings of *O. niloticus*

Palatability is defined as acceptable to the taste or sufficiently agreeable in flavour to be eaten. While it may be difficult to ascertain whether or not a fish 'likes' some flavour or not, it is certainly possible to determine differences in the amounts of feed eaten (Glencross et al., 2007). The acceptability of the diets containing PM, RB, and WB is a good indication of the palatability of the experimental diets, hence the feasibility of using these agro-industrial by-products in formulating feed for fingerlings *O. niloticus*. Similar results have been reported for *O. niloticus* by Liti et al. (2006) and Attipoe et al. (2009), and for Florida Red Tilapia, (*O. niloticus* crossed with *O. mosambicus*) by El-Dakar et al. (2008). In these reports, the acceptance of feed rations containing various agro-industrial by-products such as cocoa husk, brewery waste, RB, WB, MB and fig Jam by-product by the tilapias demonstrates the opportunistic nature of tilapia in their feeding habits. This provides an advantage to farmers because the fish can be reared in intensive systems that can be operated with lower cost feeds.

Growth performance of fingerlings *O. niloticus*

In this experiment, the results from proximate analysis demonstrate that, the test Diets (1, 2, 3 and 4) differed both in nutritional quality and efficiency in promoting the growth of fingerlings of *O. niloticus*. Fish which were fed on Diet 1 (WB-based diet), grew significantly faster than those fed on the other diets. The best performance by the fish fed on WB-based diet, as observed in this study are similar to the reports by Liti et al. (2006) and Attipoe et al. (2009) that WB-based diets improved the growth performance of tilapia compared with other brans. The good performance of fish fed on Diet 3 (RB-based diet) as observed in this study is at variance with the reported

works of Liti et al. (2005, 2006) and Solomon et al. (2007). According to these authors, rice bran-based diets gave the worse performance among other treatments. Though diet 3 (rice bran-based diet) performed well, Diet 1(WB-based diet) demonstrated the best growth because the crude fibre levels were lower. In addition, the NFE a measure of the carbohydrate level in diets (Mohanta et al., 2007; Tran-Duy et al., 2008) was higher in Diet 1 than in Diet 3 suggesting a greater the protein sparing effect (that is, the use of carbohydrate to spare the use of protein for energy purposes). This is because, more energy in the form of NFE was available in Diet 1 than it is in Diet 3 to spare the use of protein as an energy source.

Kaur and Saxena (2004) reported higher mean weight gain (MWG) in diets in which RB was replaced with brewery waste in a feed trail with *Catla catla* and *Labeo rohita*. This was attributed to better absorption and utilization by these fishes. However, in the current study, though with *O. niloticus*, the Diet 3 (rice bran-based diet) exhibited higher MWG than Diet 2 (PM-based diet). The superior growth of fish fed on the rice bran-based diet observed in this study could be a manifestation of the higher proportion of FM used in compounding Diet 3. Attipoe et al. (2009) concluded that, diets with higher proportion of FM exhibited better growth because of their excellent biological value in enhancing growth. Though the weight gain of fish fed on Diet 2 (PM-based diet) in this study did not exhibit the best growth, the higher proportion of PM in the Diet 2 suggests that, PM can enhance the growth of *O. niloticus* by reducing drastically the quantity of FM (reducing the cost of feed in the process) in diets. This is in accordance with earlier findings of Webster et al. (1992) and Wu et al. (1996). All these authors also found that, brewery waste could be used to completely or partially replace costly FM in the diets of tilapia.

According to Agbo (2008), crude fibre provides physical bulk to feed and may improve pelletability. However, fibre decreases the quantity of usable nutrients in a diet. Although the crude fibre was lowest in diet 4 (3.54%) suggesting better utilization, growth could have been impaired by the higher EE per cent (18.74%). This is consistent with several other research findings as in Manjapa et al. (2002) and Audu et al. (2008). These researchers explained that, EE levels in the diet of tilapia should range between 5 to 12%, or when in excess (above 12%), results in poor growth because of an imbalance of protein to fat ratio.

Solomon et al. (2007) using different grain sources (maize, wheat, rice, sorghum and millet) to assess the growth of tilapia in out-door hapas, reported that the best MWG for fish fed with maize grain > wheat grain > sorghum grain > millet grain > rice grain. However, as observed in this study, PM-based diet recorded a lower MWG than rice bran-based diet. This can be attributed to the high crude fibre (8.07%) levels in PM-based diet as compared to the low crude fibre (6.06%) content of the rice bran-based diet. This assertion agrees with several other researchers' findings (Gonzalez and Allan, 2007; El-

Dakar et al., 2008) that high fibre levels in the diet of tilapia resulted in poor weight gain and nutrient utilization.

The SGR as obtained in the current study was lowest for fish fed on Diet 4 (GB based diet) and highest for fish fed on the Diet 1(WB-based diet). Higher SGR values of fish fed on the WB-based diet is indicative of the fact that, the WB-based diet was better utilized than the other diets. This gives credence to the views of Iluyemi et al. (2010). For comparison, the SGR obtained in this study (1.17 to 1.49) are higher than the SGR values (0.43 to 0.53) reported by Attipoe et al. (2009).

Chatzifotis et al. (2010) demonstrated that growth can be reduced and PER lowered if fishes were fed with excessive dietary lipid levels. This is because, excessive lipids not only spare proteins, but can also result in higher fat deposits and impaired growth performance. Although PER values were highest for fish fed on the rice bran-based diet and lowest for fish fed on the diet 4 (groundnut bran-based diet) (Table 3), yet statistically, there were no significant (P > 0.05) differences among the fish fed on all the diets. However, this was at variance with the finding of Wu et al. (2000) that, diets with higher proportion of FM exhibited higher PER because of more efficient utilization of FM by fish, the results obtained in this study can be explained according to Drew et al. (2007) who reported that, digestion of plant materials by fish resulted in a lower PER because most plant materials had lower CP levels which are used as sources of energy in the form of carbohydrates. However, an imbalance of carbohydrate and lipids ratio can suppress the uses of protein even if the proportion of FM is high in a diet.

Utilization of feed measured by the feed conversion ratio (FCR) was lowest for fish fed on Diet 1 (that is, indicating better utilization) and highest for fish feed Diet 4 (that is, indicating poor utilization). Although the FCR values were not significantly (P > 0.05) different between fish fed on Diets 1, 2 and 3, the higher FCR of fish fed on Diet 4 may be explained by Manjappa et al. (2002) who reported that, growth of fish fed with diets which have higher lipid level (above 12%) is poorer, reflecting a negative impact of dietary fat beyond the optimum level. Though crude fibre was lowest in Diet 4, utilization of the diet might have been impaired by the higher EE level. Glencros et al. (2007) supported this fact when they found that, the FCR becomes lower as the efficiency of feed utilization increases. For comparison, the FCRs in this study were better than those obtained by Wu et al. (2000) and Liti et al. (2006).

In this study, the condition factor (K) recorded for fishes fed on Diets 1, 2 and 3 were similar (P > 0.05) and differed (P < 0.05) significantly from those of fishes fed on Diet 4. Ogunji et al. (2008) reported that, the higher the K the better the physiological state of the fish. This suggests that in the present study, fish fed on Diets 1, 2 and 3 were in a better condition than those fed on Diet 4. The poor K of fish fed on Diet 4 could be attributed to the poor utilization of supplemental feed (indicated by lower MWG

and PER and a higher FCR). For comparison, the lower K in the present study than those recorded by Anene (2005) for Cichlids: *C. guntheri, C, cabrae* and *T. mariae* could be ascribed to poorer utilization of supplementary feed.

Economic benefits of different dietary treatments of fingerlings of *O. niloticus*

The cost-benefit analysis (Table 4) of using different feeds in terms of the IC turned out to be similar among all the four treatments. It therefore, points to the fact that, all the diets used in the study could be used for *O. niloticus* culture depending on the availability and prices of the agro-industrial by-products used in this study. However, it is important to note that, in areas where larger quantities of these by-products exist, the cost per kilogram of feed may be lower and significantly reduce IC. For comparison, El-Dakar et al. (2008) reported an IC of 2.26 for diets containing up to 50% fig jam by-product (FJB) replacing WB in the diet of red tilapia and recommended it as an economically efficient diet for red tilapia fry. However, the IC of fingerlings of *O. niloticus* fed on Diet 1 (WB-based diet) in this study was higher. This might be because, WB was the major source of carbohydrates and the use of FJB as a less expensive feed ingredient in their study reduced the cost of the diet hence lower IC reported.

In another study Attipoe et al. (2009) in evaluating three test diets prepared from agro-industrial by-products, fish fed on the Diet (F2) (compounded with RB and groundnut bran) were reported to be the cheapest among the other diets. On the contrary, Diet 4 (groundnut bran-based diet) was the most expensive in terms of IC in the current study. The variance in results might be because, Diet 4, was prepared with FM as the major source of protein which in itself is expensive (Falaye and Jauncey, 1999; Abu et al., 2010). Coupled with that, the poor performance of the fish fed on Diet 4, did not compensate for the rather high cost of feed in this study.

The profit index (PI) was highest in Diet 2 though not significantly different from Diet 1. It may be economical and beneficial to use Diet 2 for the culture of fingerlings of *O. niloticus*. However, due to better feed utilization, growth was better for fish fed on Diet 1. Comparing the performance of fish fed on Diet 1 (WB-based diet) in this study to the report in a closely related study by Liti et al. (2006) on the economic performance of *O. niloticus*, WB-based diet gave higher returns than the other test brans for example rice bran. This is because WB had lower fibre and higher levels of carbohydrates, NFE and than RB, hence, proving more energy to spare dietary protein for growth.

Conclusion

In this study, fish fed on Diet 1 produced the fastest

growth of *O. niloticus* fingerlings among the other test diets and the worse growth was recorded for fish fed on Diet 4. The IC was lowest for fish fed on PM based diet (Diet 2) corresponding to a higher PI.

ACKNOWLEGMENTS

The authors wish to thank the technical staff of the Aquaculture Research and Development Centre (ARDEC) and the Animal Nutrition Laboratory of the School of Agriculture of the University of Cape Coast for their assistance during the experiment. Our appreciation also goes to Mr. Elliot Alhassan, the Head of Department of the Fisheries and Aquatic Resources Management of the Faculty of Renewable Natural Resources of the University for Development Studies for taken the pain to read through this work to help fine-tune it.

REFERENCES

Abu OMG, Sanni LO, Erondu ES, Akinrotimi OA (2010). Economicviability of replacing maize with whole cassava root meal in the diet of Hybrid Cat-fish. Agric. J. 1:1-5.

Adewolu MA (2008). Potentials of sweet potato (*Ipomoea batatas*) leaf meal as dietary ingredient for *Tilapia zilli* fingerlings. Pak. J. Nutr. 7(3):444-449.

Agbo NW (2008). *Oilseed meals as dietary protein sources for juvenile Oreochromis niloticus*. Unpublished doctoral thesis, Institute of Aquaculture, University of Stirling, UK.

Anene A (2005). Condition factor of four cichlid species of a man-made Lake in Imo State, Southeastern Nigeria. Turk. J. Fisher. Aquat. Sci. 5:43-47.

Association of Official Analytical Chemists (1990). Official Methods of Analysis of the Official Association of Analytical Chemists. 15th edn, Arlington, Virginia, USA: Association of the Official Analytical Chemists.

Attipoe FYK, Nelson FK, Abbran EK (2009). Evaluation of three diets formulated from local agro-industrial by-products from production of *Oreochromis niloticus* in earthen ponds. Ghana J. Agric. Sci. 42:185-191.

Audu BS, Adamu KM, Binga SA (2008). The effect of substituting fishmeal diets with varying quantities of ensiled parboiled Beniseed (*Sesamun indicum*) and raw African locust bean (*Parkia biglobosa*) on the growth response and food utilization of the Nile tilapia, *Oreochromis niloticus*. Int. J. Zool. Res. 4(1):42-47.

Charo-Karisa H, Komen H, Reynolds S, Rezk MA, Ponzoni RW, Bovenhuis H (2006). Genetic and environmental factors affecting growth of Nile tilapia (*Oreochromis niloticus*) juveniles: Modeling spatial correlations between hapas. Aquaculture 255:586-596.

Chatzifotis S, Panagiotidou M, Papaioannou N, Pavlidis M, Nengas I, Mylonas CC (2010). Effect of dietary lipid levels on growth, feed utilization, body composition and serum metabolites of Meagre (*Argyrosomus regius*) juveniles. Aquaculture 307:65-70.

Drew MD, Borgeson TL, Thiessen DL (2007). A review of processing of feed ingredients to enhance diet digestibility in finfish. Anim. Feed Sci. Technol. 138:118-136.

El-Dakar AY, Abd-Elmonem AI, Shalaby SMM (2008). Evaluation of Fig Jam by-product as an energy source in Florida Red Tilapia, *Oreochromis, niloticus x Oreochromis mosambicus* diets. Mediterranean Aquac. J. 1(1):27-34.

Falaye AE, Jauncey K (1999). Acceptability and digestibility by tilapia *Oreochromis niloticus* of feeds containing cocoa husk. Aquac. Nutr. 5:157-161.

Gabriel UU, Akinrotimi OA, Bekibele DO, Onunkwo DN, Anyanwu PE (2007). Locally produced fish feed: potentials for aquaculture development in Subsaharan Africa. Afr. J. Agric. Res. 2(7):287-295.

Glencross BD, Booth M, Allan GL (2007). A feed is only as good as its Ingredients. A review of ingredient evaluation strategies for aquaculture feeds. Aquac. Nutr. 13:17-34.

Gonzalez C, Allan G (2007). Preparing farm-made fish feed. NSW, Department of Primary Industries. P. 21.

Iluyemi FB, Hanafi MM, Radziah O, Kamarudin MS (2010). Nutrition evaluation of fermented palm kernel cake using red tilapia. Afr. J. Biotechnol. 9(4):502-507.

Kaur VI, Saxena PK (2004). Incorporation of brewery waste in supplementary feed and its impact on growth in some carps. Bioresour. Technol. 91:101-104.

Liti D, Cherop L, Munguti J, Chhorn L (2005). Growth and economic performance of Nile tilapia (*Oreochromis niloticus* L.) fed on two formulated diets and two locally available feeds in fertilized ponds. Aquac. Res. 36:746-752.

Liti DM, Mugo RM, Munguti JM, Waidbacher H (2006). Growth and economic performance of Nile tilapia (*Oreochromis niloticus* L.) fed on three brans (maize, wheat and rice) in fertilized ponds. Aquac. Nutr. 12:239-245.

Manjappa K, Keshavanath P, Gangadhara B (2002). Growth performance of common carp, *Cyprinus carpio* fed varying lipid levels through low protein diet, with a note on carcass composition and digestive enzyme activity. Acta Ichthyologica Et Piscatoria 32(2):145-155.

Mohanta KN, Mohanty SN, Jena JK (2007). Protein-sparing effect of carbohydrate in Silver barb, *Puntius gonionotus* fry. Aquac. Nutr. 13:311-317.

Ogunji J, Toor R, Schulz C, Kloas W (2008). Growth performance, nutrient utilization of Nile tilapia *Oreochromis niloticus* fed housefly maggot meal (Magmeal) diets. Turk. J. Fisher. Aquat. Sci. 8:141-147.

Sawhney S, Gandotra R (2010). Growth response and feed conversion efficiency of *Tor putitora* fry at varying dietary protein levels. Pak. J. Nutri. 9(1):86-90.

Solomon SG, Tiamiyu LO, Agaba UJ (2007). Effect of feeding different grain sources on the growth performance and body composition of tilapia, (*Oreochromis niloticus*) fingerlings fed in out-door hapas. Pak. J. Nutr. 6(3):271-27.

Tran-Duy A, Smit B, Dam AAV, Schrama JW (2008). Effects of dietary starch and energy levels on maximum feed intake, growth and metabolism of Nile tilapia, *Oreochromis niloticus*. Aquaculture 277:213-219.

Webster CD, Yancey DH, Tidwell JH (1992). Effect of partially or totally replacing fish meal with soybean meal on growth of blue catfish (*Ictalurus furcatus*). Aquaculture 103:141-152.

Wu YV, Rosati R, Brown PB (1996). Effect of diets containing various levels of protein and ethanol co-products from corn on growth of Tilapia fry. Journal of Agricultural and Food Chemistry, 44, 1491-1493.

Wu YV, Rosati R, Warner K, Brown P (2000). Growth, feed conversion, protein utilization, and sensory evaluation of Nile tilapia fed on diets containing corn gluten meal, full-fat soy, and synthetic amino acids. J. Aquat. Food Prod. Technol. 9(1):78-87.

25

Farmers' risk perceptions and adaptation to climate change in Lichinga and Sussundenga, Mozambique

Chichongue O. J.[1,2], Karuku G. N.[1], Mwala A. K.[2], Onyango C. M.[2] and Magalhaes A. M.[2]

[1]University Of Nairobi, Kenya.
[2]Agricultural Research Institute of Mozambique (IIAM), Mozambique.

In Africa, climate change exerts significant pressure on the agricultural sector. Current changes in climate for most parts of Mozambique have resulted in increased frequency of droughts, dry spells and uncertain rainfall. This has resulted in loss of food production and smallholder farmers are most vulnerable to these climatic disasters as they affect the food security status of the household. Despite an increased number of country level case studies, knowledge gaps continue to exist at the level of impact analysis. In addition, while adaptation and coping strategies with climate change and variability have become key themes in current global climate discussions and policy initiatives, literature on adaptation in Mozambique appears to be limited. The objective of this study was to assess the perception of smallholder farmers to climate change and adaptation strategies in Lichinga and Sussundenga districts of Mozambique. Using data obtained from a survey carried out in Lichinga and Sussundenga districts in Mozambique descriptive statistics analysis was undertaken using SPSS software to characterize the households, in terms of perceptions and coping strategies of the household to climate change. The farmers from both districts sited rainfall variability and higher temperatures to have severely affected maize production. Due to the late onset of rains, in Lichinga the planting period has changed from November (47.5%) to December (70%) while in Sussundenga the planting period has changed from September/October (40%) to November (62.5%). The rain seasons have become shorter and dry seasons are longer. Some farmers have switched from growing maize to growing drought tolerant crops, such as cassava, sweet potato and cultivation of horticultural crops in wetlands as strategies to cope with the climate change.

Key words: Climate variability, farmer's perceptions, adaptation strategies.

INTRODUCTION

Climate variability and droughts are already important stress factors in Africa, where rural households have adapted to such factors for decades (Mortimore and Adams, 2001). Historical data shows that the continent is already undergoing climate change. The continent is becoming warmer and drier. Rainfall is becoming less predictable. In Mali, Lacy et al. (2006) revealed a tendency for a shortening of the rainy season to induce farmers to shift some of their sorghum production to a variety with a shorter cycle than the traditional one. In a study from Burkina Faso, Nielsen and Reenberg (2010) found rainfed cereal production to be declining due to a change in climate and a shift towards a higher level of dependence on migration, livestock, small scale commerce and gardens. In the Sub-Saharan El Niño rains cause floods and destruction, while in the recent years droughts have also had catastrophic impacts (Nielsen and Reenberg, 2010). Recent events, such as the poor rains in Southern Africa 2001 and 2003, demonstrate that communities may already be suffering the consequences of less predictable weather patterns (Wiggins, 2005).

As the poorest country in Southern Africa, a region that is projected to become substantially hotter and drier, Mozambique is likely to feel the impacts of climate change more than most (Ehrhart and Twena, 2006). The most striking impacts of climate change over south-eastern Africa are expected to be an increase in the frequency and severity of extreme events such as droughts, floods, and cyclones Ribeiro and Chaúque (2010).

Climate variability directly affects agricultural production since agriculture is inherently sensitive to climatic conditions and is one of the most vulnerable sectors to the risks and impacts of global climate change. Climate change will affect food security by reducing livelihood productivity and opportunities. The impacts will be mostly negative in Mozambique (Ehrhart and Twena, 2006). Research by the Government of Mozambique suggests that mean air temperatures will raise by at least 1.8 to 3.2°C nationwide by 2075 (MICOA, 2007). Precipitation is predicted to fall by 2 to 9%, which will take greatest effect between November and May. As this coincides with the growing season, it will have an especially pronounced impact on crop yields (Ehrhart and Twena, 2006). Harvest failure and incidents of food insecurity in Africa have become regular events occurring at least once or twice every decade (Eriksen, 2005).

Most countries in Sub-Saharan Africa (SSA) rely heavily on agriculture for employment and food security for their economies. The sector also has large numbers of smallholder farmers, most of who produce under unfavourable conditions characterized by low and erratic rainfall and poor soils (Mutsvangwa, 2011).

Most households succeed in protecting their short term consumption from the full effects of income shocks, but in the long term these shocks have consequences for low income households, which are forced to reduce their investment in children's health and schooling, or sell productive assets in order to maintain consumption (Trærup, 2010). Over time, rural households develop a range of coping strategies as a buffer against uncertainties in their rural production induced by annual variations in rainfall combined with socio-economic drivers of change (Cooper et al., 2008). Coping strategies may be preventive strategies such as altering planting dates, introducing other crops and making investments of water equipment, or may be in-season adjustments in the

form of management responses (Trærup, 2010). Farmers can adapt to climate change by modifying the set of crops planted and their agronomic practices (Blanc, 2011). The latter most often include consumption smoothing, the sale of assets such as livestock, remittances from family members outside the household and income from casual employment (Niimi et al., 2009; Kinsey et al., 1998).

While extensive research on the impacts of climate change has tended to focus on impacts on country level, less effort has been directed at household level in developing countries, and little has been done on the farmer risk perception and adaptation. There is thus need to investigate the farmer risk perception and adaptation to climate change on agriculture in Mozambique, at the household level, considering that agriculture remains the backbone of the country's economy. This study seeks to contribute to the body of research on climate change by investigating the vulnerability of smallholder farmers in Mozambique.

METHODOLOGY

Study sites

The study was conducted at Lichinga and Sussundenga Research Stations in Mozambique. Mozambique's economy is essentially agricultural; its agriculture is predominantly subsistence, characterized by low levels of production and productivity. In 2009, it contributed 24% of GDP (INE). In addition, the sector employs 90% of the country's female labour force and 70% of the male labour force. This means that 80% of the active population is employed in the agriculture sector.

Lichinga Research Station is located in Lichinga district to the west of the Niassa province and lies 12° 30' to 13° 27' S; 34°50' to 35°30' E. The site receives unimodal rainfall between November and April ranging from 900 to 2000 mm per annum. The temperature ranges from 16.1°C to 32.9°C with an annual average of 22.9°C (MAE, 2005a). The agricultural production is predominantly rain fed (MAE, 2005a). The soils are ferralsols according to FAO (2006) soil classification system. The soils are red in colour with compacted structure (Geurts and Tembe, 1997). The soil fertility is poor with CEC of 27.80 cmol kg^{-1}, low base saturation ranging from 1.75 to 9.63 cmol kg^{-1} and soil organic matter of 6.78%. These soils are deficient in nitrogen (0.69%) but have high level of phosphorus (1750 ppm).

Sussundenga Agrarian Research Station is located in Manica province, Central Mozambique and lies 19° 20' S; 33° 14' E, with an altitude of 620 m. The area has a unimodal rainfall occurring between November and April with average annual rainfall of 1,155 mm (MAE, 2005b). The average minimum temperature is 9.5°C in the month of July and average maximum is 29.1°C in the month of January (Famba, 2011), giving an annual average of 23.0°C (MAE, 2005b). The agricultural production is predominantly rain fed (MAE, 2005b). The soils consist of ferralsols, lixisols and acrisols (FAO, 2006). The soil fertility is poor with CEC of 25.80 cmol kg^{-1}, low base saturation ranging from 0.18 to 0.88 cmol kg^{-1} and soil organic matter content of 7.70%. These soils are deficient in nitrogen (0.78%) but have high phosphorus (1875 ppm) content.

Data collection procedures

A survey was conducted in Lichinga and Sussundenga districts of Mozambique and two villages were randomly sampled from each of

Table 1. Farmers' awareness of climate change over the last 10 years.

District	Rainfall		Temperature	Unusual weather conditions experienced		Noted changes	
	Changed (%)	Unchanged (%)	Changed (Hot) (%)	Drought (%)	Heavy rains (%)	Longer rain periods (%)	Shorter rain periods (%)
Lichinga	87.5	12.5	100	0	40	42.5	57.5
Sussundenga	90	10	100	45	0	0	100

N = 40 each district.

the selected Districts. The study sites are located within 20 km radius of Lichinga and Sussundenga research stations. The objective of the survey was to evaluate the farmers' risk perception and adaptation to climate change. The survey was carried out at Lichinga District from 16 to 17 February and at Sussundenga from 20 to 21 February, 2012 using questionnaire with open-ended and closed questions. The survey included face-to-face interviews of 80 farmers. Forty (40) farmers were selected in each district of which 20 came from one village. Selection of respondents was based on farmer's willingness to participate in the research. During the data collection process, the participants were told the objective of the study as well as its confidentiality. Interviews were done at farmers' homesteads. Respondents were household heads and in their absence, any member of the household was interviewed. In each district, a lead farmer was identified, contacted and met to make arrangements to meet other farmers and an interpreter was used where necessary.

The first phase of the survey was to collect data to assess the factors influencing farmers' decisions making on fertilizer use. Besides general household information, the questionnaire contained modules on agricultural productivity, types of organic and inorganic fertilizer use. Data on cropping systems, land use and maize production was also collected.

The second phase of the survey was to collect data to evaluate the farmers' risk perception and adaptation to climate change. In order to understand how farmers perceive climate risk events data on weather/climate change, weather event severity, weather event effects, indicators of change in crops operation and coping strategies was collected.

The data were analysed using the Statistical Package for Social Sciences (SPSS) version 16 (SPSS 16.0 for Windows, Release 16.0.0.2007. Chicago: SPSS Inc). Descriptive statistics, means, frequencies, percentages and cross tabulations were used to present the outcome of the research.

RESULTS AND DISCUSSION

Perception about climate change

Results in Table 1 show that 87.5% of respondents in Lichinga and 90% in Sussundenga were aware of climate variability and change. Farmers reported to have noticed significant changes in rainfall and temperature over the past ten years. The higher likelihood of insights on climate change with increasing age of the head of the household is associated with experience which lets farmers observe changes over time and compare such changes with current climatic conditions. Maddison (2006) reported farmer perception of climate change

through noticing an increase in temperature and a decrease in precipitation. Similar results have been reported by Nhemachena and Hassan (2007) and Mubaya et al. (2008) who reported that majority of farmers across Southern Africa perceive warming and drying of climate and low unpredictable rainfall as indicators of climate change. Studies by McSweeney et al. (2012), Queface and Tadross (2009) and INGC (2009) indicated that in Mozambique the mean annual temperature have increased by 0.6°C and the mean annual rainfall decreased at an average rate of 2.5 mm per month between 1960 and 2006.

The results also showed that 40% of respondents in Lichinga have noticed heavy rains while 45% of respondents in Sussundenga have noticed drought in the last 10 years. 57.5 and 100% of respondents in Lichinga and Sussundenga respectively, believe that there is shift in the beginning of the rain season in both short and long rain seasons. Rains that would normally start in October and stretch up to April are now starting late in November and in most cases ending in February as indicated in Table 2. These results are supported by Usman and Reason (2004), who reported that in different parts of Southern Africa countries a significant increase in the number of heavy rainfall events have been observed and that MICOA (2007) and INGCC (2009), noted that farmers in the Central zone of Mozambique (Sussundenga) are the most likely to experience increased risk of droughts. A study by Ribeiro and Chaúque (2010) in Mozambique, revealed that farmers faced prolonged drought over the last few years causing a decrease in agricultural productivity.

The findings of this study showed that 40% of small holder farmers in the past ten years used to plant in November and but presently over 63% of farmers plant in November. This increase may be explained by the shift in the start of the rain season from October to November. These results are in agreement with those of Mary and Majule (2009) and Mortimore and Adams (2001) who found that the onset of rainfall has shifted from October to November.

Adaptation to climate change

Results in Table 3 show that coping strategies to climate

Table 2. Changes in planting dates in the last 10 years in percentage.

District	Farm operations dates changed (%)		Planting date for maize 10 years ago (%)				Planting date for maize now (%)			
	Yes	No	September	October	November	December	September	October	November	December
Lichinga	90	10	0	22.5	47.5	30	0	7.5	22.5	70
Sussundenga	75	25	5	30	40	25	2.5	12.5	62.5	22.5

N = 40 each district.

Table 3. Mitigation strategies to climate change effects.

District	Mitigation strategies (%)			Mitigations crops (%)		
	Change crop variety	Kitchen garden	Off farm job	Cassava and sweet potato	Cabbage, Onion and Tomatoes	Adopting improved maize variety
Lichinga	15	75	10	47.5	2.5	10
Sussundenga	35	32.5	32.5	65	5	17.5

N = 40 each district.

change employed by most households include change of crop variety, kitchen gardening and seeking for off farm job. With increased frequency of droughts the results showed that changing crops varieties was a strategy in which 15 and 35% of respondents in Lichinga and Sussundenga were using by growing drought tolerant crops. These results are similar to those of Mutsvangwa (2009), who indicated planting drought tolerant crops as the most common adaptive strategy among Gweru and Lupane districts in Zimbabwe. In Lichinga 47.5 and 65% in Sussundenga of the respondents were planting cassava and sweet potato. However in Lichinga 90% of respondents and 82.5% in Sussundenga reported not to be using drought tolerant maize variety. This result are similar to those of Cavane (2011), who reported that improved maize varieties, whose traits have been selected for drought tolerant, were not yet widely adopted. International Fertilizer Development Center (2012) reported that in Mozambique only five percent of smallholders use improved seed varieties. The results confirms with those reported in Zambia by Mubaya et al. (2008) that farmers do crop diversification to cope with low rainfall.

The results also indicated that 75% farmers in Lichinga and 32.5% of respondents in Sussundenga are engaged in kitchen gardening. Kitchen gardening is also intensified in both districts and this could be due to the fact that farmers take advantage of the fact that wetlands remain charged for a long time after the rains and they therefore grow crops throughout the year. This finding are consistent with studies by (Mubaya et al., 2008), which indicate kitchen gardening is a strategy adopted in Zambia and Mozambique to cope with climate change. The study indicated that farmers in Lichinga (10%) and Sussundenga (32.5%) in times of low rainfall concentrated more on off farm activities. Ziervogel and

Taylor (2008) and Maddison (2007) indicated that due to low rainfall, farmers have moved towards non farming activities. The workers mentioned that off farm activities were considered to contribute significantly to the income of rural households.

The results also indicated that 47.5% farmers in Lichinga and 65% of respondents in Sussundenga are cultivating cassava and sweet potato as mitigation crop to cope with climate change effects. The study also showed the use of horticultural crops such as cabbage, onions and tomatoes as mitigations crops. Studies by Aggarwal et al. (2004); Easterling et al. (2007) and Maddison (2007), suggested that changes in temperature and precipitation call for changes in crop varieties more adapted to mitigate the effects of climate. Study done in Ghana by Acquah (2011), showed that farmers were using different crop varieties as major methods to cope with climate change. The respondents in Lichinga (10%) and Sussundenga (17.5%) reported not using improved variety such as resistant to drought due to high cost of seeds purchase. Enete and Achike (2008) and Cavane (2011) indicated that undercapitalized farmers fail to adopt the required level of agricultural technologies that will ensure profitable return.

CONCLUSION AND RECOMMENDATION

This study established that rainfall and temperature in study area has been decreasing and increasing, respectively, thus negatively affecting the production and management of crops. Different forms of changes on rainfall have been identified including shrinking of rain seasons due to late onset of rainfall period shifting from October to November or even December. A combination of strategies to adapt; such as proper timing of agricultural operations, crop diversification, use of different crop and diversifying from farm to non–farm activities were applied.

Consequently the following recommendations have been proposed on the basis of the study:

i) Farmers should be encouraged and enabled to use crop diversification as adaptation coping strategy to guard against crop failure in times of adverse climatic conditions.

ii) All effective adaptation options that farmers have applied in the study area should be widely disseminated to others farmers.

Conflict of Interest

The authors have not declared any conflict of interest.

REFERENCES

Acquah HD (2011). Farmers perception and adaptation to climate change: A willingness to pay analysis. Clarion, Pennsylvania. J. Sustain. Dev. Afr. 13:150-161.

Aggarwal PK, Ingram JS, Gupta RP (2004). Adapting food systems of the Indo-Gangetic plains to global environment change: Key information needs to improve policy formulation. Environ. Sci. Pol. P. 7.

Blanc É (2011). The impact of climate change on crop production in Sub-Saharan Africa. Dunedin, New Zealand : University of Otago.

Cavane E (2011). Farmers' Attitude and Adoption of Improved Maize Varieties and Chemical Fertilizers in Mozambique. Maputo, Mozambique : Indian Res. J. Exten. Educ. 11:1.

Easterling W, Aggarwal P, Batima P, Brander K, Erda L, Howden S (2007). Food, fibre and forest products. Climate Change 2007: Impacts, adaptation and vulnerability. contribution of Working Group II to the Fourth Assessment Report of the Intergovernmental Panel on Climate Change, M.L. Parry, O.F. Canziani, J.P. Palutikof, P.J. Cambridge, UK: Cambridge University Press pp. 273-313.

Ehrhart C, Twena M (2006). Climate Change and Poverty in Mozambique. Maputo, Mozambique : CARE International Poverty - Climate Change Initiative.

Enete AA, Achike IA (2008). Urban agriculture and food insecurity/poverty in Nigeria; The case of Ohafia - Southeast Nigeria. Outlook Agric. 37(2):131-134.

Eriksen S (2005). The role of indigenous plants in household adaptation to climate change: the Kenyan experience. Climate Change and Africa, P.S Low. Cambridge University Press, Cambridge pp. 248-259.

Famba SI (2011). The challenges of conservation agriculture to increase maize yield in vulnerable production systems in Central Mozambique. Vienna: University of Natural Resources and Life Sciences Vienna.

FAO (Food and Agriculture Organization) (2006). Guidelines for soil description. Fourth Edition. Rome. Food and Agriculture Organization (FAO). P. 97.

Geurts PMH, Tembe AF (1997). Fertilidade dos Solos no Planalto de lichinga. Maputo, Mocambique : INIA.

IFDC (International Fertilizer Development Center) (2007). Fertilizer supply and costs in Africa. Mozambique: Bill and Melinda Gates Foundation.

INGC (National Institute for Disaster Management) (2009). Main report: INGC Climate change Report: Study on the impact of climate change on disaster risk in Mozambique. [Asante, K., Brito, R., Brundrit, G., Epstein, P., Fernandes, A., Marques, M.R., Mavume, A, Metzger, M., Patt, A., Queface, A., Sanchez R. V.]. Mozambique: INGC.

Kinsey B, Burger K, Gunning JW (1998). Coping with drought in Zimbabwe: survey evidence on responses of rural households to risk. World Dev. 26:89–110.

Maddison D (2006). The perception of and adaptation to climate change in Africa. Africa : Centre for Environmental Economics and Policy in Africa (CEEPA) Discussion Paper, Special Series on Climate Change and Agriculture in Africa.

Maddison D (2007). The perception of and adaptation to climate change in Africa. Pretoria, South Africa: The World Bank: Policy Res. Work P. 4308.

MAE (2005a). Perfil Do Distrito de Lichinga. Maputo : Minesterio da Administratracao Estatal.

MAE (2005b). Perfil do Distrito de Sussudenga. Maputo: Minesterio da Administratracao Estatal.

Mary AL, Majule AE (2009). Impacts of climate change, variability and adaptation strategies on agriculture in semi arid areas of Tanzania: The case of Manyoni District in Singida Region, Tanzania. Dares salaam, Tanzania. Afr. J. Environ. Sci. Technol. 3(8):206-218.

McSweeney C, New M, Lizcano G (2012). UNDP climate change country profiles: Mozambique. Maputo, Mozambique: United Nations Development Programme (UNDP), 2012.

MICOA (Ministry of Coordination of Environmental Affairs) (2007). National Adaptation Programme of Action (NAPA). Maputo, Mozambique: MICOA.

Mortimore MJ, Adams WM (2001). Farmer adaptation, change and crisis in the Sahel. Sahel: Glob. Environ. Change 11:49-57.

Mubaya CP, Njuki J, Paidamoyo ME (2008). Results of a baseline survey: for Building Adaptive Capacity to Cope with Increasing Vulnerability due to Climatic Change. Gweru, Zimbabwe: Unpublished report to International Development Research Centre (IDRC).

Mutsvangwa EP (2009). Climate change and vulnerability to food insecurity among smallholder farmers: A case study of Gweru and Lupane Districts in Zimbabwe. Bloemfontein, South Africa: University of Free State.

Mutsvangwa EP (2011). Climate change and vulnerability to food insecurity among Smallholder farmers: A case study of Gweru and Lupane Districts in Zimbabwe. Bloemfontein, South Africa: University of Free State.

Nhemachena C, Hassan R (2007). Micro - level analysis of farmers adaption to climate change in Southern Africa. Southern Africa: IFPRI Discussion P. 00714.

Nielsen JO, Reenberg A (2010). Cultural barriers to climate change adaptation: A case study from northern Burkina Faso. Global Environmental Change– Hum. Pol. Dimen. 20:142–152.

Niimi Y, Pham TH, Reilly B (2009). Determinates of remittances: Recent evidence usingdata on internal migrants in Vietnam. Asian Econ. J. 23(1):19-39.

Queface A, Tadross M (2009). Main report: INGC Climate Change Report: Study on the impact of climate change on disaster risk in Mozambique. Mozambique: National Institute for Disaster Management (INGC).

Ribeiro N, Chaúque A (2010). Gender and Climate change: Mozambique Case Study. Southern Africa: Heinrich Böll Foundation Southern Africa.

Trærup SLM (2010). Ensuring sustainable development within a changing climate. Copenhagen: University Of Copenhagen.

Usman MT, Reason CJC (2004). Dry spell frequencies and their variability over southern Africa. S. Afr. Clim. Res. 26:199-211.

Ziervogel G, Taylor A (2008). Feeling stressed: Integrating climate adaptation with other priorities in South Africa. S. Afr. Environment 50:33-41.

Wiggins D (2005). An idea we cannot do without: what difference will it make (e.g. to moral, political and environmental philosophy) to recognize and put to use a substantial conception of need?" Royal Institute of Philosophy Supplement 80(57):25-50.

26

Ichthyofaunal diversity of mountain streams in the Tongboshan Nature Reserve, China

Mao-Lin Hu[1], Zhi-Qiang Wu[2], Shan Ouyang[1] and Xiao-Ping Wu[1]

[1]School of Life Sciences, Nanchang University, Nanchang, Jiangxi Province, China.
[2]College of Environmental Science and Engineering, Guilin University of Technology, Guilin, Guangxi Zhuang Autonmous Region, China.

Tongboshan Nature Reserve (between 28°03′30″ - 28°10′33″N and 118°12′00″ - 118°21′36″E) is located in the northeast of the Wuyi Mountain Range in the eastern Jiangxi Province. The fish fauna of mountain streams in the nature reserve was investigated seasonally during 2012. A total of 442 samples were collected and classified into four orders, eight families and 22 species. None of them collected in the nature reserve was exotic species. Among them, *Zacco platypus* was the most abundant fish species collected, followed by *Onychostoma barbatulu* and *Acrossocheilus parallens*. A total of 10 species were found to be endemic to China. Current threats to conservation of fishes in the nature reserve were identified and management solutions are suggested.

Key words: Tongboshan Nature Reserve, Mountain streams, ichthyofauna, diversity, conservation.

INTRODUCTION

Jiangxi Province (between 24°29′14″ - 30°04′41″N and 113°34′36″ - 118°28′36″E) is located in southern China, to the south of the middle and lower reaches of the Yangtze River. Poyang Lake, the largest freshwater body in China, is located in the north of Jiangxi Province. The area immediately surrounding Poyang Lake consists of low-lying alluvial plains prone to flooding. Mountains close to the boundaries of Jiangxi Province surround this region and all the five major rivers in the province (Ganjiang, Xinjiang, Fuhe, Raohe and Xiuhe Rivers) flow into the Poyang Lake. The drainage to Poyang Lake is a narrow outlet named Hukou, which flows into the Yangtze River and marks the northern border of the province. The sources of the rivers in Jiangxi Province are located in the surrounding mountains. Of a total of 220 recorded freshwater fish species throughout Jiangxi Province, about 131 species (59.5%) are believed to be endemic, many present in the mountainous regions (Huang et al., 2011). Protected areas such as nature reserves could play an important role in conservation of freshwater fishes within Jiangxi Province, but there is a need to better identify the conservation value of these areas in relation to biogeographical diversity of fishes and the factors impacting on fish communities.

Worldwide, freshwater fishes are the most diverse of all vertebrate groups, but are also the most threatened group of vertebrates after amphibians (Moyle and Leidy, 1992; Bruton, 1995; Duncan and Lockwood, 2001). Most

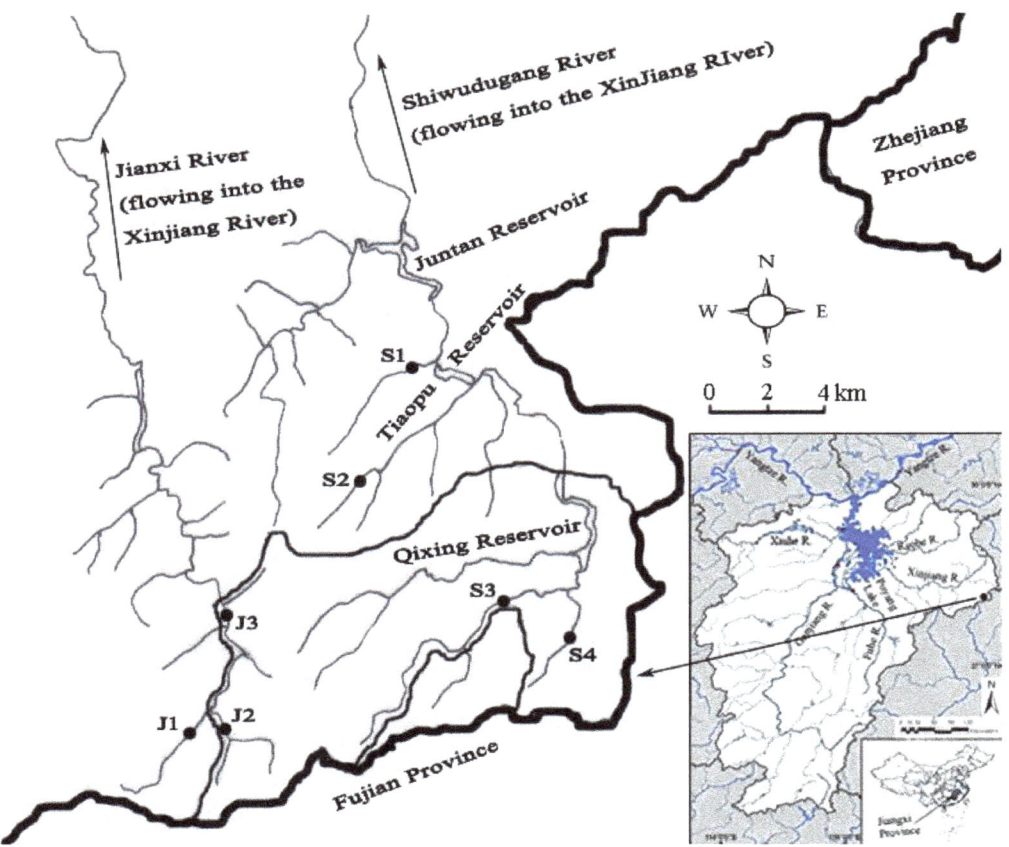

Figure 1. Map showing location of the TNR and sampling sites in the TNR.

mountain streams in the Tongboshan Nature Reserve (TNR) are shallow and the hydrology of most headwater streams has been modified by farming and irrigation of surrounding land. Recently, numerous anthropogenic disturbances, such as clear-cuts, small dams, road construction, fires and mining, have triggered physico-chemical alterations in the mountain streams (Tu et al., 2009; 2012).

At present, there have been several notable surveys of the flora and fauna within the nature reserve (Tu et al., 2009; 2012). However, until this work there have been no studies on the distribution and abundance of fish species in the nature reserve. The aims of the present study are: (1) To characterize the species composition of the fish fauna and their distribution in the nature reserve; (2) To review the main threats over fish biodiversity, and (3) To establish some recommendations to the conservation of the fish fauna.

MATERIALS AND METHODS

Study area

The TNR (total area: 108 km², altitude: 1535 m, between 28°03′30″ - 28°10′33″N and 118°12′00″ - 118°21′36″E) is located in the

northeast of the Wuyi Mountain Range in the eastern Jiangxi Province (Figure 1). The nature reserve presents humid subtropical climate and belongs to the forest ecological nature reserve for the conservation of evergreen broad-leaved forest ecological system and biodiversity. The annual precipitation is 1626.9 mm, annual temperature is 17.9°C, and forest coverage rate is up to 98% (Tu et al., 2009; 2012). Most mountain streams in the TNR flow into the Jianxi and Shiwudugang River which drain into the Xinjiang River (Figure 1).

Study site

Seven sampling sites were established on Jianxi and Shiwudugang River within the TNR (Figure 1). Sampling site selections were based on the representative habitat types present and accessibility during the study period. At each sampling site, the GPS position and altitude were recorded using a Garmin GPS map 76Cx. And water temperature, dissolved oxygen were measured with a hand-held YSI multi-meter. In addition, stream width and water depth were measured at each site.

Fish survey

Seasonally, samples were made at seven sites in the TNR during 2012. At each site, samples were collected using an electrofishing device consisting of two copper electrodes on wooden handles, powered by a 500-watt portable AC generator. Stunned fish were

Table 1. Characteristics of sampling sites within Jianxi River and Shiwudugang River, TNR, China.

Sampling sites		Altitude (m)	Depth (m)	Width (m)	Water temperature (°C)	Dissolved oxygen (mg/L)	Habitat description
Jianxi River	J1	715	0.2-1.5	1-4	9.6-17.2	9.6-10.8	Fast flowing and clear water, gravel and pebble substrate, shaded by forest canopy
	J2	796	0.1-1.2	1-3	9.0-16.8	9.8-11.2	Fast flowing and clear water, rocky and boulder substrate, shaded by forest canopy
	J3	670	0.2-2.0	2-5	10.2-17.6	9.5-10.6	Fast flowing and clear water, rocky and gravel substrate, river shaded by forest canopy
Shiwudugang River	S1	267	0.3-2.0	3-10	12.9-18.8	7.9-9.9	Slow flowing and slightly turbid water, gravel and sandy substrate, shaded by riparian vegetation
	S2	375	0.2-1.5	3-8	11.9-18.3	8.2-10.5	Slow flowing and clear water, gravel and sandy substrate, shaded by riparian vegetation
	S3	460	0.1-1.0	2-6	12.2-19.0	7.8-10.2	Slow flowing and clear water, gravel and boulder substrate, shaded by riparian vegetation
	S4	330	0.5-3.5	3-15	12.6-19.8	8.9-10.3	Slow flowing and slightly turbid water, gravel and sandy substrate, shaded by riparian vegetation

collected using dip nets or caught by hand. A cast net (mesh 5×5 mm; $\pi \times 0.6^2$ m = 1.13 m^2) was also used within shallow pools of the stream system to collect fish. Approximately 100 m of stream segment, typically comprising pool, run and riffle habitats, was sampled at each site. Collected specimens that could not be identified in the field were fixed in 10% formalin solution for accurate taxonomic verification. All specimens were identified according to Zhu (1995), Chen (1998), Chu et al. (1999) and Yue (2000).

Data analysis

The relative abundance of each species was estimated by:

$$P_j = N_j / N$$

where N_j = the number of species j collected in the TNR; N = the total number of all fish collected in the TNR. The Margalef index (D) and Shannon-Wiener index (H) were used to calculate fish species richness for each site (Peet, 1974; Magurran, 1988):

$$D = (S - 1) / \ln N \text{ and } H_k = -\sum P_j \ln P_j,$$

Where S = the total number of species collected in the TNR.

RESULTS

Stream characteristics and physicochemical parameters

The physical characteristics of each site are described in Table 1. Physico-chemical characteristics were similar among all studied sites in the TNR. Most of surveyed sampling sites were composed of sandy, gravel and pebbles substrates and the banks were lined by boulders and rocks. Shallow pools and riffles alternated in the segments studied. Generally, most mountain streams had clear water and were shaded by riparian vegetation or forest canopy. This appearance is typical of undisturbed forest stream at higher altitudes. All sampling sites were fully saturated with dissolved oxygen (mean ± SE, 9.6 ± 1.2 mg·L^{-1}). And water temperature ranged from 9.0 to 19.8°C. The high dissolved oxygen could be attributed to low water temperature and high water speed.

Fish fauna

A total of 442 specimens were collected and classified into 22 species and eight families in the TNR (Table 2). Cyprinidae (11 species, 50.00% of the total number of fish species collected) was the dominant family followed by Homalopteridae (three species, 13.64%), Bagridae and Gobiidae (two species respectively) while Cobitidae, Siluridae, Amblycipitidae and Synbranchidae were represented by only one specie respectively. The dominancy of fish species in the TNR was *Zacco platypus* (102 specimens, 23.08% of the total specimens collected), followed by *Onychostoma barbatulu* (17.87%) and *Acrossocheilus parallens* (11.99%).

Table 2. Composition and distribution of fish species in the TNR, Jiangxi, China.

Family/species	Jianxi River			Shiwudugang River			
	J1	J2	J3	S1	S2	S3	S4
Cyprinidae							
Acrossocheilus parallens (Nichols, 1931)*				11	41		1
Onychostoma barbatulum (Pellegrin, 1908)*			1	2	43		33
Opsariichthys bidens Günther, 1873							16
Zacco platypus (Temminck and Schlegel, 1846)	35	31	32			2	2
Gnathopogon imberbis (Sauvage and Dabry de Thiersant, 1874)*							30
Chanodichthys erythropterus (Basilewsky, 1855)							1
Culter alburnus (Basilewsky, 1855)							3
Hemiculter leucisculus (Basilewsky, 1855)							5
Megalobrama amblycephala (Yih, 1955)							2
Sinibrama macrops (Günther, 1868)*							2
Rhynchocypris oxycephalus (Sauvage and Dabry de Thiersant, 1874)	10	11					
Cobitidae							
Misgurnus anguillicaudatus (Cantor, 1842)						2	21
Homalopteridae							
Formosania davidi (Sauvage, 1878)*					3		
Pseudogastromyzon changtingensis tungpeiensis (Chen and Liang, 1949)*		1		17	8		12
Vanmanenia stenosoma (Boulenger, 1901)*	1			1			
Siluridae							
Silurus asotus Linnaeus, 1758							2
Bagridae							
Pseudobagrus taiwanensis (Oshima, 1919)*	15	9	15				
Pseudobagrus medianalis (Regan, 1904)*						1	
Amblycipitidae							
Liobagrus anguillicauda (Nichols, 1926)*	1	1		1			
Gobiidae							
Rhinogobius cliffordpopei (Nichols, 1925)					3		3
Rhinogobius giurinus (Rutter, 1897)				1	6		1
Synbranchidae							
Monopterus albus (Zuiew, 1793)					1	1	1

*Endemic to China (Huang et al., 2011, FishBase: www.fishbase.org).

Overall, 10 species (45.45% of the total number of fish species collected) were found to be endemic to China in the TNR. Endemic fishes were classified into four families. The dominant family of endemic fishes was Cyprinidae (four species) and the subdominant families were Homalopteridae (three species), Bagridae (two species) and Amblycipitidae (one specie). The most common endemic species to China was *Onychostoma barbatulum* (79 specimens, 17.87% of the total specimens collected), followed in order of abundance by *Acrossocheilus parallens* (53 specimens, 11.99%), *Pseudobagrus taiwanensis* (39 specimens, 8.82%), *Pseudogastromyzon changtingensis tungpeiensis* (38 specimens, 8.60%), *Gnathopogon imberbis* (30 specimens, 6.79%), *Formosania davidi* and *Liobagrus anguillicauda* (3 specimens, 0.68% respectively), *Sinibrama macrops* and *Vanmanenia stenosoma* (2 specimens, 0.45% respectively), *Pseudobagrus medianalis* (1 specimen, 0.23%) in the TNR.

General distribution of fish species collected from the seven sampling sites in the TNR was shown in Table 2. Meanwhile, the ecological indices for two rivers in the TNR, Shiwudugang River compared to Jianxi River may be because the fish habitats in the Shiwudugang River have comparatively higher species richness and diversity (Table 3).

DISCUSSION

Factors favoring diversity and endemism

The results of the present field studies on the TNR showed that a total of 22 native species (10.00% of all

Table 3. Comparison of fish species diversity between Jianxi River and Shiwudugang River, TNR, China.

Mountain stream	Total number of species (S)	Total number of individuals (N)	Margalef diversity index (D)	Shannon-Wiener diversity index (H)
Jianxi River	8	164	1.37	1.57
Shiwudugang River	20	278	3.38	3.17

Jiangxi Province freshwater species) were collected or found to be distributed in mountain streams. For example, *Zacco platypus* (23.08% of the total specimens collected), *Onychostoma barbatulu* (17.87%), *Acrossocheilus parallens* (11.99%), *Pseudobagrus taiwanensis* (8.82%), *Pseudogastromyzon changtingensis tungpeiensis* (8.60%), *Rhinogobius giurinus* (1.81%), *Rhinogobius cliffordpopei* (1.36%) and *Liobagrus anguillicauda* (0.68%) are anatomically well adapted to live in fast flowing current with clear water and relatively higher dissolved oxygen concentration. Generally, they feed on algae growing on the rock as well as detritus and insects. Overall, ten endemic species (250 specimens, 56.56% of the total specimens collected) in the TNR represented 7.63% of total endemic species in Jiangxi Province (131 endemic species; Huang et al., 2011).

This study suggests that mountain streams in the TNR are very important for freshwater fish diversity and conservation in Jiangxi Province, especially for the endemic species. The more abundant or endemic species collected in the TNR may be partially due to habitat stability and lack of disturbances, such as introduction of exotic species. The riparian zones of streams in the TNR are well forested so that stream temperatures rarely reached 20°C even during the summer and dissolved oxygen levels were high at all sites, providing suitable environmental conditions for these fishes. Such as *Rhynchocypris oxycephalus* (21 specimens, 4.75% of the total specimens collected), a representative cold water species of the Holarctic Region in China, tend to be distributed in the north of China. The alternating Quaternary glacial and interglacial periods had the effect of moving *Rhynchocypris oxycephalus* south, where it survived in the small mountain streams where the water is cold (Zhang and Chen, 1997).

It is interesting to note that the fish diversity was comparatively higher in Shiwudugang River than in Jianxi River. The habitats such as water depth and current, shoreline slopes and bottom substrates were relatively different. The substrate in Shiwudugang River was formed mainly of sandy-gravel, whereas in Jianxi River the substrate consisted mainly of rocky-pebbles which are very unstable. According to Zakaria et al. (1999) this condition could be a more suitable habitat for higher species diversity and richness. And most fishes were recorded in a channel stream part of a wide river where

the water is deeper and slower. Some species such as *Chanodichthys erythropterus*, *Culter alburnus*, *Hemiculter leucisculus*, *Megalobrama amblycephala*, *Sinibrama macrops* and *Silurus asotus* were only collected at site S4.

Current threats and conservation

During recent decades, streams and rivers in China have been drastically modified because of agricultural activities, drinking water supplies and the construction of multi-purpose dams, artificial reservoirs, levees, and weirs. These physical alterations and other human influences, such as road construction and deforestation have accelerated eutrophication (Fu et al., 2003). For example, Juntan Reservoir (closed on April 1985), Tiaopu Reservoir (completed in the 1980's) and Qixing Reservoir (closed on December, 1991) were built on Shiwudugang River. These factors strongly diminished effective migration for those species moving between different stream habitats. Small and fast-flowing streams have often been changed to large, slow-flowing streams. This change would cause that the organisms become restricted to mountainous areas and to be replaced by other beings adapted to slow-flowing streams (Hu et al., 2009).

In addition, some people go fishing as a source of food in the mountain streams of the TNR using rotenone and other poisons which usually are used to exterminate snails. This kind of fishing not only contributes to reduce fish biodiversity but is also harmful to human health.

Therefore, the primary objective for successful conservation of the freshwater ichthyofaunal diversity in the TNR must be to develop effective controls and management practices that enable life cycle success, dispersal and population maintenance within stream systems. It is necessary to improve effective fish passage facilities in order to enhance the connectivity of streams for fish dispersal and migration. Fishing activities in the TNR, especially using rotenone and other poisons must be strictly prohibited. The present work agrees with the statement that "long-term management and conservation of the fish fauna of nature reserves and other protected areas in Jiangxi Province will require good bench-mark sites and a long-term monitoring protocol" (Jang et al., 2003).

Conflict of Interest

The authors declared that they have no conflict of interest.

ACKNOWLEDGMENTS

This work was supported by the National Natural Science Foundation of China (No. 31360118), Natural Science Foundation of Jiangxi Province (No. 20122BAB214020), and Education Foundation of Jiangxi Province (No. GJJ13090). The authors would like to thank the staff of the TNR management station for their help provided during the survey.

REFERENCES

Bruton MN (1995). Have fishes had their chip? The dilemma of threatened fishes. Environ. Biol. Fish. 43:1-27.

Chen YY (1998). Fauna Sinica: Osteichthyes Cypriniformes II. Science Press, Beijing pp. 1-531.

Chu XL, Zheng BS, Dai DY (1999). Fauna Sinica: Osteichthyes Siluriformes. Science Press, Beijing pp. 1-230.

Duncan JR, Lockwood JL (2001). Extinction in a field of bullets: a search for the causes in the decline of the world's freshwater fishes. Biol. Conserv. 102:97-105.

Fu CZ, Wu JH, Chen JK, Wu QH, Lei GC (2003). Freshwater fish biodiversity in the Yangtze River basin of China: patterns, threats and conservation. Biodivers. Conserv. 12:1649-1685.

Hu ML, Wu ZQ, Liu YL (2009). The fish fauna of mountain streams in the Guanshan National Nature Reserve, Jiangxi, China. Environ. Biol. Fish. 86:23-27.

Huang LL, Wu ZQ, Li JH (2011). Fish fauna, biogeography and conservation of freshwater fish in Poyang Lake Basin, China. Environ. Biol. Fish. DOI: 10.1007/s10641-011-9806-2.

Jang MH, Martyn CL, Joo GJ (2003). The fish fauna of mountain streams in South Korean national parks and its significance to conservation of regional freshwater fish biodiversity. Biol. Conserv. 114:115-126.

Magurran AE (1988). Ecological diversity and its measurement. Cambridge University Press, London.

Moyle PB, Leidy RA (1992). Loss of biodiversity in ecosystems: evidence from fish faunas. In: Fiedler PL, Jain SK (eds.), Conservation biology: the theory and practice of nature conservation, preservation and management. Chapman and Hall, New York, pp. 127-169.

Peet RK (1974). Measurement of species diversity. Ann. Rev. Ecol. Syst. 5:285-307.

Tu YG, Huang XF, Lin CY, Tan CM, Liu YZ, Lin XY (2009). Investigation on animal and plant resources of Tongboshan Nature Reserve in Jiangxi. Jiangxi For. Sci. Technol. 2:36-38.

Tu YG, Yu NF, Wu NL, Tan CM, Jin MX, Liu YZ (2012). A preliminary study on the flora of seed plants of vegetation in Tongboshan Nature Reserve of Jiangxi Province. Acta Agriculturae Universitatis Jiangxiensis 34:754-761.

Yue PQ (2000). Fauna Sinica: Osteichthyes Cypriniformes III. Science Press, Beijing pp. 1-661.

Zakaria R, Mansor M, Ali AB (1999). Swamp-riverine tropical fish population: a comparative study of two spatially isolated freshwater ecosystems in Peninsular Malaysia. Wetl. Ecol. Manage. 6:261-268.

Zhang E, Chen YY (1997). Fish fauna in Northeastern Jiangxi province with a discussion on the zoogeographical division of east China. Acta Hydrobiol. Sin. 21(3):254-261.

Zhu SQ (1995). The synopsis of freshwater fishes of China. Jiangsu Science and Technology Press, Nanjing pp. 1-549.

27

An assessment of farm-to-market link of Indonesian dried seaweeds: Contribution of middlemen toward sustainable livelihood of small-scale fishermen in Laikang Bay

Achmad Zamroni[1,2] and Masahiro Yamao[1]

[1]Graduate School of Biosphere Science, Hiroshima University, Hiroshima, Japan.
[2]Research Center for Marine and Fisheries Socio Economics, Ministry for Marine Affairs and Fisheries, Jakarta, Indonesia.

For many years, sustainable seaweed farming has been practiced for the revitalization of village-level economy of small-scale fishermen in Laikang Bay. This study sought to identify the characteristics of small-scale seaweed farming in South Sulawesi Province in Indonesia, to assess the roles of middlemen in supporting seaweed production and marketing, and to investigate the marketing channels of dried seaweed products there in. This study adopted structured and semi-structured questionnaires as qualitative approach, and interviews were conducted with 220 seaweed farmers in Takalar and Jeneponto Districts. Interviewing with trader and middleman were also conducted. The data obtained were evaluated using descriptive statistics to summarize and compare the data set. Findings showed that most of the respondents (seaweed farmers) had two main livelihood activities, namely seaweed farming mainly *Eucheuma cottonii* and sustenance fishing activity with seaweed farming giving bigger contributions to household income. They were usually supported by particular middlemen to meet the capital and marketing requirements of their production of dried seaweed. Fishermen could usually get capital investment and daily operational funds through a quick process from the middlemen, without any interest payment. Thus, the middlemen occupy a crucial position in the production and marketing dynamics of the seaweed trade in the study area.

Key words: Sustainable livelihood, marketing channel, patron-client, middlemen, fishery.

INTRODUCTION

In Indonesia, the land area with aquaculture potential is estimated to be around 11.81 million ha, of which 8.36 million ha have marine culture potential (MMAF and JICA, 2009). Aquaculture production showed a growth rate of 20.14% within 5 years from 2001 to 200 (Nurdjana, 2006). The production of farmed seaweed in Indonesia gradually increased every year reaching 1,728,475 tonnes in 2007 (Dahuri, 2004; MMAF and JICA, 2009). According to Mira et al. (2006), there are many benefits realizable from seaweed farming such as: (1) being an environmentally friendly activity, (2) opening job opportunities, (3) improvement of fishermen's income and (4) contributing to foreign exchange revenue.

The Indonesian manufacturing industry can benefit enormously from the industrialization of carrageenan which is the principal chemical extract obtained from the farmed seaweed, Eucheuma cottonii (Tjahjana, 2010). The development of a viable seaweed industry can support the national program for job creation, reducing unemployment and contributing to national economic growth. Development can focus on the various types of seaweed available locally which in turn can support the production of carrageenan, agar and alginate. The local carrageenan industry producing semi-refined carrageenan products grew rapidly after 1990. However, it declined due to lack of raw materials. This can be attributed to the fact that the manufacturing industry cannot compete with exporters of dried unprocessed seaweed in the purchase of raw materials. On the other hand, the agar or gelatin industry has been around since the 1940s, but its industrialization started only in the 1990s, when the alginate industry also began at the same time. Both agar and alginate industries, however, are too small compared to the size of the carrageenan industry. In the cultivation of seaweed, fishermen have used various ways to address the problem of financial capital. Besides formal financial institutions that are rarely tapped, fishermen usually borrow money from the family, relatives, friends and even brokers (middlemen) in the village. This frequently happens because small-scale fishers still have problems in accessing capital from formal financial institutions such as commercial banks. The patron-client relationship within seaweed farming scheme is often referred to as punggawa (middlemen) – sawi (farmer) system. In this study, the patron called middleman, can be defined as the person who provides the capital and lending to fishermen/seaweed farmers to plant seaweed. Meanwhile, client is a person/fisher planting the seaweed, and called seaweed farmer.

This paper seeks to identify the socio-economic characteristics of seaweed farmers in Laikang Bay; to assess the role of middlemen in sustaining local seaweed cultivation activities and to describe the pattern of the local seaweed procurement chain.

MATERIALS AND METHODS

Study area

The study sites were located around Laikang Bay, specifically within Laikang Village, Takalar District, South Sulawesi Province, Indonesia (Figure 1).

Takalar district

South Sulawesi Province is located in the southernmost part of Sulawesi Island (formerly Celebes) between 0°12' to 8' S and from 116°48' E up to 122°36' E. Makassar City, the provincial capital is located between from 5°30' 18" S to 5° 14' 49" S and from 119° 18' 97" E to 119° 32' 3" E. The average temperature in Makassar is between 22 and 33°C (South Sulawesi Province, 2010).

Takalar District is located on the south side of South Sulawesi Province. It has a land area of 566.51 km2 and is located between 5°3' S and 5°38' S and from 119°22' E to 119°39' E. It is bounded by Gowa District on the north, Gowa District and Jeneponto District

on the east, Flores Sea to the south and Makassar Strait to the west (Dinas Kelautan dan Perikanan Takalar -Takalar Marine and Fisheries Service Office-TMFSO and Narayana Adicipta Persero, 2007). It is about 42 km south of Makassar City. Interviews and data collection were conducted in Laikang Village, in Managarabombang Sub-district in Takalar District, 16 km from the central district and 63 km from Makassar City.

Laikang Village is one of the 12 villages of Mangarabombang Sub-District. It has an area of 19.6 km2, comprising about 19.57% of Mangarabombang Sub-District (±100.14 km2). The population is approximately 4,139, or 12% of the total population of the sub-district (35,526 people) with a population density of about 211 people/km^2.

Most of the people work in fisheries, and some work in the agricultural sector. Laikang Village is rich in natural resources like fisheries, agriculture and tourism sectors which largely contribute to the economic development of the village. However, poorly developed fisheries infrastructure, telecommunication and public transportation, further hamper the economic development of the village.

Jeneponto district

Jeneponto District is located in the western part of South Sulawesi Province, and it has a coastal area that stretches for about 95 km in the southern part, covering an area of 74,979 ha or 749.79 km^2. It is bounded by Gowa District on the north, the Flores Sea on the south, Takalar District on the west and Bantaeng on the east. Jeneponto District consists of 9 sub-districts and 105 villages, and the population in 2004 was 324,927, consisting of 158,043 men and 166,884 women. There were 18,943 fishermen, fish farmers and seaweed farmers. Similar to other coastal areas, it has two dominant fishing activities namely capture fisheries and aquaculture including seaweed farming (Figure 2).

Figure 2 shows that during 2003 to 2005, production of seaweed increased sharply compared to the other fishery sectors. It clearly indicates the potential of seaweeds to contribute to the fishery sector in Jeneponto District.

Data collection and analysis

The current investigation involved surveys, direct observations and interviews. Total number of respondents was 220 seaweed farmers selected by the appropriate sampling method with data obtained through direct interview and focus group discussion. A structured questionnaire was prepared and used for direct interview, while semi structured questionnaires were used as guides in the focus group discussion. The structured questionnaire covered socio-economic information of seaweed farmers.

Key informants were selected purposively including middlemen, staff or researchers from government offices, research centers, universities, local government officers, community leaders (tokoh masyarakat), heads of the villages (kepala desa), religious leaders, who understand well the social and economic conditions of the village. Secondary data were obtained mainly from statistical data and scientific journals. Descriptive analysis focused on socio economic conditions of respondents and the study areas.

RESULTS

Fishermen activity in seaweed farming

The survey shows that most seaweed farmers in Laikang Village are 26 to 40 years old.

An assessment of farm-to-market link of Indonesian dried seaweeds: Contribution of middlemen toward sustainable...

175

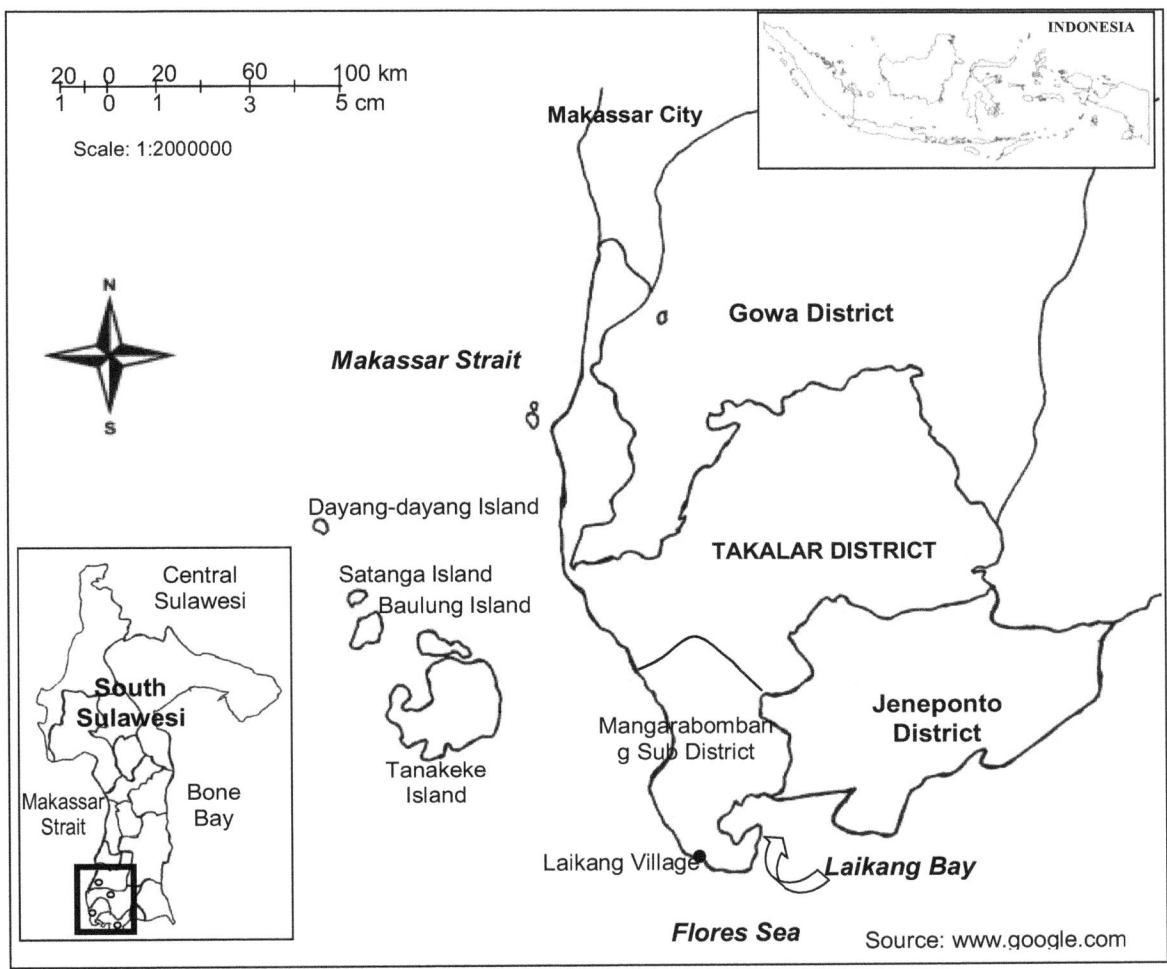

Figure 1. Map of Laikang Bay in Laikang Village, Takalar District, South Sulawesi Province.

They have households composed of 2 to 5 persons in each family. Most of them are of poor level of education having graduated only from elementary school. The income of respondents was from 2 main activities, capture artisanal fisheries and seaweed farming. Both activities were conducted by people in Laikang, Garassikang, Lurah Pantai Bahari (LPB) and Ujunga Villages. The fishers who derive more income from seaweed farming than compared with other forms of culture fisheries obviously preferred to give higher priority to seaweed culture as their main income source. Most of respondents (70.5%) have income of less than 1 million Indonesian rupiah (IDR) per month (Table 1). This amount is still below the Regional Minimum Wage (UMR) 2010 in South Sulawesi Province (IDR 1,049, 321) (BPS, 2010).

The main activities in Laikang Bay are seaweed farming and capture fisheries. Approximately 10 years ago, people were mainly engaged in fishing by using simple technologies, such as nets and fishing rods. Their boats were very simple, either with 5 to 15 Horse Power (HP) engine or without an engine. Thus, the areas of their operations were limited to 1 to 3 nautical miles from the shore with operation time limited to only one day fishing. Fishermen usually went to the sea in the afternoon at about 4 to 5 pm to set up their fishing nets in the fishing ground. In the morning between the hours of 4 and 5 am, they returned to sea to retrieve the nets. Depending on the volume of catch, they sold to markets or consume at home. Certain valuable species of fish would be sold either to a local trader or collector or in a traditional market, while others would be retained for domestic consumption.

Figure 3 shows the size of one unit of seaweed farm as 100 m long and 30 m wide developed at water depths of less than 10 m. Styrofoam material are used for large buoys, while 500 ml and 1000 ml empty bottles of mineral water are also used. Seaweed farmers assisted by their wives and other family members prepare the seedlings to be attached on the stretch of rope. The average length of

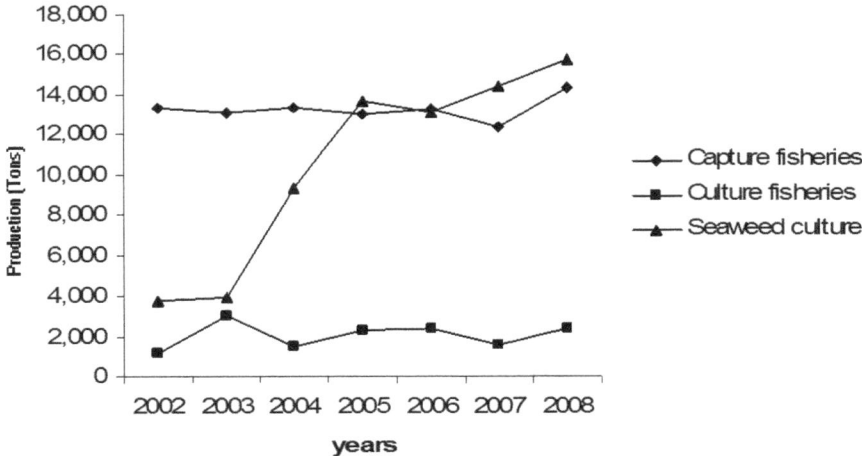

Figure 2. Production growth of three fishery activities in Jeneponto District. Source: Statistics of Jeneponto Fisheries office, various years.

Table 1. Demographic characteristics of respondents.

Variable (*N=220 respondents*)	Frequency	%	Mean	S.D
Age (years)			36.92	9.39
≤ 25	16	7.27		
26 - 40	132	60		
41 - 60	72	32.72		
Gender (male/female)			1	0
Male	220	100		
Female	0	0		
Education			2.23	1.97
Elementary school	136	61.8		
Junior High School	33	15		
Senior High School	6	2.7		
None	45	20.5		
Main income generating activity			3.29	2.87
Seaweed culture	132	60		
Capture fisheries	0	0		
Seaweed culture + capture fisheries	54	24.5		
Seaweed culture + public officer	4	1.8		
Seaweed culture + non fisheries	30	13.6		
Number of family member (persons)			1.9	0.29
≤ 2	7	17.5		
3 - 5	31	77.5		
≥ 6	2	5		
Income value per month			1.35	0.46
≤ 1000,000	155	70.5		
>1000,000	65	29.5		
Number of seaweed plots (1 plot=100 m × 30 m)			1.35	0.48
≤2	142	64.5		
≥3	78	35.5		

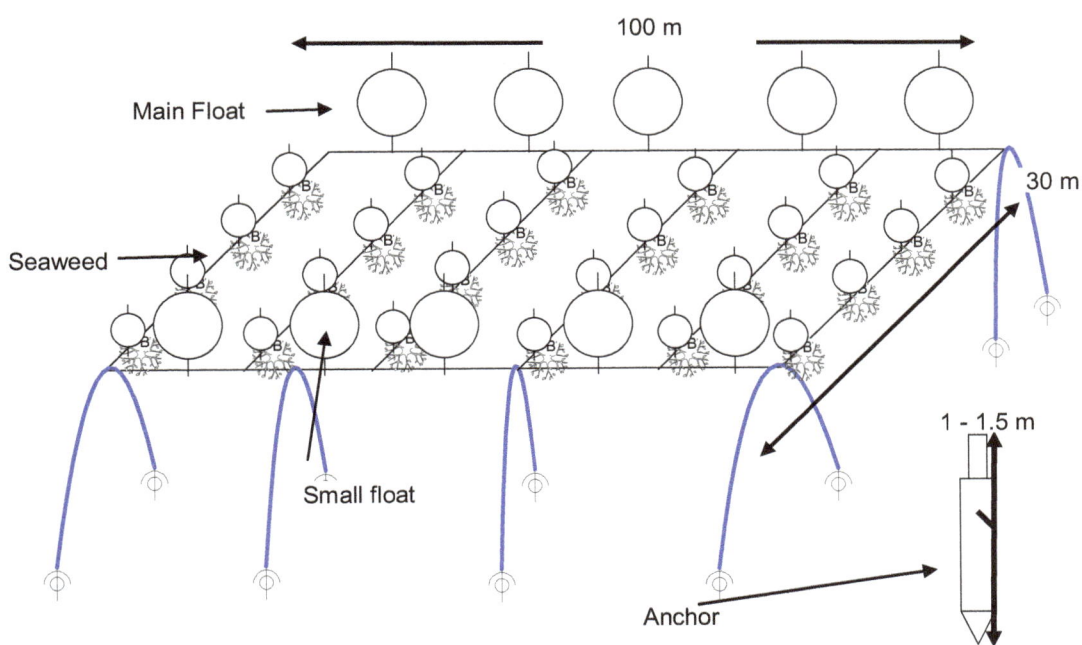

Figure 3. The construction of seaweed farm (floating method). Source: Field observation, 2009 and 2010.

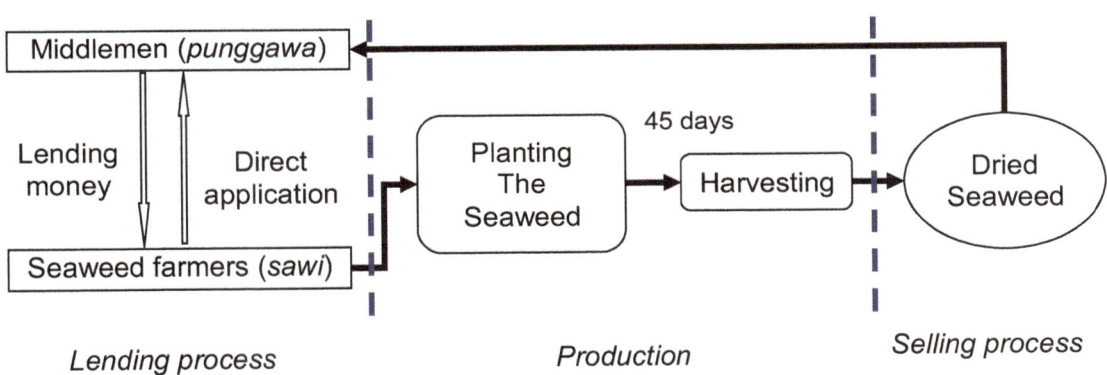

Figure 4. Relationship of seaweed trader and seaweed farmers. Source: Field observation, 2009 and 2010

seaweed growth is 45 days, but could be shorter in case disease affects the seaweed that can result to damage or death of these plants. In this case, the seaweed can still be sold but at a price lower than usual.

Roles of middlemen in sustaining seaweed farming

Over the years, farmers have sold dried seaweeds to middlemen. The findings of this study showed that some seaweed farmers are closely affiliated with the traders/middleman in many ways particularly related with financial capital and product sales. As a consequence, they sold their product exclusively to a particular middleman (Figure 4).

Figure 4 describes the mutual relationships between farmers and middlemen. Farmers came to the middleman directly to borrow money. This money was used for restarting the planting cycle and allocated for purchasing of seaweed seed stock and/or repairing equipments in seaweed farm. The loan is done without any collateral and usually with minimal or no documentation. In return, farmers pledge to sell their products exclusively to the middleman. Middlemen do not fix the duration of repaying debt and even the amount of loan repayment. However, dried seaweed that is bought from farmers is treated as cash paid immediately back to the middleman, not as delayed payment as experienced by other seaweed farmers (Namudu and Pickering, 2006). The presence of middlemen in Laikang Village has been documented

Table 2. The two patron-client systems in capture fisheries and seaweed farming activities.

| Instruments | Fishing activity | | Seaweed farming | |
	Punggawa-middlemen (patron)	*Sawi*-fishermen (client)	Middlemen (patron)	Seaweed Farmer (client)
Role	Owner of fishing equipment	Worker	Moneylender Buyer	Farmer Borrower
Products or service provided	Fuel, boat, fishing gears	Manpower	Funds/money	Dried seaweed
Benefits	Profit from business/activity	Receives a salary	Easy to get dried seaweed products	Get capital money Easy to sell dried seaweed
Organizational form	Group	Group	Individual	Individual

Source: Field observation 2010 supported by Arif (2007) (unpublished).

since the start of seaweed farming there. At present, middlemen mainly function as providers of capital lent out to particular seaweed farmers and as a buyer of raw dried seaweed from them. The middlemen's financial sources are independently different from regular institutional sources. The same punggawa-sawi relationship exists in the traditional fishery system but it is different from that in the seaweed farming system. Table 2 compares the two different patron-client systems in capture fisheries and seaweed farming activities.

According to the survey, a business relationship between the middleman and seaweed farmer is based solely on the viability and continuity of supply and demand trends. Middleman will get the assurance of steady availability of dried seaweed from farmers who borrowed capital from them. Thus, the business relationship between the both have each other implicitly "tied" to one another for mutual benefit. Day (2000) emphasized the value of relationship building when there are a few valuable customers or suppliers transacting lucrative business. However, the relationship between the middleman and fishermen in the long term tends to evolve into a monopolistic trade at the local level. Ogawa et al. (2006) argued that middlemen as the monopolist can act as intermediaries and have better control in adjusting the prices to obtain maximum profits for himself and at times for the suppliers as well because farmers are often heavily dependent on the middleman. Middlemen often refuse seaweed farmers from paying off their debts. They preferred the debt to accumulate, unpaid, in order to maintain exclusive business link with the farmers. This is somewhat restrictive to seaweed farmers because they have little freedom to sell their products to other traders who may offer better prices than the middlemen who lent them money. Nevertheless, the price of seaweed offered by middlemen is usually in accordance with prevailing market prices which the respondents perceive as reasonable.

Seaweed farmers actually have another alternative for obtaining needed capital. Government projects related to

community empowerment and economic development in coastal areas have provided various credit schemes in the form of revolving loan funds. However, the lack of sustainability of these programs has led seaweed farmers back to the middlemen. The failure of such income-generating projects could be related to the lack of financial capital, technical training and market access (Marais and Botes, 2006). This study found that although there is minimal interest on government loans, there are also inefficient repayment and capital recycling which weaken the programs, limiting availability of revolving funds to future borrowers. Hence, government credit program contributes very little towards supporting community-level seaweed farming (Table 3).

Marketing channel of seaweed: some preliminary investigation

The seaweed farmers in Takalar sold the seaweed in dried form to traders/middlemen coming to the village, who then sell to the trader at the district level and finally to wholesalers who have warehouses in Makassar/Ujung Pandang or to processing companies there. The typical marketing channels of seaweeds in Takalar can be seen in Figure 5.

Fishers culturing seaweeds at the village level can also serve as middlemen, while some the company's agent referred to in the marketing chain is a person who is a paid employee of processing companies and who is tasked by the company in purchasing raw materials. Other types of middlemen are free-lance buyers and money lenders who sell dried seaweeds mainly to exporters. Some exporters are also engaged in seaweed processing aside from being exporters of dried seaweed. There are particular exporters such as Semi Refined Carrageenan (SRC) and Alkali Treatment Cottonii (ATC) producing value-added products.

Bulk shipment of seaweed out of the villages is done about once a month, or when the accumulated seaweed

Table 3. Comparative roles of two financial sources supporting the activities of seaweed farmers.

Instruments	Lenders	
	Middleman	**Financial institution**
Application for credit or loan	Non-formal Based on trust	By document By contract
Collateral	- No collateral	Mostly collateral required
Considerations	Trust Relationships	Requirements Track record Trust
Durations	Flexible time limit	Fixed time limit
Interest	No interest	With interest
Role	Creditors Agent/marketing/buyer	Creditors

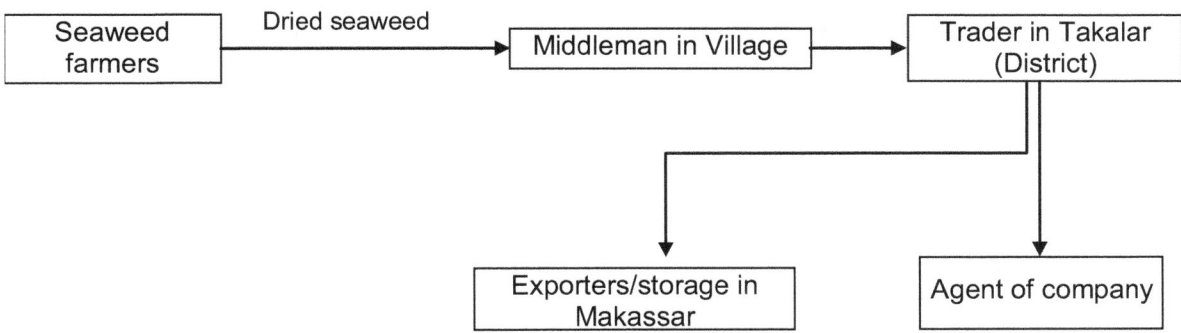

Figure 5. Marketing channel of seaweed in Laikang Village. Source: Field observation 2010 and supported by Zamroni et al., 2006.

seaweed from the same traders who have become frequent and reliable suppliers over time thus reinforcing the strong ties between wholesalers and traders. Personal trust plays an important role in the success of their business. According to the survey, wholesalers require technical specifications of dried seaweed such as water content of seaweed is less than 37% and pH less than 12. These requirements are often demanded from farmers and middlemen but sometimes not followed.

Recently, the most critical problem of Indonesian seaweed industry development in is related to marketing and breeding. Seaweed farmers have not received much economic benefit from the current marketing set up than they had expected. In general, these marketing problems are regarded as institutional aspects, complicated marketing network, and communication gap between producers and consumers or production output is often not in accordance with standards established by the processing industry or the export market of seaweed. This becomes a compelling reason for industry to buy raw seaweeds at lower prices from production sites, and then to process by themselves.

Farmers also encounter problems associated with declining seaweed quality. Seaweed are often affected by

seasonal disease outbreaks which result to premature harvesting. Some farmers observed slower growth of seaweeds which can be attributed to inferior genetic quality. This can partly be addressed by identifying new seaweed varieties or strains which can grow faster or are more disease resistant. Clearly, more research and development addressing declining seaweed seed stocks are needed.

DISCUSSION

The Government of Indonesia (GoI) has defined three main policies, three core programs and six support programs to achieve and realize aquaculture development (Nurdjana, 2006). Until 2008, South Sulawesi has been the highest producer of seaweed (690,385 tons) followed by East Nusa Tenggara (566,495 tons), Central Sulawesi (208,040 tons), and Bali (170,860 tons) and other regions (304,220 tons) (Figure 6) (Sudartanto, 2010). The overall national seaweed production still shows a positive trend, likewise in the local level (Figure 7).

Coastal fisheries in Asia need relevant strategies to

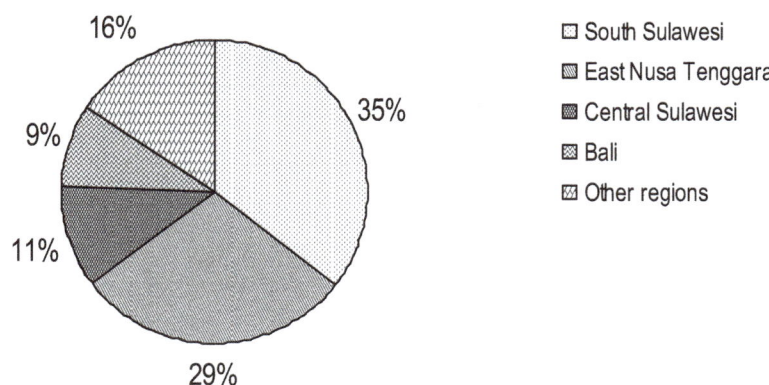

Figure 6. Indonesian Wet Seaweed Production by Province in 2008. Source: DGCF, 2008.

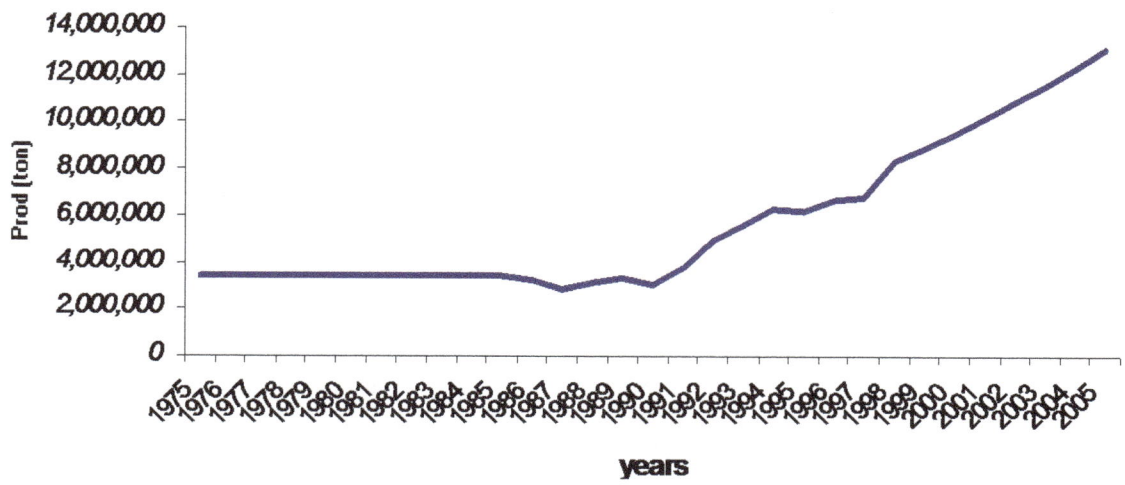

Figure 7. Trend production of seaweed in Indonesia (BPS. Various years).

ensure future fish production and to improve the livelihoods or standard of living of coastal communities (Stobutzki et al., 2006; Hurtado and Agbayani, 2002; Munoz et al., 2004). Recently, seaweeds appear to play important roles to achieve these strategies. Seaweed farming has also reduced some forms of local fishing effort and environmental pressure (Sievanen et al., 2005). Since there were several projects related to coastal community development and poverty reduction, people had begun to develop seaweed farming since 1982 by culturing the economic seaweed *E. cottonii* using the long line method. This technology uses rope of various sizes, which are arranged transversely forming a rectangle and allowed to float away from the bottom but anchored to the problem by ropes. Seaweed cultivation is maintained individually by fishermen with 50 to 100 long

lines per respondent (Figure 3). This method is strongly recommended for cultivating *E. cottonii* in Tanzania compared to the off-bottom method (Msuya et al., 2007). However, Thirumaran and Anantharaman (2009) mentioned that seedling density, monsoon season, water temperature, nutrients, and water movement strongly influenced the growth rate of seaweeds. Any big changes in any or all of these parameters would affect seaweed growth and overall farm output.

Cooke (2004) and Crawford (2006) emphasized that the success of seaweed farming would depend on the extent of the roles of farmer's wives and family members who help in the drying and the preparation of seaweed plantations. They also help in the maintenance and weeding of any stretch of culture lines (line) from dirt and debris as well as tying of seedlings unto the ropes and sorting of harvested seaweeds. The supplemental labor provided by family members can contribute towards cutting down farming expense.

The present study has found that the observed positive impact on the relationship between seaweed farmers and middlemen can be explained in several points; (1) seaweed farmers could get loans through quicker process, (2) there was no interest collected on loans, (3) the farmers were assured of selling their harvest and (4) the farmers got cash payment for their produce which actually represents additional debt. Middlemen may not be the best buyers, but they can provide some social benefits in the long run (Masters, 2008) by seeking the best market prices whose profits can sometimes be passed on to the farmers (Ju et al., 2010; Rust and Hall, 2003; Shevchenko, 2004; Li, 1998). Indeed, middlemen have two essential roles as direct links to the external market and as provider of credit to fishermen (Crona et al., 2010). Characteristics of seaweed marketing in Takalar seem typical of the trends seen in this part of Indonesia. In Gorontalo southeast Sulawesi, dried seaweeds were sold to a local trader, then purchased by wholesalers/processing company (Neish, 2007).

According to Ju et al. (2010), intermediaries begin by making capacity choice then buy goods from producers and sell them to consumers effectively bridging the production-consumption gap (Gadde and Snehota, 2001). Johri and Leach (2002) and Vesala (2008) argued that the adverse selection problems in the trade of goods of different quality may be alleviated through a middleman. Middlemen can act as an alternative and advantageous way to reduce market frictions (Masters, 2007). Meanwhile, wholesalers made the products available, bringing an assortment of conveniences essential for bulk-breaking, providing credit and finance, performing customer service functions, as well as providing advice and technical support (Samali and El-Ansary, 2007). Wholesalers exist because they are able to provide the most effective and efficient distribution process than all other channel participants (Rosenbloom, 2007; Torii and Nariu, 2004). In the Japanese system, for example, to improve wholesaler's performance involves: Promoting strong collaboration, offering trustworthy information, accommodating a variety of needs, supporting the mission, and determining the price (Rawwas et al., 2008). Thus, accessibility and risks of the product-market chain depended on market structure, size, expected demand levels and the nature of competition (Roberts and Stekoll, 1993). The encounter of any problems generally indicates lack of cooperation among stakeholders.

Conclusions

Seaweed farming activity in Laikang Bay is mainly conducted using the long mono line method by seaweed

farmers who were formerly exclusively small-scale fishermen who found seaweed farming as a more lucrative livelihood. At present, seaweed farming has become the main income source for these small-scale fishermen because it provides substantial additional household income besides the usual income from occasional fishing activity.

From both farming and marketing perspectives, middlemen play important roles to support the fishermen; (1) to provide the financial capital expeditiously and (2) to collect/buy the dried seaweed from seaweed produce farmers. The existence of middlemen is crucial in the dried seaweed supply chain in Laikang Bay, as long as the local/central government could not implement a better and effective market chain for seaweeds at the local level. Because most farmers have low educational background or are sometimes illiterate, borrowing from middlemen eliminates the need for financial institutions which require complicated documentation and strict repayment schemes. Middlemen can offer more flexible repayment terms in kind or goods. This works better for farmers because of the flexibility it offers and assured disposal of their dried products even if the middlemen-farmer relationship is perceived by some respondents as exploitative and unfair. This is because the seaweed buying price is mostly set by middlemen and most farmers usually cannot sell to other traders who may be offering higher buying prices. Indeed, this traditionally disadvantageous relationship will be maintained in the absence of government intervention and big industry players which can offer more equitable business terms to further encourage seaweed farming. The acute supply of raw materials these days should prompt processors to set up more vigorous procurement efforts by putting up buying programs characterized by higher prices and easier credit extension.

The growing dependence of many fishing communities to seaweed farming as a main income source will allow seaweed farming to develop more quickly potentially resulting to greater prosperity in the coastal areas. However, the density of seaweed plots and the unclear definition of farm ownerships are prone to lead to conflicting claim among interested parties. The issues about foreshore claims should be addressed by village and government leadership to avert a socio-economic crisis in the future. Furthermore, ecological studies should be conducted on the carrying capacity of the coastal environment in Laikang Bay and how seaweed farming can impact on the environment in an effort to strike a balance of social acceptability and positive ecological effects of this particular activity.

ACKNOWLEDGEMENTS

First of all, the authors wish to acknowledge their gratitude to the anonymous reviewers who gave freely time and effort, constructive recommendations that enhanced the value of this manuscript. The authors also would like to express their deepest thanks and appreciation to all the seaweed farmers who participated in the interview research and Dr. Lawrence Liao, visiting professor of the Graduate School of Biosphere Science, Hiroshima University for his academic suggestions. Specials thanks go to the Monbukagakusho Scholarship for supporting the studies of the first author that enabled him to conduct this research.

REFERENCES

BPS (2010). South Sulawesi in Figure 2008. Nat. Stat. agency - South Sulawesi Province.

Cooke FM (2004). Symbolic and social dimensions in the economic production of seaweed. Asia Pasific Viewpoint 45(3):387-400.

Crawford B (2006). Seaweed farming: An alternative livelihood for small-scale fishers? Working paper. Coastal Resources Center. Univ. of Rhode Island, available at http://www.crc.uri.edu/download/Alt_Livelihood.pdf (accessed 19 July 2010).

Crona B, Nystrom M, Folke C, Jiddawi N (2010). Middleman, a critical social-ecological link in coastal communities of Kenya and Zanzibar. Mar. Pol. 34(4):761-771.

Dahuri (2004). Indonesia berpotensi menjadi produsen ikan terbesar (Indonesia has the potential to become the largest fish producer). Mina Bahari 2(10):16-17.

Day GS (2000). Managing market relationships. J. Acad. Mark. Sci. 28(1):24-30.

Gadde LE, Snehota I (2001). Rethinking the role of middlemen. Paper for IMP 2001, BI, Oslo, 9 - 11 September, available at http://www.impgroup.org/uploads/papers/182.pdf (accessed 9 December 2010).

Hurtado AQ, Agbayani RF (2002). Deep-sea farming of Kappaphycus using the multiple raft, long-Line method. Bot. Mar. 45(5):438-444.

Johri A, Leach J (2002). Middlemen and the allocation of heterogeneous goods. Int. Econ. Rev. 43(2):347-361.

Ju J, Linn SC, Zhu Z (2010). Middlemen and oligopolistic market makers. J. Econ. Manag. Strateg. 19(1):1-23.

Li Y (1998). Middlemen and private information. J. Monet. Econ. 42:131-159.

Marais L, Botes L (2006). Income generation, local economic development and community development: paying the price for lacking business skills? Com. Dev. J. 42(3):379-395.

Masters A (2007). Middlemen in search equilibrium. Int. Econ. Rev. 48(1):343-362.

Masters A (2008). Unpleasant middlemen. J. Econ. Behav. Organiz. 68 (1): 73-86.

Mira, Mulyawan I, Zamroni A (2006). Analisis Keunggulan Kompetitif Usaha Budidaya Rumput Laut di Indonesia (Competitive Advantage Analysis of Seaweed Farming in Indonesia), in Semnaskan UGM 2006 proceedings of the 3rd annual meeting of marine and fisheries research in Yogyakarta, Indonesia, July 27, 2006, UGM, Gajah Mada University, Yogyakarta, Indonesia. pp. 655 - 662.

MMAF, JICA (2009). Indonesian Fisheries Book 2009. Ministry of Marine Affairs and Fisheries and Japan International Cooperation Agency (JICA). P. 83.

Msuya FE, Shalli MS, Sullivan K, Crawford B, Tobey J, Mmochi AJ (2007). A Comparative Economic Analysis of Two Seaweed Farming Method in Tanzania. The Sustainable Coastal Communities and Ecosystems Program. Coastal Resource Center, University of Rhode Island and the Western Indian Ocean Marine Science Association. P 27.

Munoz L, Freile-Pelegrin Y, Robledo D (2004). Mariculture of Kappaphycus alvarezii (Rhodophyta, Solieriaceae) color strains in tropical water of Yucatan, Mexico. Aquaculture 239(1-4):161-177.

Namudu MT, Pickering TD (2006). Rapid survey technique using socio-economic indicators to assess the suitability of Pacific island rural communities for Kappaphycus seaweed farming development. J. Appl. Phycol. 18:241-249.

Neish IC (2007). Assessment of the seaweed value chain in Indonesia. USAID, available at http://www.amarta.net/amarta/ConsultancyReport/EN/AMARTA%20Value%20Chain%20Assessment%20Seaweed.pdf (accessed 24 June 2010).

Nurdjana ML (2006). Indonesian aquaculture development. Directorate General for Aquaculture, MMAF, available at http://www.agnet.org/library/bc/55007/ (accessed 1st December 2010).

Ogawa K, Koyama Y, Oda SH (2006). A Middleman in an ambiguous situation-experimental evidence. J. Soc. Econ. 35(3):412-439.

Rawwas MYA, Konishi K, Kamise S, Al-Khatib, J (2008). Japanese distribution system: The impact of newly designed collaborations on wholesalers' performance. Ind. Market. Manag. 37(1):104-115.

Roberts WA, Stekoll MS (1993). Commercial potential of seaweed from St Lawrence Island, Alaska: Evaluation of market opportunity. J. Appl. Phycol. 5(2):167-173.

Rosenbloom B (2007). The wholesaler's role in the marketing channel: disintermediation vs reintermediation. Int. Rev. Retail Dist. Cons. Res. 17(4):327-339.

Rust J, Hall G (2003). Middlemen versus market makers: A theory of competitive exchange. J. Polit. Econ. 111(2):353-403.

Samali AC, El-Ansary Al (2007). The role of wholesalers in developing countries. Int. Rev. Retail, Dist. Cons. Research 17(4):353 - 358.

Shevchenko A (2004). Middlemen. Int. Econ. Rev. 45(1):1-4.

Sievanen L, Crawford B, Pollnac R, Lowe C (2005). Weeding through assumptions of livelihood approaches in ICM: seaweed farming in the Philippines and Indonesia. Oce. Coast. Manag. 48(3-6):297-313.

South Sulawesi Province (2010), Sekilas tentang Provinsi Sulawesi Selatan (overview of the South Sulawesi Province). Available at http://www.sulsel.go.id/indonesia/media.php?module=peta, (accessed 17 January 2011).

Stobutzki IC, Silvestre GT, Garces LR (2006). Key issues in coastal fisheries in South and Southeast Asia, Outcomes of a Regional Initiative. Fish. Res. 78(1-3):109-118.

Sudartanto E (2010). Penyuluh dan bibit sebagai kendala rumput laut (Extension staff and seed as obstacles in seaweed culture), available at http://www.dkp.go.id (accessed 19 April 2010).

Dinas Kelautan dan Perikanan Takalar -Takalar Marine and Fisheries Service Office-TMFSO and Narayana Adicipta Persero, 2007. Final report of Small Scale Natural Resource Management in Takalar. South Sulawesi (Unpublished).

Thirumaran G, Anantharaman P (2009). Daily growth rate of field farming seaweed Kappaphycus alvarezii (Doty) Doty ex P. Silva in Vellar Estuary. World J. Fish. Mar. Sci. 1(3):144-153.

Tjahjana (2010). Produksi rumput laut Indonesia terbesar di dunia (Indonesia is largest producer of seaweed in the world), available at http://www.bipnewsroom.info/?_link=loadnews.php&newsid=60849. (accessed 23 April 2010).

Torii A, Nariu T (2004). On the length of wholesale marketing channels in Japan. Japan. Econ. 32(3):5-26.

Vesala T (2008). Middlemen and the adverse selection problem. Bull. Econ. Res. 60(1):1-11.

28

Information sources of knowledge based economic development for fisheries in Turkey

Ahmet AYDIN[1] and Guchgeldi BYASHIMOV[2]

[1]Department of Fisheries and Aquaculture, Finike Vocational School, Akdeniz University, 07740 Antalya, Turkey.
[2]Departmant of Business Administration, Nigde University, 51240 Nigde, Turkey.

Fishery is a very important sector in Turkey due to economic, geographic, traditional, and cultural conditions. The sector has being one of the four sub-sectors of the agriculture till recent year when the Ministry of Food, Agriculture, and Livestock separate Fishery as an independent sector. The coastal areas, the amounts of lakes and rivers have been very important for fishery supplies in Turkey. Although Turkey has big potential fishery sector, the production commonly survive as traditional practices. Information sources of fishery sector, water resources, and ecological issues come from master-apprentice connections. The objective of this study is to share information sources of knowledge, science, and technology where stakeholders can reach or create Knowledge Based Economic Development (KBED) in the sector. As a result of the study, the stakeholders of fishery in Turkey as well as in the other countries obtain information sources of fishery from the study where knowledge is shared.

Key words: Fishery, knowledge sources, development, Turkey.

INTRODUCTION

Fishery sector has always been very important activity in coastal areas due to economic, geographic, traditional, and cultural factors. Therefore, in Turkey, as in many other countries, the most intensive users of coastal zone have been fishermen (Unal, 2006). Fishery being one of the four sub-sectors of the agriculture in Turkey. It has been vital importance in contributing beneficial nutrition for human beings, providing raw material for the industrial sector, creating the employment possibilities and high potential for export. Turkey, with its favorable geographic

position between the Black Sea and Mediterranean Sea, has access to the fish resources of both water bodies. The country is also endowed with rich inland waters and rivers with significant capture fishery and aquaculture potential (Anonymous, 2008a). Turkey has high potential about catching and growing aquaculture products. But it has not been successful to raise the value of the potential.

Turkey is a country that's surrounded with seas three size of it such as Black Sea, Mediterranean Sea, Aegean

Sea and Marmara Sea. In addition that Marmara Sea has an inside sea property. Also the amounts of lakes and rivers have been very important for fishery supplies (Tasdan et al., 2010). Total fish production of Turkey is 653,080 tons according to the 2012 fishery statistics. Of the total, 68.2 % is obtained from the marine fisheries, 6.2% from island and 25.6% comes from aquaculture (Anonymous, 2011). Although Turkey has big potential fishery sector, the production commonly survive as traditional practices. Knowledge Based Economic Development (KBED) of the sector has not been created yet. Information sources of fishery sector, water resources, and ecological issues come from master-apprentice connections. The information sources and connection within them are weak to trust in. The objective of this study is to share information sources of knowledge, science, and technology where stakeholders can reach or create KBED in the sector. As result of study, the stakeholders of fishery in Turkey as well as in the other countries obtain information sources of fishery from the study where knowledge is shared.

STAKEHOLDERS OF FISHERIES IN TURKEY

Public

Ministry of Food, Agriculture and Livestock

The Ministry of Food, Agriculture and Livestock (MFAL) is the main state organization responsible for fisheries (including aquaculture) administration, regulation, protection, and promotion and technical assistance through Directorate General of Fisheries and Aquaculture (DFA). "According to the issue No. 27958 of the Official Gazette published on 8 of June in 2011, the Ministry of Agriculture and Rural Affairs (MARA) has been reconstituted as the MFAL while the department of fisheries, previously named for directorate-general, has been reconstituted as the DFA" (Can and Demirci, 2012). All activities in fisheries and aquaculture are based on the Fisheries Law No. 1380, enacted in 1971. With this law, and its related bureaucracy, definitions were codified. Based on this law, regulations and circulars are prepared to regulate fisheries. The Fisheries Law No. 1380 of 1971 is amended by law 3288 of 1986. According to Laws 1380 and 3288 and Continental Waters Law No. 2674 of 1982, foreigners are not allowed to take part in commercial fishing activities (Anonymous, 2008b). The main duties of DFA on fisheries are to:

(1) Determine and to promote the main issues of fishing and aquaculture both in marine and inland water systems for sustainable fishing and aquaculture,
(2) Make up issues to establish, to operate and to control fishing ports and fishing infrastructures,

(3) Protect fisheries resource, to determine marine protected, production and aquaculture areas, and making provisions to protect those areas from harmful activities,
(4) Set up legislative issues for import and export fisheries product,
(5) Conduct facilities that are aimed to improve and to enhance of fisheries and aquatic resources.
(6) Make provisions for inputs that are needed to improve the production of fishing and aquaculture resources.
(7) Set up the legislative issues that are connected to areas where fishing and aquaculture activities are held. To determined the characteristics and conditions of the limits of production tools and their usage and renting bases.
(8) Prepare and implement research projects relating to improving of fishing and aquaculture production,
(9) Set up an information network related to fisheries, fishing and aquaculture activities (Can and Demirci, 2012).

Ministry of Development

The objectives of fisheries management policy in Turkey used to be set up by State Planning Organization (DPT) till 2011 when the Ministry of Development established in Turkey. Managing fishery resources in a sustainable way is the main objective of the fisheries policy. Therefore, region-based preliminary fisheries plans have been designed. The objectives set out for these plans include rebuilding of depleted stocks, long-term resource management, introduction of fishing rights and sustainability of fishing opportunities for fishermen. The Ministry of Development prepares long-term development plans and annual programs conforming to the targets determined by the government. To this end, it coordinates the activities of ministries and public institutions concerning economic, social and cultural policies in order to ensure efficient implementation and advices the government on fishery policy issues. A special Committee for each sector including the stakeholders from fishery and aquaculture gathers every year during the preparation of the Development Plans, and serves as a platform for The Ministry of Development to consult. During the development of the latest development plan, the 9th five year development plan covering the period between 2007 and 2013, the Committee for Fisheries mainly focused on sustainable exploitation of resources, integration of environmental considerations to fishery and institutional restructure for the adoption of the common fishery policy (Can and Demirci, 2012).

Fishery port offices

Turkey has a large number of fishing ports and the MFAL has traditionally not had a presence (office) in any of these ports. The need to strengthen fisheries protection and control was identified as an important priority. For this purpose, Fishery Port Offices were established in 2006 with 30 ports. All of them will also eventually be linked through the computer-based FIS to Provincial Offices and Headquarters of the Ministry of Food, Agriculture and Livestock in Ankara. The port offices have been operating since the beginning of 2007. These offices are an important aspect of improved fisheries management in Turkey. They will serve to improve the collection, checking and the use of information on the quantities and species of fish landed, and helping to improve compliance with national regulations related to grading and marketing of fish and fishing vessel licensing. A permanent presence in these ports will also help the Ministry of Food, Agriculture and Livestock to investigate fishing offences and take appropriate enforcement action. There will also be an additional benefit for the industry in facilitating easier communication with government officials, enabling more effective communication of industry concerns and needs (Can and Demirci, 2012).

Other organizations

Fisheries production data are gathered and evaluated by the State Statistics Institute in collaboration with the MFAL. The institute uses a complete questionnaire method for large scale fishermen, and sub-sampling for small scale fishermen. The Under Secretariat of Foreign Trade of the Prime Ministry is the other public organization which regulates fish exports and imports regime. The Agricultural Bank of Republic of Turkey and Under Secretariat of the Treasury operate credit and incentive schemes to support the fisheries and aquaculture sectors. The Scientific and Technical Research Council also plays an important role organizing and subsidizing research activities. The Export Promotion Centre of Turkey, which is the only public organization in this field, acts as an intermediary in establishing business contacts between foreign importers and Turkish exporters to develop and to promote Turkish fisheries exports (Duzgunes and Erdogan, 2008).

Universities

Nationally, there are many universities with a fisheries faculty or vocational school, or a fisheries department within an agriculture faculty (Table 1). All fisheries faculties and schools consist of three main departments: aquaculture, capture fisheries and processing, and science.

Cooperatives

Cooperatives have great importance in countries, where small holdings prevail. Since small holdings prevail in Turkish agriculture and fisheries sectors, cooperatives are of vital importance. Most of the cooperatives in Turkey are operating in their small and local markets (Unal et al., 2009). Cooperatives in the small-scale fisheries sector are a way of maximizing long-term community benefits to deal with the threats of fisheries mismanagement, livelihood insecurity and poverty-harsh realities for many of the world's small-scale fishers. Communities with successful community-based organizations are better off than those without (Ostrom, 1990). Cooperatives can: (i) increase fishers' price-negotiating power with market intermediaries, help stabilize markets, improve post-harvest practices and facilities, provide marketing logistics and information, and facilitate investment in shared structures such as ice plants and fish processing facilities; (ii) increase market competition by setting up auctioning systems; (iii) use their greater negotiating power to make cost-saving bulk purchases of fishing gear, engines, equipment and fuel and to advocate with government; (iv) facilitate microcredit schemes for fishers, to reduce their dependency on intermediaries and give them greater freedom in selecting buyers (Anonymous, 2009a). In 1980, there were 229 fishery cooperatives with 14,750 members in Turkey, and 96 of these cooperatives were on the sea coast (Hazar, 1990). In 1992, while there were 8,020 agricultural cooperatives, only 262 fishery cooperatives existed in Turkey (Cikin and Kizildag, 1997). Both number of cooperatives and members have increased in 2009. According to MFAL, there are 528 fishery cooperatives with a total of 28,455 members and 12 fishery cooperatives associations and 1 central union of fishery cooperatives associations (Anonymous, 2012) (Table 2).

Information sources for the development of fisheries sector in Turkey

Information is a basic and fundamentally important element in any development activity. The value of information lies in its ability to affect a behavior, decision, or outcome. Information is an essential ingredient in agricultural development programs (Ozowa, 1995). Fisheries and aquaculture information is produced in different institutions. Much of information is published by commercial publishers and the cost is high and increasing. Efforts such as Access to Global Online Research in Agriculture (AGORA), Health Inter Network Access to Research (HINARI) and Online Access to Research in the Environment (OARE) alleviate the prohibitive costs of access for developing countries (Anonymous, 2009b). Much regional fisheries information is also published as grey literature by intergovernmental organizations such as the Caribbean Regional Fisheries Mechanism, Pacific Islands Forum Fisheries Agency, Secretariat of the Pacific Community, Network of

Table 1. Academic institutions of fisheries or marine science faculties.

Name of University	Name of Faculty	Web Address
Adıyaman University	Kahta Vocational School Department of Fisheries	www.adiyaman.edu.tr
Adnan Menderes University	Faculty of Agriculture Department of Fisheries Engineering	www.adu.edu.tr
Adnan Menderes University	Bozdogan Vocational School Department of Fisheries	www.adu.edu.tr
Akdeniz University	Faculty of Fisheries	www.sufak.akdeniz.edu.tr
Akdeniz University	Finike Vocational School Department of Fisheries	www.finike.akdeniz.edu.tr
Ankara University	Faculty of Agriculture Department of Fisheries Engineering	www.agri.ankara.edu.tr
Atatürk University	Faculty of Fisheries	www.atauni.edu.tr
Atatürk University	Hınıs Vocational School Department of Fisheries	www.atauni.edu.tr
Atatürk University	İspir Hamza Polat Vocational School Department of Fisheries	www.atauni.edu.tr
Bingöl University	Faculty of Agriculture Department of Fisheries	www.bingol.edu.tr
Bingöl University	Genc Vocational School Department of Fisheries	www.bingol.edu.tr
Cumhuriyet University	Gurun Vocational School Department of Fisheries	www.cumhuriyet.edu.tr
Cumhuriyet University	Suşehri Timur Karabal Vocational School Department of Fisheries	www.cumhuriyet.edu.tr
Çanakkale Onsekiz Mart University	Faculty of Marine Sciences and Technology	www.denbiltek.comu.edu.tr
Çanakkale Onsekiz Mart University	Bayramic Vocational School Department of Fisheries	www.bmyo.comu.edu.tr
Çukurova University	Faculty of Fisheries	www.suurunleri.cu.edu.tr
Çukurova University	Feke Vocational School Department of Fisheries	www.fekemyo.cu.edu.tr
Çukurova University	İmamoglu Vocational School, Department of Fisheries	www.imamoglumyo.cu.edu.tr
Dokuz Eylül University	Institute of Marine Sciences and Technology	www.imst.deu.edu.tr
Ege University	Faculty of Fisheries	www.egefish.ege.edu.tr
Ege University	Research and Application Center of Underwater	www.saum.ege.edu.tr
Ege University	Ege Vocational School Department of Fisheries	www.egemyo.ege.edu.tr
Erzincan University	Kemaliye Hacı Ali Akın Vocational School Department of Fisheries	www.erzincan.edu.tr
Erzincan University	Tercan Vocational School Department of Fisheries	www.erzincan.edu.tr
Fırat University	Faculty of Fisheries	www.firat.edu.tr
Fırat University	Keban Vocational School Department of Fisheries	www.firat.edu.tr
Gaziosmanpaşa University	Faculty of Agriculture Department of Fisheries Engineering	www.ziraat.gop.edu.tr
Gaziosmanpaşa University	Almus Vocational School Department of Fisheries	www.almusmyo.gop.edu.tr
Giresun University	Tirebolu Mehmet Bayrak Vocational School Department of Fisheries	www.tmyo.giresun.edu.tr
İstanbul University	Institute of Marine Sciences and Management	www.istanbul.edu.tr
İstanbul University	Faculty of Fisheries	www.suurunleri.istanbul.edu.tr
Kahramanmaraş Sutcu Imam University	Faculty of Agriculture Department of Fisheries	www.su.ksu.edu.tr
Karadeniz Teknik University	Surmene Faculty of Marine Science	www.deniz.ktu.edu.tr
Karadeniz Teknik University	Macka Vocational School Department of Fisheries	www.ktu.edu.tr
Kastamonu University	Faculty of Fisheries	www.su.kastamonu.edu.tr
Kocaeli University	Gazanfer Bilge Vocational School Department of Fisheries	www.gazanferbilge.kocaeli.edu.tr

Table 1. Contd.

Mersin University	Faculty of Fisheries	www.mersin.edu.tr
Mugla Sitki Kocman University	Faculty of Fisheries	www.mu.edu.tr
Mugla Sitki Kocman University	Underwater Practice and Research Center	www.mu.edu.tr
Muğla Sitki Kocman University	Ortaca Vocational School Department of Fisheries	www.mu.edu.tr
Mustafa Kemal University	Faculty of Marine Sciences and Technology	www.mku.edu.tr
Mustafa Kemal University	Dörtyol Vocational School Department of Fisheries	www.mku.edu.tr
Mustafa Kemal University	Samandağ Vocational School Department of Fisheries	www.mku.edu.tr
Middle East Technical University	Institute of Marine Sciences	www.ims.metu.edu.tr
Recep Tayyip Erdogan University	Faculty of Fisheries	www.suf.rize.edu.tr/tr
Suleyman Demirel University	Egirdir Fisheries Faculty	www.esuf.sdu.edu.tr
Süleyman Demirel University	Water Institute	www.sue.sdu.edu.tr
Sinop University	Faculty of Fisheries	www.sinop.edu.tr
Sinop University	Vocational School Department of Fisheries	www.sinop.edu.tr
Tunceli University	Faculty of Fisheries	www.tunceli.edu.tr
Yalova University	Armutlu Vocational School Department of Fisheries	www.yalova.edu.tr
Yuzuncu Yil University	Faculty of Fisheries	www.yyu.edu.tr

Table 2. Regional unions of fisheries cooperatives in Turkey.

Regions	Number of cooperatives	Total numbers of partners
Adana Regional Association	9	590
Balıkesir Regional Association	12	674
Çanakkale Regional Association	22	567
Hatay Regional Association	11	xx
Istanbul Regional Association	32	2240
Izmir Regional Association	22	1227
Kocaeli Regional Association	10	607
Marmara Regional Association	14	1529
Mersin Regional Association	9	389
Muğla Regional Association	14	567
Sinop Regional Association	9	394
Tekirdağ Regional Association	8	317

Source: MFAL, 2012.

Aquaculture Centres in Asia-Pacific, Southeast Asian Fisheries Development Center and by regional fishery science bodies such as the North Pacific Marine Science Organization and the International Council for the Exploration of the Sea (Anonymous, 2009b), Tables 3

Table 3. International available information sources about fisheries.

Information sources	Published by authority	Web address
Publications of Fisheries and Aquaculture	Food and Agriculture Organization of the United States	http://www.fao.org/fishery/en
Publications of Fisheries	U.N. FAO Globefish	http://www.eurofish.dk/
Publications of Fisheries	Eurofish International Organization	http://www.eurofish.dk/
Publications of Fisheries Unit	European Commission	http://ec.europa.eu/fisheries/index_en.htm
Publications of Fishery	Asia Pacific Fishery Commission	http://www.apfic.org/modules/wfdownloads
National Agricultural Statistics Service	United States Department of Agriculture	http://www.nass.usda.gov/
Publications of Fisheries Management	Australian Government Australian Fisheries Management Authority	http://www.afma.gov.au/
Publications of FisheriesandOceanography	Russian Federal Research Institute of Fisheries and Oceanography	http://www.vniro.ru/en/
Journal of Northwest Atlantic Fishery Science	Northwest Atlantic Fisheries Organization	http://journal.nafo.int/index.html
American Fisheries Society Journals	American Fisheries Society	http://afsjournals.org/
Scientia Marina	Spanish National Research Council	http://www.icm.csic.es/scimar/index.php
Pan-American Journal of Aquatic Sciences	Panamjas	http://www.panamjas.org/index.htm
Fisheries Journal	www.fisheriessciences.com	http://www.fisheriessciences.com/
Fishery Technology	Society of Fisheries Technologists	http://epubs.icar.org.in/ejournal/index.php/FT
Elsevier Science Fisheris Journals	Elsevier Publishing	http://www.elsevier.com/journals/title/a
The Asian Fisheries Science Journal	Asian Fisheries Society	http://www.asianfisheriessociety.org/publication/index.php
Journal of Taiwan Fisheries Research	Fisheries Research Institute, Taiwan	http://www.tfrin.gov.tw/
African Journals	AJOL	http://www.ajol.info/index.php
Journal of Natural Resources and Life Sciences Education	American Society of Agronomy	http://www.jnrlse.org/
Egyptian Journal of Aquatic Biology and Fisheries	Ain Shams University Faculty of Science Department of Zoology	http://www.ejabf.eg.net/index.html
International Aquatic Research	Islamic Azad University	http://www.intelaquares.com/
Fish farming and fisheries	Panorama Publisher	http://panor.ru/journals/fish
The Journal of Shellfish Research	National Shellfisheries Association	http://www.shellfish.org/jsr-public

and 4 show lists of some international and national available information sources on fisheries, aquaculture and aquatic sciences.

CONCLUSIONS

The Fishery sector has always had a very important activity in coastal areas of Turkey.

Therefore, in Turkey, as in many other countries, the most intensive users of coastal zone have been fishery industry sources. The country is also endowed with rich inland waters and rivers with significant capture fishery and aquaculture potential. Although Turkey has big potential fishery sector, the production commonly survive as traditional practices. Knowledge Based Economic Development (KBED) of the sector has not been created yet. The stakeholders of fishery

have to reach the source of development for fishery where KBED may create. MFAL, Ministry of Development, Fishery Port Office, Cooperatives, Universities, and Companies are main stakeholders of the sector. Also, national and international available information sources about fisheries as web site are very good source of information. It is necessary to convert information data to knowledge where KBED created in the sector.

Table 4. National available information sources about fisheries.

Information source	Published by authority	Web address
Fisheries Statistics Data General Directorate of Fisheries and Seafood Products	General Directorate of Fisheries and Seafood Products	http://www.bsgm.gov.tr/
Fisheries Statistics Service	Turkish Statistical Institute	http://www.turkstat.gov.tr/
Publications of Fisheries	General Directorate of Agricultural, Researchand Policy	www.sumae.gov.tr/
Publications of Fisheries Research	Mediterranean Fisheries Research Productionand Training Institute	http://www.akdenizsuurunleri.gov.tr/index_en.asp
Publications of Fisheries Research	Elaziğ Fisheries Research Station	http://www.elazigsuurunleri.gov.tr/default.asp?content=main&lang=en
Publications of Marine Research	Turkish Marine Research Foundation	http://www.tudav.org/index.php?lang=en
Journal of Fisheries	E.U. Fisheries Faculty	www.egejfas.org
Water World Magazine	Monthly Journal of Fisheries	www.sudunyasidergisi.com
Egirdir Journal of Fisheries Faculty	S.D.U. Fisheries Faculty	edergi.sdu.edu.tr
K.U.Journal of Fisheries	K.U. Fisheries Faculty	su.kastamonu.edu.tr
Journal of the Association of Fisheries Engineering	Association of the Fisheries Engineering	www.suurunleri.org.tr
Turkish Journal of Fisheries and Aquatic Sciences	Central Fisheries Research Institute	www.trjfas.org
Dolphin Research Bulletin	Central Research Institute for Fisheries	www.sumae.gov.tr/yunus
Academic Journals of The Scientific and Technological Research Council of Turkey	The Scientific and Technological Research Council of Turkey	http://journals.tubitak.gov.tr/

Conflict of Interest

The authors have not declared any conflict of interest.

REFERENCES

Anonymous (2008a). Fisheries Management Systems in OECD countries, Organization for Economic Co-operation and Development, http://www.oed.org

Anonymous (2008b). Fishery Country Profile: The Republic of Turkey, Food and Agriculture Organization of the United States, ftp://ftp.fao.org/fi/document/fcp/en/FI_CP_TR.pdf

Anonymous (2009a). Report of the Global Conference on Small-scale Fisheries: Securing sustainable small-scale fisheries: Bringing together responsible fisheries and social development. Fisheries and Aquaculture Report Rome. P. 911.

Anonymous (2009b). Information and Knowledge Sharing. FAO Fisheries Technical Guidelines for Responsible Fisheries. Rome, FAO. 2009. P. 12.

Anonymous (2011). Fishery Statistics, Turkish Statistical Institute, Printing Division.

Anonymous (2012). The Ministry of Food, Agriculture and Livestock General Directorate of Fisheries and Aquaculture, www.bsgm.gov.tr.

Can MF, Demirci A (2012). Fisheries Management in Turkey. Int. J. Aquac. 2(8):48-58.

Cikin A, Kizildag N (1997). Agricultural cooperative movement in Turkey and European Union, (in Turkish) Bulletin. Cooperative Special Issue, Chamber of Agricultural Engineers Branch of Izmir.

Duzgunes E, Erdogan N (2008). Fisheries Management in the Black Sea Countries. Turk. J. Fisher. Aquat. Sci. 8:181-192.

Hazar N (1990). Cooperative History (in Turkish), Turkish Cooperative Education Foundation Publications, Ankara.

Ostrom E (1990). Governing the Commons: The Evolution of Institutions for Collective Action, Cambridge University Press, New York. http://dx.doi.org/10.1017/CBO9780511807763

Ozowa VN (1995). Information Needs of Small Scale Farmers in Africa: The Nigerian Example. Q. Bull. Int. Assoc. Agric. Inform. Specialists 40:1.

Tasdan K, Celiker SA, Arisoy H, Ataseven Y, Donmez D, Gul U, Demir A (2010). Socio-Economic Analysis of the Fisheries Enterprises in the Mediterranean Region, (in Turkish), Agricultural Economics Research Institute Publication, P. 179, Ankara.

Unal V (2006). Profile of Fishery Cooperatives and Estimation of Socio-Economic Indicators in Marine Small-Scale Fisheries; Case Studies in Turkey. M. Sc. Thesis on Fisheries Economics and Management, University of Barcelona, Barcelona, Spain.

Unal V, Yercan M, Guclusoy H, Goncuoglu H (2009). A Better Understanding of Fishery Cooperatives in the Aegean, Turkey. J. Anim. Veter. Advan. 8(7):1361-1366

Permissions

All chapters in this book were first published in AJAR, by Academic Journals; hereby published with permission under the Creative Commons Attribution License or equivalent. Every chapter published in this book has been scrutinized by our experts. Their significance has been extensively debated. The topics covered herein carry significant findings which will fuel the growth of the discipline. They may even be implemented as practical applications or may be referred to as a beginning point for another development.

The contributors of this book come from diverse backgrounds, making this book a truly international effort. This book will bring forth new frontiers with its revolutionizing research information and detailed analysis of the nascent developments around the world.

We would like to thank all the contributing authors for lending their expertise to make the book truly unique. They have played a crucial role in the development of this book. Without their invaluable contributions this book wouldn't have been possible. They have made vital efforts to compile up to date information on the varied aspects of this subject to make this book a valuable addition to the collection of many professionals and students.

This book was conceptualized with the vision of imparting up-to-date information and advanced data in this field. To ensure the same, a matchless editorial board was set up. Every individual on the board went through rigorous rounds of assessment to prove their worth. After which they invested a large part of their time researching and compiling the most relevant data for our readers.

The editorial board has been involved in producing this book since its inception. They have spent rigorous hours researching and exploring the diverse topics which have resulted in the successful publishing of this book. They have passed on their knowledge of decades through this book. To expedite this challenging task, the publisher supported the team at every step. A small team of assistant editors was also appointed to further simplify the editing procedure and attain best results for the readers.

Apart from the editorial board, the designing team has also invested a significant amount of their time in understanding the subject and creating the most relevant covers. They scrutinized every image to scout for the most suitable representation of the subject and create an appropriate cover for the book.

The publishing team has been an ardent support to the editorial, designing and production team. Their endless efforts to recruit the best for this project, has resulted in the accomplishment of this book. They are a veteran in the field of academics and their pool of knowledge is as vast as their experience in printing. Their expertise and guidance has proved useful at every step. Their uncompromising quality standards have made this book an exceptional effort. Their encouragement from time to time has been an inspiration for everyone.

The publisher and the editorial board hope that this book will prove to be a valuable piece of knowledge for researchers, students, practitioners and scholars across the globe.

List of Contributors

Junping Liu
Research Center of Fluid Machinery Engineering and Technology, Jiangsu University, P. R. China, 212013

Xingye Zhu
Research Center of Fluid Machinery Engineering and Technology, Jiangsu University, P. R. China, 212013
Key Laboratory of Modern Agricultural Equipment and Technology, Ministry of Education and Jiangsu Province, Jiangsu University, P. R. China, 212013

Shouqi Yuan
Research Center of Fluid Machinery Engineering and Technology, Jiangsu University, P. R. China, 212013

Megha Aziz
Department of Marine Biology, Microbiology and Biochemistry, Cochin University of Science and Technology,
Fine Arts Avenue, Cochin 682016, Kerala, India

V. Ambily
Department of Marine Biology, Microbiology and Biochemistry, Cochin University of Science and Technology,
Fine Arts Avenue, Cochin 682016, Kerala, India

S. Bijoy Nandan
Department of Marine Biology, Microbiology and Biochemistry, Cochin University of Science and Technology,
Fine Arts Avenue, Cochin 682016, Kerala, India

Muhammad Luqman
University College of Agriculture, University of Sargodha, Pakistan

Tahir Munir Butt
University of Agriculture, Faisalabad, Sub-Campus Toba Tek Singh-Pakistan

Ayub Tanvir
Department of Forestry, University of Agriculture, Faisalabad-Pakistan

Muhammad Atiq
Department of Plant Pathology, University of Agriculture, Faisalabad-Pakistan

Muhammad Zakaria Yousuf Hussan
Department of Agriculture, Government of the Punjab, Pakistan

Muhammad Yaseen
University College of Agriculture, University of Sargodha, Pakistan

Siamak Bagheri
Inland Waters Aquaculture Institute, Iranian Fisheries Research Organization, 66 Anzali, Iran
School of Biological Sciences, Universiti Sains Malaysia, 11800 USM, Penang, Malaysia

Mashhor Mansor
School of Biological Sciences, Universiti Sains Malaysia, 11800 USM, Penang, Malaysia

Azemat Ghandi
Inland Waters Aquaculture Institute, Iranian Fisheries Research Organization, 66 Anzali, Iran

Esmaeil Yosefzad
Inland Waters Aquaculture Institute, Iranian Fisheries Research Organization, 66 Anzali, Iran

S. Matsiori
Department of Ichthyology and Aquatic Environment, School of Agricultural Sciences, University of Thessaly, Nea Ionia, Magnesia, Greece

Z. Stamkopoulos
Department of Ichthyology and Aquatic Environment, School of Agricultural Sciences, University of Thessaly, Nea Ionia, Magnesia, Greece

S. Aggelopoulos
Department of Agricultural Development and Agribusiness Management, Alexander Technological Educational
Institute of Thessaloniki, 57400, Sindos Thessaloniki, Greece

K. Soutsas
Department of Forestry and Management of the Environment and Natural Resources, Democritus University of Thrace, N. Orestiada PC 68200, Greece

Ch. Neofitou
Department of Ichthyology and Aquatic Environment, School of Agricultural Sciences, University of Thessaly, Nea Ionia, Magnesia, Greece

D. Vafidis
Department of Ichthyology and Aquatic Environment, School of Agricultural Sciences, University of Thessaly, Nea Ionia, Magnesia, Greece

E. Y. Kpoclou
Department of Nutrition and Food Science, Laboratory of Food Microbiology and Biotechnology, Faculty of Agronomic Sciences, University of Abomey-Calavi, 01 BP 526, Cotonou, Benin

V. B. Anihouvi
Department of Nutrition and Food Science, Laboratory of Food Microbiology and Biotechnology, Faculty of Agronomic Sciences, University of Abomey-Calavi, 01 BP 526, Cotonou, Benin

M. L. Scippo
Departement of Food Science, Laboratory of Food Analysis, Faculty of Veterinary Medicine, Centre of Analytical Research and Technology (CART), University of Liège, B43b, Boulevard de colonster 20, Star-tilman, B-4000 Liège, Belgium

J. D. Hounhouigan
Department of Nutrition and Food Science, Laboratory of Food Microbiology and Biotechnology, Faculty of Agronomic Sciences, University of Abomey-Calavi, 01 BP 526, Cotonou, Benin

E. A. Eze
Department of Microbiology, University of Nigeria, Nsukka, Enugu State, Nigeria

Chijioke N. Eze
Department of Microbiology, University of Nigeria, Nsukka, Enugu State, Nigeria

V. O. Amaeze
Department of Microbiology, University of Nigeria, Nsukka, Enugu State, Nigeria

Chibuzor N. Eze
Department of Microbiology, University of Nigeria, Nsukka, Enugu State, Nigeria

O. K. Gbadamosi
Department of Fisheries and Aquaculture Technology, P. M. B 704, Akure, Ondo State, Nigeria

E. A. Fasakin
Department of Fisheries and Aquaculture Technology, P. M. B 704, Akure, Ondo State, Nigeria

O. T. Adebayo
Department of Fisheries and Aquaculture Technology, P. M. B 704, Akure, Ondo State, Nigeria

I. Bayaso
Ministry of Agriculture headquarters P. M. B. 2079, Yola, Adamawa State, Nigeria

H. Nahunnaro
Department of Crop Production and Horticulture, Federal University of Technology P. M. B. 2076, Yola, Adamawa State, Nigeria

D. M. Gwary
Department of Crop Protection, Faculty of Agriculture, University of Maiduguri, P. M. B. 1069, Maiduguri, Borno State, Nigeria

Hui Fang
Chinese Academy of Fishery Sciences, Beijing 100141, China

Ying Jing
Chinese Academy of Fishery Sciences, Beijing 100141, China

Gang Han
Chinese Academy of Fishery Sciences, Beijing 100141, China

Yingren Li
Chinese Academy of Fishery Sciences, Beijing 100141, China

Theophilus Miebi Gbigbi
Department of Agricultural Services, Ministry of Agriculture and Natural Resources, Asaba, Delta State, Nigeria

V. C. Asogwa
Department of Agricultural Education, University Agriculture, P. M. B. 2373 Makurdi, Nigeria

D. O. Onu
Department of Agricultural Education, University Agriculture, P. M. B. 2373 Makurdi, Nigeria

B. N. Egbo
Department of Agricultural Education, Enugu State College of Education (Technical), Enugu, Nigeria

S. K. Odetola
Department of Agricultural Economics, University of Ibadan, Nigeria

T. T. Awoyemi
Department of Agricultural Economics, University of Ibadan, Nigeria

S. Ajijola
Institute of Agricultural Research and Training, Moor Plantation, Ibadan, Nigeria

Emad M. El-Kholie
Research Center, College of Science, King Saud University, P. O. Box 2455, Riyadh11451, Kingdom of Saudi Arabia
Department Nutrition and Food Science, Faculty of Home Economics, Menufiya University, Egypt

Mohammed A. T. Abdelreheem
Department of Biochemistry, Faculty of Agriculture, Ain Shams University, Egypt

Seham A. Khader
Department Nutrition and Food Science, Faculty of Home Economics, Menufiya University, Egypt

Naim Saglam
Department of Aquaculture and Fish Diseases, Faculty of Fisheries, University of Fırat, 23119, Elazığ-Turkey

Jin Han Bae
Division of Marine Environment and Bioscience, Korea Maritime University, Busan 606-791, Korea

Sun Young Lim
Division of Marine Environment and Bioscience, Korea Maritime University, Busan 606-791, Korea

Xueqin Jiang
School of Life Science, East China Normal University, Shanghai, 200062 China

Liqiao Chen
School of Life Science, East China Normal University, Shanghai, 200062 China

Jianguang Qin
School of Biological Sciences, Flinders University, Adelaide, SA 5001, Australia

Chuanjie Qin
School of Life Science, East China Normal University, Shanghai, 200062 China

Haibo Jiang
School of Life Science, East China Normal University, Shanghai, 200062 China

Erchao Li
School of Life Science, East China Normal University, Shanghai, 200062 China

Tasaduq Hussain Shah
Faculty of Fisheries, Sher-e-.Kashmir University of Agricultural Sciences and Technology of Kashmir, Rangil, Ganderbal, C/o Shuhama Campus, Alusteng, Srinagar, Kashmir, Jammu and Kashmir, 190006, India

Masood Ul Hassan Balkhi
Faculty of Fisheries, Sher-e-.Kashmir University of Agricultural Sciences and Technology of Kashmir, Rangil, Ganderbal, C/o Shuhama Campus, Alusteng, Srinagar, Kashmir, Jammu and Kashmir, 190006, India

Oyas Ahmad Asimi
Faculty of Fisheries, Sher-e-.Kashmir University of Agricultural Sciences and Technology of Kashmir, Rangil, Ganderbal, C/o Shuhama Campus, Alusteng, Srinagar, Kashmir, Jammu and Kashmir, 190006, India

Imran Khan
Faculty of Fisheries, Sher-e-.Kashmir University of Agricultural Sciences and Technology of Kashmir, Rangil, Ganderbal, C/o Shuhama Campus, Alusteng, Srinagar, Kashmir, Jammu and Kashmir, 190006, India

Wei Song
Key Laboratory of East China Sea and Oceanic Fishery Resources Exploitation, Ministry of Agriculture, East China Jiangsu Engineering Laboratory for Breeding of Special Aquatic Organisms, Huaiyin Normal University, 111 West Changjiang Road, 223300, Jiangsu, China

Zhiqiang Wu
College of Environmental Science and Engineering, Guilin University of Technology, 12 Jiangan Road, 541004, Guilin, China

Nan Wu
College of Life Science, Huaiyin Normal University, 111 West Changjiang Road, 223300, Jiangsu, China
Jiangsu Engineering Laboratory for Breeding of Special Aquatic Organisms, Huaiyin Normal University, 111 West Changjiang Road, 223300, Jiangsu, China

Lingbo Ma
Key Laboratory of East China Sea and Oceanic Fishery Resources Exploitation, Ministry of Agriculture, East China Sea Fisheries Research Institute, Chinese Academy of Fishery Science, 200090, Shanghai, China

Uttam Kumar Baruah
Krishi Vigyan Kendra Goalpara, National Research Centre on Pig, Indian Council of Agricultural Research, Dudhnoi – 783124, Assam, India

Jyotish Barman
Krishi Vigyan Kendra Goalpara, National Research Centre on Pig, Indian Council of Agricultural Research, Dudhnoi – 783124, Assam, India

Hitu Choudhury
Krishi Vigyan Kendra Goalpara, National Research Centre on Pig, Indian Council of Agricultural Research, Dudhnoi – 783124, Assam, India

Popiha Bordoloi
Krishi Vigyan Kendra Goalpara, National Research Centre on Pig, Indian Council of Agricultural Research, Dudhnoi – 783124, Assam, India

Mohammad Abul Mansur
Department of Fisheries Technology, Faculty of Fisheries, Bangladesh Agricultural University, Mymensingh, Bangladesh

Shafiqur Rahman
Faculty of Earth Science, University Malaysia Kelantan, Jeli Campus, 17600 Jeli, Malaysia

Mohammad Nurul Absar Khan
Faculty of Fisheries, Chittagong Veterinary and Animal Science University, Chittagong, Bangladesh

Md. Shaheed Reza
Department of Fisheries Technology, Faculty of Fisheries, Bangladesh Agricultural University, Mymensingh-2202, Bangladesh

Kamrunnahar
Department of Fisheries Technology, Faculty of Fisheries, Bangladesh Agricultural University, Mymensingh-2202, Bangladesh

Shoji Uga
Department of Parasitology, Graduate School of Health Sciences, Kobe University, Japan

C. O Olaniyi
Department of Animal Production and Health, Ladoke Akintola University of Technology, Ogbomosho, Oyo State, Nigeria

B. R. Salau
Department of Animal Production and Health, Ladoke Akintola University of Technology, Ogbomosho, Oyo State, Nigeria

A. S. Oyekale
Department of Agricultural Economics and Extension, North-West University Mafikeng Campus, Mmabatho, 2735 South Africa

O. I. Oladele
Department of Agricultural Economics and Extension, North-West University Mafikeng Campus, Mmabatho, 2735 South Africa

F. Mukela
Department of Agricultural Economics, University of Ibadan, Ibadan, Oyo State, Nigeria

Emmanuel D. Abarike
Department of Fisheries and Aquatic Resources Management, University for Development Studies, P. O. Box TL 1882, Tamale-Ghana

Edward A. Obodai
Department of Fisheries and Aquatic Sciences, University of Cape Coast, Ghana

Felix Y. K. Attipoe
Aquaculture Research and Development Centre (ARDEC), Council for Scientific and Industrial Research (CSIR), Akosombo, Ghana

O. J. Chichongue
University Of Nairobi, Kenya Agricultural Research Institute of Mozambique (IIAM), Mozambique

G. N. Karuku
Agricultural Research Institute of Mozambique (IIAM), Mozambique

A. K. Mwala
University Of Nairobi, Kenya

C. M. Onyango
Agricultural Research Institute of Mozambique (IIAM), Mozambique

A. M. Magalhaes
Agricultural Research Institute of Mozambique (IIAM), Mozambique

Mao-Lin Hu
School of Life Sciences, Nanchang University, Nanchang, Jiangxi Province, China

Zhi-Qiang Wu
College of Environmental Science and Engineering, Guilin University of Technology, Guilin, Guangxi Zhuang Autonmous Region, China

Shan Ouyang
School of Life Sciences, Nanchang University, Nanchang, Jiangxi Province, China

Xiao-Ping Wu
School of Life Sciences, Nanchang University, Nanchang, Jiangxi Province, China

Achmad Zamroni
Graduate School of Biosphere Science, Hiroshima University, Hiroshima, Japan
Research Center for Marine and Fisheries Socio Economics, Ministry for Marine Affairs and Fisheries, Jakarta, Indonesia